# Mapping of Nervous System Diseases via MicroRNAs

# FRONTIERS IN NEUROTHERAPEUTICS SERIES

## Series Editors
Diana Amantea, Laura Berliocchi, and Rossella Russo

**Rational Basis for Clinical Translation in Stroke Therapy**
*Giuseppe Micieli*, IRCCS, Pavia, Italy
*Diana Amantea*, University of Calabria, Rende, Italy

**Mapping of Nervous System Diseases via MicroRNAs**
*Christian Barbato*, Institute of Cell Biology and Neurobiology (IBCN), Rome, Italy
*Francesca Ruberti*, Institute of Cell Biology and Neurobiology (IBCN), Rome, Italy

# Mapping of Nervous System Diseases via MicroRNAs

Edited by
## Christian Barbato

National Research Council (CNR)
Institute of Cell Biology and Neurobiology (IBCN)
Rome, Italy

## Francesca Ruberti

National Research Council (CNR)
Institute of Cell Biology and Neurobiology (IBCN)
Rome, Italy

**CRC Press**
Taylor & Francis Group
Boca Raton London New York

CRC Press is an imprint of the
Taylor & Francis Group, an **informa** business

CRC Press
Taylor & Francis Group
6000 Broken Sound Parkway NW, Suite 300
Boca Raton, FL 33487-2742

First issued in paperback 2019

© 2016 by Taylor & Francis Group, LLC
CRC Press is an imprint of Taylor & Francis Group, an Informa business

No claim to original U.S. Government works

ISBN-13: 978-1-4822-6352-7 (hbk)
ISBN-13: 978-0-367-87115-4 (pbk)

**Visit the Taylor & Francis Web site at**
**http://www.taylorandfrancis.com**

**and the CRC Press Web site at**
**http://www.crcpress.com**

# Contents

Preface..................................................................................................................vii
Editors...................................................................................................................ix
Contributors ..........................................................................................................xi

## SECTION I    MicroRNAs in the Nervous System and Neurological Diseases (Basic): MicroRNAs in the Nervous System

**Chapter 1**    MicroRNA Biology and Function in the Nervous System...................3

*Francesca Ruberti and Christian Barbato*

## SECTION II    MicroRNAs in the Nervous System and Neurological Diseases (Basic): MicroRNAs in Neuropsychiatric Diseases

**Chapter 2**    MicroRNAs in Mood and Anxiety Disorders.....................................21

*Karen M. Ryan, Erik Kolshus, and Declan M. McLoughlin*

**Chapter 3**    MicroRNA Dysregulation in Schizophrenia
and Functional Consequences ...........................................................59

*Liliana Laskaris, Ting Ting Lee, Piers Gillett,
and Gursharan Chana*

**Chapter 4**    Impact of MicroRNAs in Synaptic Plasticity, Major Affective
Disorders, and Suicidal Behavior.....................................................101

*Gianluca Serafini, Yogesh Dwivedi, and Mario Amore*

## SECTION III    MicroRNAs in the Nervous System and Neurological Diseases (Basic): MicroRNAs in Neurological Diseases

**Chapter 5**    MicroRNAs in Prion Diseases ...........................................................139

*Daniela Zimbardi and Tiago Campos Pereira*

**Chapter 6**  MicroRNAs in Epileptogenesis and Epilepsy ................................. 153

*Eva M. Jimenez-Mateos, Tobias Engel, Catherine Mooney, and David C. Henshall*

**Chapter 7**  MicroRNAs and Pain ....................................................................... 183

*Hjalte H. Andersen and Parisa Gazerani*

## SECTION IV   MicroRNA Biomarkers in Neurological Diseases (Applications)

**Chapter 8**  Circulating Cell-Free MicroRNAs as Biomarkers for Neurodegenerative Diseases ....................................................... 207

*Margherita Grasso, Francesca Fontana, and Michela A. Denti*

**Chapter 9**  Circulating Cell-Free MicroRNAs as Biomarkers for Neural Development and Their Importance in Fetal Programing for Postnatal Disease .............................................................................. 237

*Mario Lamadrid-Romero, Néstor Fabián Díaz Martínez, and Anayansi Molina-Hernández*

**Index** .................................................................................................................. 255

# Preface

Molecular and cellular neurobiological studies of microRNA (miRNA)-mediated gene silencing in the nervous system represent the exploration of a new frontier of miRNA biology and the potential development of new diagnostic tests and genetic therapies for neurological disease. Over recent years our understanding of microRNA (miRNA) biogenesis, molecular mechanisms by which miRNAs regulate gene expression, and the functional roles of miRNAs has been expanded. MiRNAs are ≈22 nucleotide-long double-stranded RNAs. One strand of these small noncoding RNA molecules operates as guide for RISC (RNA-induced silencing complex) to produce either the block of translation or the decay of target mRNAs. In mammalian cells, miRNA action is mediated by an imperfect pairing between the 3′UTRs of the mRNA targets and nucleotides 2–8 from the 5′ of the miRNA.

Recently, microRNAs are emerging as important players in posttranscriptional regulation in the brain. Several studies have shown spatially and/or temporally restricted distribution of miRNAs, suggesting that they may control the fine-tuning regulation of neuronal gene expression.

Individual microRNAs can reduce the production of a hundred proteins and miRNA-mediated posttranscriptional regulation is involved in neuronal differentiation, dendritic spine development, and synaptic plasticity.

Increasing evidence suggests that miRNAs are dysregulated in several neurological disorders. The collection of data on the association between human brain diseases and miRNAs has focused on expression profiles of miRNAs and their quantitative modulation (i.e., upregulation versus downregulation), according to age, gender, phase of the disease, and specific brain area.

Depending on the role of the miRNA, the goal of the treatment will be to either increase or reduce miRNA function. Given the importance that miRNAs might play in neuropathology, several strategies to manipulate miRNA activity and expression are being pursued. Two main strategies may be applied to target miRNA expression in the brain: directly, by using oligonucleotides or virus-based constructs, and indirectly, by using drugs to modulate miRNA expression at the transcriptional and/or processing level. To date, all delivery strategies have been important for identifying a suitable way to generate microRNA-based therapies for neurological diseases related to the perturbation of miRNAs. The main challenge for miRNA therapeutics in neurology, beyond stability and safety, is delivery to the appropriate tissue and neurons. Molecules modulating miRNA action must reach the cells and must function at the site of the disease.

Due to the ever-expanding knowledge of miRNAs as fine tuners of gene expression in all aspects of biology and medicine, and to the emerging impact of sequence-specific posttranscriptional gene silencing mediated by miRNA as a potential therapeutic approach directed to the nervous system, we believe that this book in the series Frontiers in Neurotherapeutics will be of great interest to a broad scientific audience. This book, titled *Mapping of Nervous System Disease via MicroRNAs*, consists of nine chapters, and opens a window on our current understanding of the microRNAs involved in neurological diseases. This is an exciting work because the collection of selected chapters provides insight into the full range of concepts and a snapshot of the current status of this dynamic field.

The book is divided into four sections. Section I gives an overview of the landscape of miRNAs biology and function in the nervous system. This is followed by Chapters 2 through 4 under Section II, which focuses on discovery, regulatory functions, and molecular mechanism of miRNAs associated with neuropsychiatric disease. These chapters provide an excellent account of the miRNAs in anxiety and depression, and in mental disorders characterized by abnormal social behavior, such as schizophrenia, or important affective disorders with a high risk for suicidal behavior. These chapters are particularly important because they offer a new point of view to identify novel insights into brain/behavioral diseases.

Section III provides the state of the art in neurological diseases, which have been little explored from a "microRNA research view." The authors of these chapters illustrate how microRNAs could be involved in the prion pathogenesis, or in molecular mechanisms of epilepsy and pain processing and conditions, showing a new frontier of neuroscience research.

The "Applications" Section IV includes Chapters 8 and 9, which analyze the advancement of the study of miRNAs in biofluids in both physiological and pathological conditions. In addition, miRNAs are described as biomarkers for different neurodegenerative diseases, such as Alzheimer's, Parkinson's, amyotrophic lateral sclerosis or for CNS development and during postnatal life, an emerging field of research.

*Mapping of Nervous System Diseases via MicroRNAs* serves as an introduction to the miRNAs in the nervous system, and their increasingly important role in neurological diseases. In addition, in each chapter of the book, readers and scholars will find material on the future diagnostic and therapeutic advances of microRNA in selected neurological diseases.

We thank all the authors who have contributed excellent chapters to this book and reviewers for their critical comments to improve the quality and integrity of the chapters. We are grateful to Laura Berliocchi for the invitation to initiate this book in the Frontiers in Neurotherapeutics Series and for her continuing support and commitment in making this book a reality and to staff members involved in the production of the book.

**Christian Barbato**
**Francesca Ruberti**

# Editors

**Christian Barbato**, MD, PhD is a researcher at the National Research Council (CNR) at the Institute of Cell Biology and Neurobiology (IBCN) in Rome, Italy. He earned his degree in medicine and surgery from the Sapienza University of Rome and his PhD in Neuroscience in 2001 from Tor Vergata University of Rome. From 2002 to 2005, he worked as a postdoc at the Institute of Neurobiology and Molecular Medicine (INMM), National Research Council (CNR), Rome, Italy. Then from 2005 to 2009 he was an associate researcher at the Fondazione EBRI Rita Levi-Montalcini, European Brain Research Institute, MicroRNAs in the Nervous System Unit, Rome, Italy. Since 2009, he has been appointed as a researcher at the Institute of Cell Biology and Neurobiology (IBCN), CNR, Italy. His recent work aims to the characterization of cellular and molecular neurobiology of RNA-induced silencing complex (RISC) and noncoding RNA in synaptic plasticity, learning and memory. Dr. Barbato's recent research interest focused on the function of microRNAs and their gene targets implicated in physiological and neuropathological processes of neuronal cells.

**Francesca Ruberti** earned her PhD in biophysics in 1996 from the International School of Advanced Studies, Trieste, Italy. From 1998 to 2000 she was a postdoctoral fellow of the European Community at the laboratory of Dr. C.G. Dotti at the European Molecular Biology Laboratory (EMBL), Heidelberg, Germany. In 2000, she had a collaboration contract on "Methodologies of molecular biology applied to the study of intracellular traffic in neurons" at the Biophysics sector, International School for Advanced Studies, Trieste, Italy. In September 2001, Ruberti worked as a researcher at the National Research Council, Institute of Cell Biology and Neurobiology, CNR, Italy. Her previous research activity has been directed to studying the role of nerve growth factor (NGF) on specific neuronal populations of the central nervous system (CNS) and on synaptic plasticity in the hippocampus, and on the molecular mechanisms involved in the cellular localization of specific neuronal proteins or mRNAs. Dr. Ruberti's recent research interest focused on the study of microRNAs and their gene targets implicated in the physiological and pathological (neurodegenerative diseases, Alzheimer's disease) processes of neurons.

# Contributors

**Mario Amore**
Department of Neuroscience
University of Genoa
Genoa, Italy

**Hjalte H. Andersen**
Department of Health Science and
    Technology
Aalborg University
Aalborg, Denmark

**Christian Barbato**
Institute of Cell Biology and
    Neurobiology (IBCN)
National Research Council (CNR)
Rome, Italy

**Gursharan Chana**
Department of Electrical and Electronic
    Engineering
Centre for Neural Engineering
University of Melbourne
Victoria, Australia

**Michela A. Denti**
Centre for Integrative Biology
University of Trento
Trento, Italy

**Yogesh Dwivedi**
Department of Psychiatry and
    Behavioral Neurobiology
University of Alabama
Birmingham, Alabama

**Tobias Engel**
Department of Physiology and Medical
    Physics
Royal College of Surgeons in Ireland
Dublin, Ireland

**Francesca Fontana**
Centre for Integrative Biology
University of Trento
Trento, Italy

**Parisa Gazerani**
Department of Health Science
    and Technology
Aalborg University
Aalborg, Denmark

**Piers Gillett**
Department of Electrical and Electronic
    Engineering
Centre for Neural Engineering
and
Department of Psychiatry
University of Melbourne
Victoria, Australia

**Margherita Grasso**
Centre for Integrative Biology
University of Trento
Trento, Italy

**David C. Henshall**
Department of Physiology and Medical
    Physics
Royal College of Surgeons
    in Ireland
Dublin, Ireland

**Eva M. Jimenez-Mateos**
Department of Physiology and Medical
    Physics
Royal College of Surgeons
    in Ireland
Dublin, Ireland

**Erik Kolshus**
Institute of Neuroscience
and
Department of Psychiatry
St Patrick's University Hospital
Trinity College
Dublin, Ireland

**Mario Lamadrid-Romero**
Cell Biology Department
Instituto Nacional de Perinatología
Isidro Espinosa de los Reyes
and
School of Science-UNAM
Mexico City, Mexico

**Liliana Laskaris**
Department of Electrical and Electronic
Engineering
Centre for Neural Engineering
and
Department of Psychiatry
University of Melbourne
Victoria, Australia

**Ting Ting Lee**
Department of Electrical and Electronic
Engineering
Centre for Neural Engineering
and
Department of Psychiatry
University of Melbourne
Victoria, Australia

**Néstor Fabián Díaz Martínez**
Cell Biology Department
Instituto Nacional de Perinatología
Isidro Espinosa de los Reyes
Mexico City, Mexico

**Declan M. McLoughlin**
Institute of Neuroscience
and
Department of Psychiatry
St Patrick's University Hospital
Trinity College
Dublin, Ireland

**Anayansi Molina-Hernández**
Cell Biology Department
Instituto Nacional de Perinatología
Isidro Espinosa de los Reyes
Mexico City, Mexico

**Catherine Mooney**
Department of Physiology and Medical
Physics
Royal College of Surgeons in Ireland
Dublin, Ireland

and

Irish Centre for Fetal and Neonatal
Translational Research
Cork, Ireland

**Tiago Campos Pereira**
Department of Biology—FFCLRP
University of São Paulo
São Paulo, Brazil

**Francesca Ruberti**
Institute of Cell Biology and
Neurobiology (IBCN)
National Research Council (CNR)
Rome, Italy

**Karen M. Ryan**
Institute of Neuroscience
and
Department of Psychiatry
St Patrick's University Hospital
Trinity College
Dublin, Ireland

**Gianluca Serafini**
Department of Neuroscience
University of Genoa
Genoa, Italy

**Daniela Zimbardi**
Department of Genetics—IBB
São Paulo State University
São Paulo, Brazil

# Section I

*MicroRNAs in the Nervous System and Neurological Diseases (Basic)*

*MicroRNAs in the Nervous System*

# 1 MicroRNA Biology and Function in the Nervous System

*Francesca Ruberti and Christian Barbato*

## CONTENTS

Abstract ........................................................................................................................... 3
1.1 Introduction ............................................................................................................ 4
1.2 MiRNA Biogenesis ................................................................................................ 4
1.3 Nomenclature of MiRNAs ..................................................................................... 6
1.4 MiRNAs in the Nervous System ........................................................................... 7
    1.4.1 MiRNAs in Neuronal Development and Differentiation .................... 7
    1.4.2 MiRNAs at the Synapse ................................................................. 9
    1.4.3 MiRNAs and Dendritic Spines .............................................................. 9
    1.4.4 RNA-Induced Silencing Complex, MiRNAs, and Cognitive
        Functions ........................................................................................ 10
    1.4.5 MiRNAs in Brain Disease ................................................................. 12
1.5 Perspectives ......................................................................................................... 12
Acknowledgments ......................................................................................................... 13
References ...................................................................................................................... 13

## ABSTRACT

MicroRNAs (miRNAs) are highly expressed in the mammalian nervous system and regulate neuronal gene expression during neurogenesis, neurodevelopment, differentiation, dendritic morphogenesis, synaptic plasticity, learning, and memory.

MiRNAs are spatially and temporally modulated in the nervous system and they exhibit context-dependent functions. MiRNAs often act through regulatory networks in specific cellular contexts and at specific times to ensure the progression through each biological state. MiRNAs are involved in local protein synthesis and contribute to synaptic plasticity by modulating dendritic mRNA translation at dendritic spines.

Studies in animal models showed that RNA-induced silencing complexes and specific miRNAs might be recruited in synaptic plasticity processes supporting learning, memory, and cognition. Significant progress has been made in our understanding of miRNAs in the nervous system and it provides an encouraging starting point to investigate miRNA pathway involvement in the development and progression of neurological and psychiatric diseases and to search future therapeutic applications. Here, we describe recent molecular and cellular neurobiological findings that highlight the role of miRNAs in the neuronal regulatory networks, which represent the exploration of a new frontier of miRNAs biology.

## 1.1   INTRODUCTION

The brain is the most complex biological structure known, and multiple neuronal cell types are connected via synapses. The synaptic circuitry is determined during development and differentiation and is achieved by multiple levels of gene regulation. The transcriptional and posttranscriptional gene regulation mechanisms drive synaptic plasticity, memory formation, and cognitive functions. MicroRNAs (miRNAs) are emerging as important players in posttranscriptional regulation in the brain, and recently it was evidenced that the miRNAs might participate in the establishment and maintenance of such brain complexity (Goldie and Cairns, 2012; Chiu et al., 2014). MiRNAs are short noncoding regulatory RNAs, double-stranded RNAs (dsRNAs) ≈22 nucleotides in length, which associate with RNA-induced silencing complex (RISC) to inhibit the target mRNA translation (miRNA) when there is an imperfect pairing between miRNAs and the 3′ untranslated regions (3′-UTRs) of the mRNA targets (Bartel, 2009). MiRNAs are fine regulators expressed at different levels and in a neuronal specific-type manner, and they are expressed in a spatially and temporally controlled manner in the nervous system (Miska et al., 2004). Bioinformatic tools predict hundreds of mRNA targets for each miRNA (Tarang and Weston, 2014), suggesting that many neuronal genes are subjected to miRNA-mediated regulation. A step forward in the understanding how miRNA work in the brain comes from studies of miRNA expression profiles from pathological and normal tissues. By support of several *in vitro* and *in vivo* studies aimed at the investigation of functional role of miRNA, it was suggested that miRNAs have multiple role in regulating neuronal fate specification and differentiation, dendritic spine development, synaptic plasticity, and local protein synthesis. Molecular and cellular neurobiological exploration of the miRNA-mediated gene silencing opens a new frontier in our understanding of the regulation of neural gene expression in physiological and pathological conditions.

## 1.2   MiRNA BIOGENESIS

Over the last 10 years, our understanding of miRNA biogenesis, cellular and molecular mechanisms by which miRNAs regulate gene expression, and the functional roles of miRNAs have been developed (Ha and Kim, 2014). MiRNAs are an evolutionarily conserved class of small ≈22-nucleotide long double-stranded noncoding RNAs that modulate gene expression at the posttranscriptional level by both inducing mRNA degradation and inhibiting the translation of target mRNA (Bartel, 2009).

In the genome, there are two classes of miRNA intergenic or residing in introns of coding or noncoding genes. The miRNA biogenesis pathway begins with the transcription, generally exerted by RNA polymerase II, of a primary miRNA transcript (pri-miRNA). Pri-miRNA extends from hundreds to thousands of nucleotides and contains a 60- to 80-nucleotide stem-loop structure (Ha and Kim, 2014; Table 1.1).

**TABLE 1.1**
**MicroRNA (miRNA) Size and Biogenesis**

| Size Ranges in Nucleotide | ncRNAs |
| --- | --- |
| 18–24 | MiRNAs |
| 70–110 | Pre-MiRNAs (miRNA small hairpin precursors) |
| >300 | Pri-MiRNAs (primary miRNA gene transcript) |

In the nucleus, the pri-miRNAs are processed by a protein complex, named microprocessor, which contains a 160-kDa ribonuclease (RNase) III-like Drosha enzyme and the dsRNA-binding protein called DiGeorgie syndrome critical region gene 8 (DGCR8), releasing the hairpin and generating the intermediate precursor miRNA (pre-miRNA). The pre-miRNA is then exported into the cytoplasm by exportin-5 (Exp-5) and processed by another RNase III enzyme Dicer, which cuts out the loop region of the hairpin, releasing the mature miRNA:miRNA* duplex (Figure 1.1). After strand separation, the guide strand or mature miRNA is incorporated into a RISC, whereas the passenger strand, miRNA*, is usually degraded (Ha and Kim, 2014; Figure 1.1). One strand of an miRNA is incorporated into the Argonaute (AGO)-containing RISC (miRISC) and drives the RISC to bind target mRNAs, leading to translational repression and/or mRNA destabilization. RISC contains a key component of AGO family protein, named AGO2, that directly binds

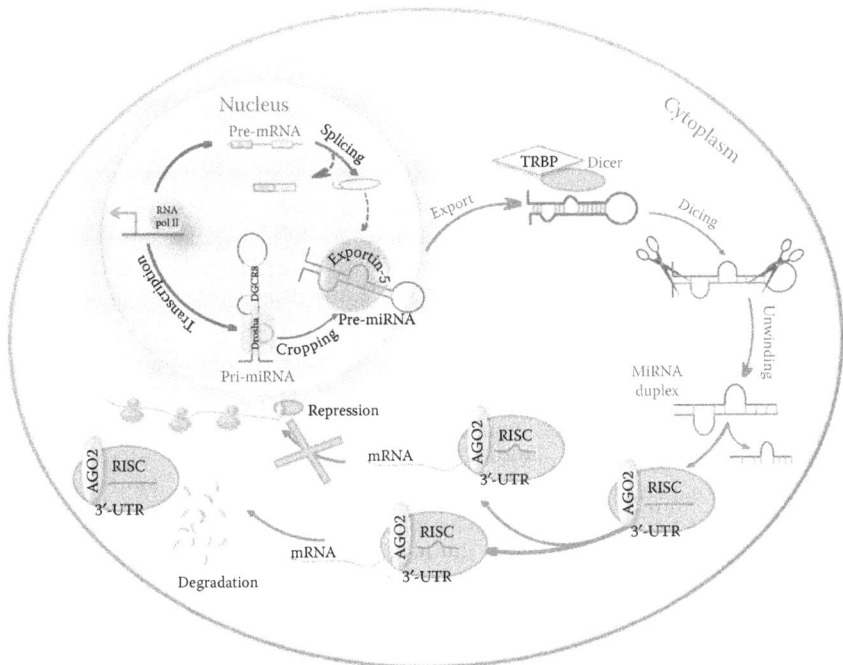

**FIGURE 1.1**   MicroRNAs biogenesis pathway.

the miRNA and the dsRNA-binding protein TRBP (Meister, 2013). The RISC core component AGO2 with trinucleotide repeat-containing gene TNRC6 A/B/C, associating with other RNA-binding protein as MOV10 and fragile X mental retardation protein (FMRP), participates in translational repression (Figure 1.1). The target mRNAs are recognized depending on the complementarities between positions 2–8 from the 5′ of miRNA (the seed sequence) and an miRNA responsive element (MRE), usually located within the 3′-UTR of mRNA (Bartel, 2009; Figure 1.1). In mammals, the interaction between miRNAs and MREs generally results in either the block of translation or the decay of the target mRNAs, which are deadenylated, decapped, and eventually degraded (Fabian et al., 2010; Fabian and Sonenberg, 2012). Emerging paradigms and mode of actions of miRNA regulatory network, as well as stepwise regulation of miRNA biogenesis, stability, and decay, have been described (Rüegger and Großhans, 2012; Ameres and Zamore, 2013; Meister, 2013).

The estimation of the number of known miRNAs is more than 1000 different miRNAs in human cells. Each miRNA has been predicted to target a large number of mRNAs, hundreds of genes for each miRNA, suggesting that a large population of the protein-coding genes may be somehow regulated by miRNAs (Friedman et al., 2013; Kozomara and Griffiths-Jones, 2014).

A single target mRNA is regulated by different miRNAs that bind on its 3′-UTR, indicating that miRNAs may act in a cooperative and/or combinatorial mode in the posttranscriptional regulation of an mRNA and in synergistic or antagonistic pathways (Ul Hussain, 2014). These features show that miRNAs are an increasing class of "small master regulatory molecules" that may add layer of complexity to the intricate regulatory networks necessary to govern complex biological processes, playing a pivotal role aimed to fine-tune gene expression.

## 1.3 NOMENCLATURE OF MiRNAs

In 2003, Ambros et al. suggested a uniform system for miRNA annotation and nomenclature. A comprehensive and searchable database, named miRNA Registry, assigning names prior to the publication of novel miRNA sequences, was published (Griffiths-Jones, 2004). Two years later, miRBase database takes over functionality from the miRNA Registry, to provide integrated interfaces to comprehensive miRNA sequence data and annotation (miRBase sequences), and links to a database of predicted gene targets. MiRBase is available at http://mirbase.org (Griffiths-Jones et al., 2006; Kozomara and Griffiths-Jones, 2014). For example, the miRNAs numbered to discover and experimentally confirm are composed of "miR" followed by a dash (miR-101). If the miRNA was discovered in humans or mouse, the hsa or mmu suffix indicates the organism (hsa-miR-101 or mmu-miR-101). The mature miRNA is denoted as miR-101 with a capitalized R while the uncapitalized mir-101 indicates the miRNA gene and the pre-miRNA. Mature miRNAs with identical sequences, which originate from discrete precursor sequences and genomic loci are indicated by a progressive number after the dash (e.g., hsa-miR-101-1 and hsa-miR-101-2), whereas the presence of one or two different nucleotides in the mature sequence is evidenced with a lettered suffix (e.g., mmu-miR-101a or mmu-miR-101b).

## 1.4 MiRNAs IN THE NERVOUS SYSTEM

MiRNAs are conserved during the evolution and abundantly expressed in the brain, where they have been found to play important roles in the regulation of brain function (Eacker et al., 2013). Neurons compartmentalize specific mRNAs in different subcellular compartments, and it was suggested that miRNAs might provide a unique system to spatially regulate neuronal gene expression (Vo et al., 2010). The temporally regulated expression of miRNAs suggests that miRNAs play an important role in brain development (Miska et al., 2004). MiRNAs are enriched in or unique to brain tissue, and selected miRNA subsets are expressed in specific brain area and neuronal and glial cell subtypes (Jovičić et al., 2013).

Given the complex architecture of miRNAs regulation in the brain, miRNAs are ideal candidates for temporally and spatially regulating several context-dependent functions in neurons, ranging from early neurogenesis and neuronal differentiation to dendritic morphogenesis and synaptic plasticity and from memory and behavior to cognition in the brain (Barry, 2014; Díaz et al., 2014; Earls et al., 2014; Follert et al., 2014; Smalheiser, 2014). The studies of miRNA expression profiles in the nervous system, comparative analysis of miRNA expression in the normal and pathologic brain represent the first step in understanding of the miRNAs regulation in the nervous system. Although there is clear evidence of the involvement of miRNAs in normal and pathologic brain, the specific role of miRNAs in different conditions, as well as the identification of the target mRNAs that are relevant to a specific phenotype, is only beginning to be defined. In addition, the complex composition of the brain, containing several neuronal and glial isotypes, while representing its main biological characteristic, is also the principal obstacle for an accurate analysis. In this chapter, we present an overview of general aspects of the involvement of miRNA in nervous system regulation in neuronal differentiation, plasticity, and cognition.

### 1.4.1 MiRNAs IN NEURONAL DEVELOPMENT AND DIFFERENTIATION

MiRNAs are essential players in the regulatory network involved in neuronal development and differentiation. Dicer knockout zebrafish resulted in asymmetric brain development and abnormal brain phenotype with severe morphological malformations, which could be rescued by the introduction of the miR-430 family of miRNAs (Giraldez et al., 2005). Brain-specific conditional Dicer knockout mouse model in neural stem cells led to a reduction of cortical structures (De Pietri Tonelli et al., 2008), suggesting that a loss of Dicer impaired neural stem cell expansion and differentiation (Andersson et al., 2010; Kawase-Koga et al., 2010). Dicer loss of function in post-mitotic neurons drives abnormal maturation characterized by reduced dendritic arborization in excitatory neurons (Davis et al., 2008), and apoptosis in dopaminergic and cerebellar Purkinje neurons (Schaefer et al., 2007). About the role of specific miRNAs in the developing nervous system, the first step of the investigation was the analysis of miRNA expression profile during different stages of neuronal development *in vitro* and *in vivo*. Among miRNAs highly enriched in the brain, miR-9 and miR-124 were evidenced. MiR-124 is the most abundant miRNA in the brain (50% of all miRNAs), and in mammals, it is encoded by three genes located on three different chromosomes (Lagos-Quintana et al., 2002; Deo et al., 2006). It is broadly expressed in all post-mitotic neurons in the adult mouse brain (Maiorano and Mallamaci, 2009; Åkerblom et al., 2012) during neuronal differentiation (Smirnova et al., 2005) and brain development (Krichevsky et al., 2006). In mammals, miR-124 controls the switch from non-neuronal to neuronal gene expression, repressing non-neuronal genes, and controlling neuronal identity. MiR-124 overexpression in human non-neuronal cells induces the downregulation of more than hundred non-neuronal mRNAs, determining a neuron-like expression profile playing a role in neuronal differentiation (Lim et al., 2005; Conaco et al., 2006). Moreover, in cortical neurons, miR-124 knockdown allows the increasing expression of non-neuronal mRNA transcripts (Conaco et al., 2006), whereas the miR-124 expression is necessary to maintain the neuronal identity during spinal cord development (Visvanathan et al., 2007). MiR-124 promotes neuronal differentiation, by silencing of non-neuronal transcripts, as CTD phosphatase-1 (SCP1), a component of REST complex and downregulates RE1-silencing transcription (REST) factor, a transcriptional repressor of neural genes in non-neuronal tissues (Visvanathan et al., 2007). Inversely, in non-neuronal cells and neuronal precursors, REST and SCP1 repress the expression of neuronal genes, among these, miR-124 is also repressed (Wu and Xie, 2006; Visvanathan et al., 2007), suggesting the presence of a negative feedback loop between REST/SCP1 and miR-124 for the transition from neural progenitors to post-mitotic neurons. Among genes controlled by miR-124 is SRY box containing gene 9 (Sox9; Cheng et al., 2009), lamin gamma 1 (LAMC1) and integrin beta 1 (ITGB1; Cao et al., 2007), Rho-associated coiled-coil-containing protein kinase 1 (ROCK1; Gu et al., 2014), early growth response gene 1 (Egr1; Wang et al., 2013), and the polypyrimidine tract-binding protein (PTBP1), which is an important splicing regulator that is substituted by its neuronal homolog, PTBP2 during neuronal differentiation (Makeyev et al., 2007). MiR-9 is one of the most abundantly expressed miRNAs in the E11.5 mouse brain during development. Shibata et al. (2008, 2011) showed that miR-9 modulates Cajal–Retzius cell differentiation by suppressing the expression of the transcription factor of the forkhead family FoxG1 in the developing mouse telencephalon. Moreover, miR-9 promotes the differentiation and fate of spinal motoneurons reducing the expression of FoXP1 (Otaegi et al., 2011; Luxenhofer et al., 2014). Foxg1 is almost absent, where miR-9 is highly expressed. In addition, miR-9-2/3 double mutants show multiple defects in telencephalon associated to alteration of Foxg1, nuclear receptor subfamily 2, group E, member 1 (Nr2e1), genomic screened homeobox 2 (Gsh2), and Meis homeobox 2 (Meis2) expression (Shibata et al., 2011). The nuclear receptor TLX is essential in the regulation of neural stem cell proliferation and promotion of differentiation. MiR-9 inhibits at the posttranscriptional level the expression of TLX, and TLX represses the expression of miR-9 by a feedback regulatory loop (Zhao et al., 2009). The orchestrated actions of miR-124 and miR-9 promote neuronal differentiation by repressing BAF53a, a subunit of the chromatin complex that is essential for post-mitotic neuronal development and dendritic morphogenesis (Yoo et al., 2009, 2011). In neuronal precursors, REST represses miR-124 and miR-9, and when REST is inactive, miR-124 and miR-9 repress their targets such as BAF53b, which results in a derepression of proneural genes (Conaco et al., 2006;

Packer et al., 2008). This triple negative switching in human fibroblasts converts them to neurons (Yoo et al., 2011).

### 1.4.2 MiRNAs at the Synapse

The experience-dependent plasticity is the process that drives the formation and persistence of memory. The synapses, the place where axons and dendrites meet, change their structure and function adapting to a changing environment. Neuronal, morphological, and biochemical changes that occur during synaptic plasticity and distinguish experienced (stimulated) from naive (unstimulated) synapses are necessary for the information storage capacity of the brain. The dendritic localization of selected mRNA and polyribosomes was suggested to be a mechanism for rapid dendritic protein synthesis, triggered by synaptic activity, aimed to long-term plasticity at the specific synapse. The presence of miRNAs at the synapse suggests that the posttranscriptional regulation of gene expression may contribute to the regulation of dendrites and spine architecture by modulating the expression of site-specific mRNAs.

### 1.4.3 MiRNAs and Dendritic Spines

The local protein synthesis occurs at dendritic spines. MiRNAs may participate in the mRNA-specific regulation of local translation by tuning gene expression at the posttranscriptional level. The profile of miRNAs content in synaptodendritic compartments was performed by analysis of synaptoneurosome, and a handful of miRNAs were associated with dendritic morphology by regulating cytoskeleton proteins, mRNA transport, and neurotransmission (Vo et al., 2005; Schratt et al., 2006; Wayman et al., 2008; Fiore et al., 2009; Siegel et al., 2009; Edbauer et al., 2010). Neuronal activity induced an increase of miR-132 (Vo et al., 2005; Wayman et al., 2008). The induction of miR-132 is regulated by cAMP response element-binding (CREB) protein during synaptogenesis, and its inhibition blocks dendritic morphogenesis (Wayman et al., 2008). The RNAi of Rho family GTPase-activating protein p250GAP, an miR-132 target, increase dendritic growth, while experiments performed with a p250GAP mutant unresponsive to miR-132 attenuates this activity (Wayman et al., 2008). MiR-132/p250GAP pathway also regulates a Rac1 activity and spine formation by the synapse-specific gene Kalirin-7 (Impey et al., 2010). Interestingly, another CREB-dependent miRNA, miRNA-212, is located 200 bases upstream from that of miR-132 and promoter regions of both miRNA contain CRE sequences (Magill et al., 2010; Remenyi et al., 2013). The deletion of the miR-212/132 locus caused a dramatic decrease in dendrite length, arborization, and spine density in newborn hippocampal neurons in young adult mice (Magill et al., 2010). Moreover, dendritic plasticity is modulated by miR-132 by controlling the expression of methyl CpG-binding protein 2 (MeCP2), a protein regulating neuronal development (Klein et al., 2007). Since brain-derived neurotrophic factor (BDNF) induces miR-212/132 expression, which targets MeCP2 that it was suggested to control BDNF expression, it is conceivable a feedback mechanism involving these molecules (Klein et al., 2007). In addition, the BDNF-induced expression of the glutamate receptors NR2A,

NR2B, and GluR1 in cortical neurons showed an increase in miR-132 expression, suggesting other route of involvement of miR-132 in synaptic functions (Kawashima et al., 2010). A recent study showed that FMRP interacts with RISC by associating with AGO in hippocampal neurons and that FMRP is linked among several miR-NAs, also with miR-132 and miR-125b (Edbauer et al., 2010). Mouse hippocampal neurons with FMRP knockdown showed that miR-132 effect on spine density was abolished, as well as the miR-125b action on spine width (Edbauer et al., 2010). MiR-125b targets NR2A, and also FMRP participates in the negative regulation of this N-methyl-D-aspartate (NMDA) receptor subunit, indicating that FMRP contributes to miRNA function during synapse development and that FMRP modulates NMDA function in mice (Pfeiffer and Huber, 2007). The involvement of miR-132 and miR-212 in memory formation and cognition is described in Section 1.4.4.

A brain-specific miRNA associated with the local protein synthesis in mammalian neurons is the miR-134 (Schratt et al., 2006). MiR-134 is localized at the synapse and negatively regulates dendritic spine size by repressing the translation of Lim-domain containing protein kinase 1 (Limk1) mRNA. The increase of miR-134 expression induces a reduction in dendritic spine size, and BDNF reduces the miR-134 repressive effect on Limk1 mRNA. Consistently BDNF treatment induced the translation of the 3′-UTR Limk1 mRNA luciferase reporter and did not affect the translation of the mutant reporter in which the miR-134 responsive element was abolished (Schratt et al., 2006). More recently, BDNF stimulation was shown to increase the transcription of miR-134 in neurons, by activation of myocyte-enhancing factor 2 (Mef2; Fiore et al., 2009). One possible explanation to reconciliate these findings is that miR-134 might to be regulated by a general and/or local synaptic mechanism. In addition, a fine regulation of miR-134 expression was evidenced for only specific cortical neurons, somatostatin and calretinin-positive interneurons (Chai et al., 2013). Searching for other miRNAs with a dendritic localization, Schratt and collaborators showed that miR-138 is abundantly expressed at the synapse, and negatively regulate spine size by posttranscriptional regulation of acyl-protein thioesterase 1 (APT1) mRNA (Siegel et al., 2009). On the other hand, Kosik's group identified, among several dendritic RISC-regulated mRNA, both Limk1 and Lypla1 (Banerjee et al., 2009).

### 1.4.4  RNA-INDUCED SILENCING COMPLEX, MiRNAs, AND COGNITIVE FUNCTIONS

Local protein synthesis at the synapse is required for synaptic plasticity and has been implicated in learning and memory. In 2006, the involvement of RISC in memory formation in *Drosophila melanogaster* was reported (Ashraf et al., 2006). In this animal model, CaMKII, a kinase required for memory, changes its expression after classical olfactory conditioning. The localization and translation of CaMKII mRNA to the antennal lobe synapses were dependent on the presence of the 3′-UTR. Putative miRNA-binding sites are present within the 3′-UTR of CaMKII. Searching for local protein synthesis component, the authors identify that the disruption of the silencing complex component Armitage leads to the removal of the miRNA-mediated repression of CaMKII. Synaptic activation drives an antiparallel expression of Armitage, which decreases, and of CaMKII, which increases. The lowering expression level of Armitage protein was due to the ubiquitin–proteasome pathway activity, suggesting a model characterized by miRNAs/CaMKII translational repression regulated by the activity-dependent proteasome-mediated degradation of Armitage (Ashraf et al., 2006).

The effect of RISC/AGO2 complex inactivation in the mouse brain was investigated (Batassa et al., 2010). Five different plasmids that express siRNA-targeting AGO2 mRNA and induce AGO2 downregulation were injected into the dorsal hippocampus of C57BL/6 mice. After 1 week of recovery, the study of treated mice engaged in hippocampus-related tasks showed that AGO2 silencing impaired both short-term memory and long-term contextual fear memory (Batassa et al., 2010). When AGO2 expression levels were rescued 3 weeks after AGO2 siRNA plasmid injection, memory recovered, indicating that the memory deficit was not due to a broad-spectrum impairment in hippocampal function. This was the first study showing a role for the RISC/AGO2 pathway in mammalian memory formation *in vivo*. The inducible neuronal deletion of Dicer in the adult mouse brain showed an improvement in several behavioral tests of learning and memory 12 weeks following the induction of Dicer deletion (Konopka et al., 2010). The Dicer deletion was associated with a progressive loss of a full set of brain-specific miRNAs. Neurons from mutated mice showed elongated filopodia-like dendritic spines and an increased expression of synaptic-related gene miRNA regulated (Konopka et al., 2010). From these data emerged that "less miRNAs" sustain a "better memory," however, the exact relationship between the miRNAs pathway and a molecular component of memory is debated (Konopka et al., 2010). Sirtuin1 (SIRT1) is essential for normal cognitive function and synaptic plasticity (Michán et al., 2010), and the SIRT1 knockout mice show a reduction in dendritic spine density in pyramidal neurons and an impairment of associative memory and spatial learning (Gao et al., 2010). In the SIRT1-KO hippocampus, a reduction in CREB protein, but not of its cognate mRNA, suggested the involvement of a posttranscriptional regulatory mechanism. MiR-134 was shown to regulate CREB expression. Significantly, an SIRT1 complex negatively regulates miR-134 transcription through its association with DNA sequences upstream of the pre-miR-134 sequence (Gao et al., 2010). Consistently, in the SIRT1-KO mice, miR-134 is upregulated and targets the CREB 3′-UTR. Since the inhibition of miR-134 in SIRT1-KO mice rescues LTP and partially rescues memory formation, the synaptic plasticity impairments observed in SIRT1 KO mice are partly due to miR-134 upregulation and the consequent inhibition of miR-134 target genes. SIRT1/miR-134 is a novel pathway regulating memory and plasticity in mammals (Gao et al., 2010).

MeCP2 is an miRNA target regulated by miR-132 (Klein et al., 2007), and the altered expression of MeCP2 has been associated with Rett syndrome, a neurodevelopmental disorder in which dendrite development and synaptogenesis are affected. MiR-132 overexpression in mice induces a decrease of MeCP2 expression in hippocampal neurons, an increase of dendritic spine density, and deficits in novel object recognition (Hansen et al., 2010). With the aim to evaluate whether miR-132 under normal physiological conditions could effectively be associated with cognition and not only the high level of transgenic miR-132 overexpression (Hansen et al., 2010), Hansen and colleagues performed *in vivo* experiments within the context of learning and memory (Hansen et al., 2013). The spatial memory tasks induced an increase of miR-132 expression in CA1, CA3, and CGL excitatory cell layers of the hippocampus. A doxycycline-regulated miR-132 transgenic mouse strain was created to permit to modulate, by varying concentrations of doxycycline added to the drinking water, several levels of miR-132 expression. Low levels of transgenic miR-132 expression were associated with enhanced cognitive capacity while supra-physiological miR-132 levels inhibited learning. These data showed that the functional involvement of miR-132 in normal memory formation depends on maintenance within a limited range of the miRNA expression levels (Hansen et al., 2013).

These findings described above focused on the importance of miRNA-mediated posttranscriptional regulation in several memory, learning, and behavioral paradigms, indicating an additional level of complexity localized at the synaptic level in neurons.

## 1.4.5   MiRNAs in Brain Disease

Significant progress on expression profile, role, and function of miRNAs has been made in neurological diseases, such as neurodegenerative and neuropsychiatric diseases, epilepsy, and brain neoplasia. It is difficult to determine whether the changes in miRNA expression detected in the brains or cerebrospinal fluid (CSF) of patients are primary or secondary events or both, as well as whether miRNAs are involved in early or late stage of the disease pathogenesis. The identification of miRNAs regulating the translation of target genes associated in brain diseases could represent the first step aimed to therapeutic applications (Cogoni et al., 2015). Nevertheless, unique patterns of miRNA expression profile in the CSF of specific neurological disease could be useful as molecular biomarkers for the diagnosis or therapy.

The identification of specific miRNA/target mRNA pathways potentially causing a specific pathology could open new therapeutic perspectives to block endogenous miRNAs or to treat with exogenous miRNAs (Table 1.2). To date either antisense oligonucleotides chemically modified or expressed sequences corresponding to multiple miRNA seed target (miRNA sponge) have been used as miRNA inhibitors (Ruberti et al., 2012). Experimental studies on small-RNA delivery methods to the central nervous system, avoiding toxicities, could be the challenge of future research (Table 1.2).

All correlations and future research between miRNA-mediated gene silencing and neurological diseases will be discussed in the next chapters of this book.

## 1.5   PERSPECTIVES

During the last years, significant progress has been made in the analysis of miRNAs expression in the nervous system. MiRNAs emerged recently in the neuronal regulation scene of gene expression, and they have helped many neurobiologists to partially understand one of the most fascinating phenomena of neuronal cell biology, the local protein synthesis. The small RNAs represent a versatile molecule that dynamically participates in a complex network of regulatory mechanisms involved during all phases of neuronal development, differentiation synaptic plasticity, memory

## TABLE 1.2
## How to Study MicroRNAs (miRNAs)

| Experimental Approaches | Toolbox |
|---|---|
| High-throughput screening of miRNA expression profile (cell specific, area specific, total brain) | • RNA sequencing<br>• MiRNA microarrays<br>• Real-time PCR |
| Bioinformatics tools for predicting miRNA targets | • Target-scan (www.targetscan.org; Lewis et al., 2005)<br>• DIANA microT (www.microrna.gr/microT; Paraskevopoulou et al., 2013)<br>• MiRanda (www.microrna.org; Enright et al., 2003) |
| MiRNA-target interaction assay | Luciferase reporter assay |
| Animal models expressing modified levels of miRNA | • Mutation of miRNA biogenesis genes pathway<br>• Transgenic mice expressing up/down levels of miRNA<br>• Virus-mediated site-specific manipulation of miRNA<br>• Administration of miRNA mimics/miRNAs inhibitor |

formation, and cognitive functions. Studies of miRNAs and the 3′-UTR of mRNAs highlight the complexity and significance of posttranscriptional regulation mediated by the 3′-UTR in mammalian gene expression (Tarang and Weston, 2014; Eulalio and Mano, 2015). A major effort and the combination of new technologies will be needed to define the role of RNA-mediated gene-silencing machinery in neurons, the neuronal miRNA targets, and specific components of RISC that are relevant in physiological and pathological conditions. The exploration of a new frontier of miRNAs biology in the nervous system likely represents just the tip of the iceberg compared to what we expect to learn in the next decade.

## ACKNOWLEDGMENTS

This work was supported by PNR-CNR Aging Program 2012–2016 (to FR) and Italian Ministry for Education, University, and Research in the framework of the Flagship Project NanoMAX (to CB).

## REFERENCES

Åkerblom M, Sachdeva R, Barde I, Verp S, Gentner B, Trono D, Jakobsson J. 2012. MicroRNA-124 is a subventricular zone neuronal fate determinant. *J Neurosci* 32:8879–89.

Ambros V, Bartel B, Bartel DP, Burge CB, Carrington JC et al. 2003. A uniform system for microRNA annotation. *RNA* 9:277–9.

Ameres SL, Zamore PD. 2013. Diversifying microRNA sequence and function. *Nat Rev Mol Cell Biol* 14:475–88.

Andersson T, Rahman S, Sansom SN, Alsiö JM, Kaneda M, Smith J, O'Carroll D, Tarakhovsky A, Livesey FJ. 2010 Reversible block of mouse neural stem cell differentiation in the absence of dicer and microRNAs. *PLoS One* 5:e13453.

Ashraf SI, McLoon AL, Sclarsic SM, Kunes S. 2006. Synaptic protein synthesis associated with memory is regulated by the RISC pathway in *Drosophila*. *Cell* 124:191–205.

Banerjee S, Neveu P, Kosik KS. 2009. A coordinated local translational control point at the synapse involving relief from silencing and MOV10 degradation. *Neuron* 64:871–84.

Barry G. 2014. Integrating the roles of long and small non-coding RNA in brain function and disease. *Mol Psychiatry* 19:410–6.

Bartel DP. 2009. MicroRNAs: Target recognition and regulatory functions. *Cell* 136:215–33.

Batassa EM, Costanzi M, Saraulli D, Scardigli R, Barbato C, Cogoni C, Cestari V. 2010. RISC activity in hippocampus is essential for contextual memory. *Neurosci Lett* 471:185–8.

Cao X, Pfaff SL, Gage FH. 2007. A functional study of miR-124 in the developing neural tube. *Genes Dev* 21:531–6.

Chai S, Cambronne XA, Eichhorn SW, Goodman RH. 2013. MicroRNA-134 activity in somatostatin interneurons regulates H-Ras localization by repressing the palmitoylation enzyme, DHHC9. *Proc Natl Acad Sci USA* 110:17898–903.

Cheng C, Pastrana E, Tavazoie M, Doetsch F. 2009. miR-124 regulates adult neurogenesis in the subventricular zone stem cell niche. *Nat Neurosci* 12:399–408.

Chiu H, Alqadah A, Chang C. 2014 The role of microRNAs in regulating neuronal connectivity. *Front Cell Neurosci* 7:283.

Cogoni C, Ruberti F, Barbato C. 2015. MicroRNA landscape in Alzheimer's disease. *CNS Neurol Disord Drug Targets* 14:168–75.

Conaco C, Otto S, Han JJ, Mandel G. 2006. Reciprocal actions of REST and a microRNA promote neuronal identity. *Proc Natl Acad Sci USA* 103:2422–7.

Davis TH, Cuellar TL, Koch SM, Barker AJ, Harfe BD, McManus MT, Ullian EM. 2008. Conditional loss of Dicer disrupts cellular and tissue morphogenesis in the cortex and hippocampus. *J Neurosci* 28:4322–30.

De Pietri Tonelli D, Pulvers JN, Haffner C, Murchison EP, Hannon GJ, Huttner WB. 2008. miRNAs are essential for survival and differentiation of newborn neurons but not for expansion of neural progenitors during early neurogenesis in the mouse embryonic neocortex. *Development* 135:3911–21.

Deo M, Yu JY, Chung KH, Tippens M, Turner DL. 2006. Detection of mammalian microRNA expression by *in situ* hybridization with RNA oligonucleotides. *Dev Dyn* 235:2538–48.

Díaz NF, Cruz-Reséndiz MS, Flores-Herrera H, García-López G, Molina-Hernández A. 2014. MicroRNAs in central nervous system development. *Rev Neurosci* 25:675–86.

Eacker SM, Dawson TM, Dawson VL. 2013. The interplay of microRNA and neuronal activity in health and disease. *Front Cell Neurosci* 7:136.

Earls LR, Westmoreland JJ, Zakharenko SS. 2014. Non-coding RNA regulation of synaptic plasticity and memory: Implications for aging. *Ageing Res Rev* 17:34–42.

Edbauer D, Neilson JR, Foster KA, Wang CF, Seeburg DP, Batterton MN, Tada T, Dolan BM, Sharp PA, Sheng M. 2010. Regulation of synaptic structure and function by FMRP-associated microRNAs miR-125b and miR-132. *Neuron* 65:373–84.

Eulalio A, Mano M. 2015. MicroRNA screening and the quest for biologically relevant targets. *J Biomol Screen* 20:1003–17.

Fabian MR, Sonenberg N. 2012. The mechanics of miRNA-mediated gene silencing: A look under the hood of miRISC. *Nat Struct Mol Biol* 19:586–93.

Fabian MR, Sonenberg N, Filipowicz W. 2010. Regulation of mRNA translation and stability by microRNAs. *Annu Rev Biochem* 79:351–79.

Fiore R, Khudayberdiev S, ChristensenM, Siegel G, Flavell SW, Kim TK, Greenberg ME, Schratt G. 2009. Mef2-mediated transcription of the miR379–410 cluster regulates activity-dependent dendritogenesis by fine-tuning Pumilio2 protein levels. *EMBO J* 28:697–710.

Follert P, Cremer H, Béclin C. 2014. MicroRNAs in brain development and function: A matter of flexibility and stability. *Front Mol Neurosci* 7:5.

Friedman Y, Balaga O, Linial M. 2013. Working together: Combinatorial regulation by microRNAs. *Adv Exp Med Biol* 774:317–37.

Gao J, Wang WY, Mao YW, Gräff J, Guan JS, Pan L, Mak G, Kim D, Su SC, Tsai LH. 2010. A novel pathway regulates memory and plasticity via SIRT1 and miR-134. *Nature* 466:1105–9.

Giraldez AJ, Cinalli RM, Glasner ME, Enright AJ, Thomson JM, Baskerville S, Hammond SM, Bartel DP, Schier AF. 2005. MicroRNAs regulate brain morphogenesis in zebrafish. *Science* 308:833–8.

Goldie BJ, Cairns MJ. 2012. Post-transcriptional trafficking and regulation of neuronal gene expression. *Mol Neurobiol* 45:99–108.

Griffiths-Jones S. 2004. The microRNA Registry. *Nucleic Acids Res* 32 (Database issue):D109–11.

Griffiths-Jones S, Grocock RJ, van Dongen S, Bateman A, Enright AJ. 2006. MiRBase: MicroRNA sequences, targets and gene nomenclature. *Nucleic Acids Res* 34 (Database issue):D140–4.

Gu X, Meng S, Liu S, Jia C, Fang Y, Li S, Fu C, Song Q, Lin L, Wang X. 2014. miR-124 represses ROCK1 expression to promote neurite elongation through activation of the PI3K/Akt signal pathway. *J Mol Neurosci* 52:156–65.

Ha M, Kim VN. 2014. Regulation of microRNA biogenesis. *Nat Rev Mol Cell Biol* 15:509–24.

Hansen KF, Karelina K, Sakamoto K, Wayman GA, Impey S, Obrietan K. 2013. miRNA-132: A dynamic regulator of cognitive capacity. *Brain Struct Funct* 218:817–31.

Hansen KF, Sakamoto K, Wayman GA, Impey S, Obrietan K. 2010. Transgenic miR132 alters neuronal spine density and impairs novel object recognition memory. *PLoS One* 5:e15497.

Impey S, Davare M, Lesiak A, Fortin D, Ando H, Varlamova O, Obrietan K, Soderling TR, Goodman RH, Wayman GA. 2010. An activity-induced microRNA controls dendritic spine formation by regulating Rac1-PAK signaling. *Mol Cell Neurosci* 43:146–56.

Jovičić A, Roshan R, Moisoi N, Pradervand S, Moser R, Pillai B, Luthi-Carter R. 2013. Comprehensive expression analyses of neural cell-type-specific miRNAs identify new determinants of the specification and maintenance of neuronal phenotypes. *J Neurosci* 33:5127–37.

Kawase-Koga Y, Low R, Otaegi G, Pollock A, Deng H, Eisenhaber F, Maurer-Stroh S, Sun T. 2010. RNAase-III enzyme Dicer maintains signaling pathways for differentiation and survival in mouse cortical neural stem cells. *J Cell Sci* 123:586–94.

Kawashima H, Numakawa T, Kumamaru E, Adachi N, Mizuno H, Ninomiya M, Kunugi H, Hashido K. 2010. Glucocorticoid attenuates brain-derived neurotrophic factor-dependent upregulation of glutamate receptors via the suppression of microRNA-132 expression. *Neuroscience* 165:1301–11.

Klein ME, Lioy DT, Ma L, Impey S, Mandel G, Goodman RH. 2007. Homeostatic regulation of MeCP2 expression by a CREB-induced microRNA. *Nat Neurosci* 10:1513–4.

Konopka W, Kiryk A, Novak M, Herwerth M, Parkitna JR et al. 2010. MicroRNA loss enhances learning and memory in mice. *J Neurosci* 30:14835–42.

Kozomara A, Griffiths-Jones S. 2014. miRBase: Annotating high confidence microRNAs using deep sequencing data. *Nucleic Acids Res* 42(Database issue):D68–73.

Krichevsky AM, Sonntag KC, Isacson O, Kosik KS. 2006. Specific microRNAs modulate embryonic stem cell-derived neurogenesis. *Stem Cells* 24:857–64.

Lagos-Quintana M, Rauhut R, Yalcin A, Meyer J, Lendeckel W et al. 2002. Identification of tissue-specific microRNAs from mouse. *Curr Biol* 12:735–9.

Lim LP, Lau NC, Garrett-Engele P, Grimson A, Schelter JM et al. 2005. Microarray analysis shows that some microRNAs downregulate large numbers of target mRNAs. *Nature* 433:769–73.

Luxenhofer G, Helmbrecht MS, Langhoff J, Giusti SA, Refojo D, Huber AB. 2014. MicroRNA-9 promotes the switch from early-born to late-born motor neuron populations by regulating Onecut transcription factor expression. *Dev Biol* 386:358–70.

Magill ST, Cambronne XA, Luikart BW, Lioy DT, Leighton BH, Westbrook GL, Mandel G, Goodman RH. 2010. MicroRNA-132 regulates dendritic growth and arborization of newborn neurons in the adult hippocampus. *Proc Natl Acad Sci USA* 107:20382–7.

Maiorano NA, Mallamaci A. 2009. Promotion of embryonic cortico-cerebral neuronogenesis by miR-124. *Neural Dev* 4:40.

Makeyev EV, Zhang J, Carrasco MA, Maniatis T. 2007. The microRNAmiR-124 promotes neuronal differentiation by triggering brain-specific alternative pre-mRNA splicing. *Mol Cell* 27:435–48.

Meister G. 2013. Argonaute proteins: Functional insights and emerging roles. *Nat Rev Genet* 14:447–59.

Michán S, Li Y, Chou MM, Parrella E, Ge H et al. 2010. SIRT1 is essential for normal cognitive function and synaptic plasticity. *J Neurosci* 30:9695–707.

Miska EA, Alvarez-Saavedra E, Townsend M, Yoshii A, Sestan N et al. 2004. Microarray analysis of microRNA expression in the developing mammalian brain. *Genome Biol* 5:R68.

Otaegi G, Pollock A, Hong J, Sun T. 2011. MicroRNA miR-9 modifies motor neuron columns by a tuning regulation of FoxP1 levels in developing spinal cords. *J Neurosci* 31:809–18.

Packer AN, Xing Y, Harper SQ, Jones L, Davidson BL. 2008. The bifunctional microRNA miR-9/miR-9* regulates REST and CoREST and is downregulated in Huntington's disease. *J Neurosci* 28:14341–6.

Pfeiffer BE, Huber KM. 2007. Fragile X mental retardation protein induces synapse loss through acute postsynaptic translational regulation. *J Neurosci* 27:3120–30.

Remenyi J, van den Bosch MW, Palygin O, Mistry RB, McKenzie C, Macdonald A, Hutvagner G, Arthur JS, Frenguelli BG, Pankratov Y. 2013. miR-132/212 knockout mice reveal roles for these miRNAs in regulating cortical synaptic transmission and plasticity. *PLoS One* 8:e62509.

Ruberti F, Barbato C, Cogoni C. 2012. Targeting microRNAs in neurons: Tools and perspectives. *Exp Neurol* 235:419–26.

Rüegger S, Großhans H. 2012. MicroRNA turnover: When, how, and why. *Trends Biochem Sci* 37:436–46.

Schaefer A, O'Carroll D, Tan CL, Hillman D, Sugimori M, Llinas R, Greengard P. 2007. Cerebellar neurodegeneration in the absence of microRNAs. *J Exp Med* 204:1553–8.

Schratt GM, Tuebing F, Nigh EA, Kane CG, Sabatini ME et al. 2006. A brain-specific microRNA regulates dendritic spine development. *Nature* 439:283–9.

Shibata M, Kurokawa D, Nakao H, Ohmura T, Aizawa S. 2008. MicroRNA-9 modulates Cajal-Retzius cell differentiation by suppressing Foxg1 expression in mouse medial pallium. *J Neurosci* 28:10415–21.

Shibata M, Nakao H, Kiyonari H, Abe T, Aizawa S. 2011. MicroRNA-9 regulates neurogenesis in mouse telencephalon by targeting multiple transcription factors. *J Neurosci* 31:3407–22.

Siegel G, Obernosterer G, Fiore R, Oehmen M, Bicker S et al. 2009. A functional screen implicates microRNA-138-dependent regulation of the depalmitoylation enzyme APT1 in dendritic spine morphogenesis. *Nat Cell Biol* 11:705–16.

Smalheiser NR. 2014. The RNA-centred view of the synapse: Non-coding RNAs and synaptic plasticity. *Philos Trans R Soc Lond B Biol Sci* 369:20130504.

Smirnova L, Grafe A, Seiler A, Schumacher S, Nitsch R, Wulczyn FG. 2005. Regulation of miRNA expression during neural cell specification. *Eur J Neurosci* 21:1469–77.

Tarang S, Weston MD. 2014. Macros in microRNA target identification: A comparative analysis of in silico, *in vitro*, and *in vivo* approaches to microRNA target identification. *RNA Biol* 11:324–33.

Ul Hussain M. 2014. Micro-RNAs (miRNAs): Genomic organisation, biogenesis and mode of action. *Cell Tissue Res* 349:405–13.

Visvanathan J, Lee S, Lee B, Lee JW, Lee SK. 2007. The microRNA miR-124 antagonizes the antineural REST/SCP1 pathway during embryonic CNS development. *Genes Dev* 21:744–9.

Vo NK, Cambronne XA, Goodman RH. 2010. MicroRNA pathways in neural development and plasticity. *Curr Opin Neurobiol* 20:457–65.

Vo N, Klein ME, Varlamova O, Keller DM, Yamamoto T et al. 2005. A cAMP-response element binding protein-induced microRNA regulates neuronal morphogenesis. *Proc Natl Acad Sci USA* 102:16426–31.

Wang X, Wang ZH, Wu YY, Tang H, Tan L et al. 2013. Melatonin attenuates scopolamine-induced memory/synaptic disorder by rescuing EPACs/miR-124/Egr1 pathway. *Mol Neurobiol* 47:373–81.

Wayman GA, Davare M, Ando H, Fortin D, Varlamova O et al. 2008. An activity-regulated microRNA controls dendritic plasticity by down-regulating p250GAP. *Proc Natl Acad Sci USA* 105:9093–8.

Wu J, Xie X. 2006. Comparative sequence analysis reveals an intricate network among REST, CREB and miRNA in mediating neuronal gene expression. *Genome Biol* 7:R85.

Yoo AS, Staahl BT, Chen L, Crabtree GR. 2009. MicroRNA-mediated switching of chromatin-remodelling complexes in neural development. *Nature* 460:642–6.

Yoo AS, Sun AX, Li L, Shcheglovitov A, Portmann T, Li Y, Lee-Messer C, Dolmetsch RE, Tsien RW, Crabtree GR. 2011. MicroRNA-mediated conversion of human fibroblasts to neurons. *Nature* 476:228–31.

Zhao C, Sun G, Li S, Shi YA. 2009. Feedback regulatory loop involving microRNA-9 and nuclear receptor TLX in neural stem cell fate determination. *Nat Struct Mol Biol* 16:365–71.

# Section II

MicroRNAs in the Nervous System and Neurological Diseases (Basic)

MicroRNAs in Neuropsychiatric Diseases

# 2 MicroRNAs in Mood and Anxiety Disorders

*Karen M. Ryan, Erik Kolshus, and*
*Declan M. McLoughlin*

## CONTENTS

2.1 Introduction to Mood and Anxiety Disorders .................................................21
2.2 Treatment of Mood and Anxiety Disorders....................................................22
2.3 Molecular Neurobiology of Mood and Anxiety Disorders ............................23
2.4 Role of MicroRNAs in the Brain....................................................................24
2.5 Systematic Review of MicroRNAs in Mood and Anxiety Disorders ............25
    2.5.1 Introduction .........................................................................................25
    2.5.2 Methods ................................................................................................25
2.6 MicroRNAs in Mood Disorders .....................................................................25
    2.6.1 Preclinical Studies ...............................................................................25
        2.6.1.1 MicroRNAs and Stress .........................................................29
        2.6.1.2 Depressive-Like Behavior and Antidepressant Therapies......31
    2.6.2 Clinical Studies....................................................................................37
        2.6.2.1 Clinical Studies in BPAD .....................................................37
        2.6.2.2 Clinical Studies in Depression..............................................40
        2.6.2.3 Postmortem Studies ..............................................................41
        2.6.2.4 Peripheral Tissue Studies......................................................45
        2.6.2.5 Genotyping and Rare Variants ..............................................46
    2.6.3 Summary ..............................................................................................47
2.7 Review of MicroRNAs and Anxiety Disorders...............................................47
    2.7.1 Preclinical Studies ...............................................................................47
    2.7.2 Clinical Studies in Anxiety Disorders.................................................49
        2.7.2.1 Genotyping and Rare Variants ..............................................50
        2.7.2.2 Expression Studies in Peripheral Blood................................50
    2.7.3 Summary ..............................................................................................52
2.8 Conclusions.....................................................................................................52
Acknowledgments......................................................................................................53
References...................................................................................................................53

## 2.1 INTRODUCTION TO MOOD AND ANXIETY DISORDERS

Mood and anxiety disorders are among the most common mental health disorders, with a huge individual, societal, and economic burden (Wittchen et al., 2011). Affective disorders are characterized by episodes of abnormally depressed or elated mood. Depression is frequently recurrent with lifetime prevalence rates in the region of 10%–20% (Kessler et al., 2012; Kessler and Bromet, 2013). It can be difficult to treat and is associated with increased suicide rates (Bostwick and Pankratz, 2000). Bipolar affective disorder (BPAD) is less common, with lifetime prevalence rates around 2.5% (Kessler et al., 2012). Anxiety disorders is an umbrella term for a number of disorders, including specific phobias, social phobias, posttraumatic stress disorder (PTSD), generalized anxiety disorder, panic disorder, agoraphobia, and obsessive-compulsive disorder (OCD). The overall lifetime prevalence of any anxiety disorder is estimated at 31.6% (Kessler et al., 2012). Both mood and anxiety disorders are often life-long conditions, which is reflected in the WHO's latest ranking of top 10 causes of Global Years Lived with Disability with depression ranking second and anxiety disorders seventh (Murray et al., 2012).

Unipolar depression is characterized by a clear period of low mood and/or anhedonia accompanied by a variable number of other symptoms, including "biological" symptoms such as disturbed sleep, appetite, activity levels and energy, as well as "psychological" symptoms such as poor concentration, feelings of guilt, worthlessness, and suicide (American Psychiatric Association, 2013). BPAD is characterized by alternating episodes of depression (as described above) and hypomanic or manic episodes that are defined by a period of persistently elevated, expansive, or irritable mood accompanied by a variable number of other symptoms, including increased self-esteem, racing thoughts, pressure of speech, distractibility, risk-taking behavior, and goal-directed behavior. Anxiety disorders all share a common theme of a heightened sense of arousal accompanied by physical and psychological symptoms that can be either generalized or linked to specific triggers. Physical symptoms include shortness of breath, palpitations, chest pain or discomfort, and choking sensations, while psychological symptoms include fear of dying, fear of losing control, and apprehension. Diagnosis of mood and anxiety disorders currently depends on clinical judgment, with no laboratory test to date having any practical utility.

There is considerable genetic influence on both mood and anxiety disorders. Heritability is ~39%–42% for unipolar depression (Flint and Kendler, 2014), 70%–80% for BPAD (Craddock and Sklar, 2013), and 30%–67% for anxiety disorders (Domschke and Deckert, 2012). Candidate genes for unipolar depression include *BDNF*, *5HTT*, *FKBP5*, *TPH2*, and *HTRA2* (Flint and Kendler, 2014). While no single gene with a large effect has been found for BPAD, genes such as *CACNA1C* and *ANK3* and genes involved in circadian rhythm have been associated with it (McCarthy et al., 2012). There has been little progress in identifying candidate genes for specific anxiety disorders. Therefore, although the evidence-base points to considerable genetic influence on both mood and anxiety disorders, current thinking is that genetic influences arise from many genes with small effect.

## 2.2   TREATMENT OF MOOD AND ANXIETY DISORDERS

Major breakthroughs in the 1950s and 1960s resulted in the development of tricyclic antidepressants (TCAs) and monoamine oxidase inhibitors (MAOIs) for depression and benzodiazepines for anxiety disorders. However, there have been few new major pharmacological breakthroughs since (Spedding et al., 2005).

The mainstay of pharmacological therapy for depression remains selective serotonin reuptake inhibitors (SSRIs), serotonin–noradrenaline reuptake inhibitors (SNRIs), TCAs, MAOIs, and others, which all aim to increase the amount of monoamines, such as serotonin or noradrenaline, at the synapse. However, the most acutely effective treatment for severe depression remains electroconvulsive therapy (ECT), which has been used since the 1930s (Group, 2003). Treatment of BPAD can involve the use of antidepressant medications during a depressive episode, but typically involves the use of mood stabilizers such as lithium and anticonvulsants for continuation therapy, and mood stabilizers and antipsychotics or benzodiazepines in the acute phase. Pharmacological treatments for anxiety disorders primarily target the gamma-aminobutyric acid (GABA) (benzodiazepines or pregabalin) or serotonergic (SSRIs, SNRIs, and TCAs) systems.

## 2.3 MOLECULAR NEUROBIOLOGY OF MOOD AND ANXIETY DISORDERS

One of the great challenges facing neuroscience is improving our understanding of the molecular basis of psychiatric disorders, including mood and anxiety disorders.

The accidental discovery that some antitubercular medications had antidepressant effects led to the development of the TCAs in the 1950s (Berton and Nestler, 2006). This laid the premise for early models of major depression, which postulated a "chemical imbalance" in monoamine neurotransmitters. Subsequent models of depression, that are also relevant in BPAD and anxiety disorders, include hypothalamic–pituitary–adrenal axis (HPA axis) dysregulation, inflammatory models, and neurotrophic models of affective disorders.

The monoamine hypothesis of depression states that depression is caused by a deficiency of monoamines in the brain and that antidepressant treatment will normalize these levels (Berton and Nestler, 2006). However, although catecholamine levels at the synapse increase acutely in response to treatment, a clinical response can take weeks or months. A number of neurotransmitters have been implicated in BPAD, chiefly serotonin and the catecholamines. Drugs that cause an increase in serotonin or catecholamines (e.g., antidepressants, amphetamine, and cocaine) can induce hypomanic or manic episodes (Manji et al., 2003). Initial pharmacological treatment for anxiety targeted the GABA system; however, SSRIs are the current first-line treatment for many anxiety disorders. Thus, monoamine neurotransmitters and their pathways clearly play a part in the pathophysiology of mood and anxiety disorders, although it has become clear they are only one part of a more complex system.

Hyperactivity of the HPA axis has long been linked to anxiety and affective disorders, particularly depression. In response to stress, the hypothalamus secretes corticotropin-releasing hormone (CRH), which acts on the anterior pituitary to release adrenocorticotropic hormone (ACTH). ACTH in turn stimulates cortisol release from the adrenal glands (Pariante and Lightman, 2008). Circulating cortisol normally triggers a negative feedback loop, inhibiting further release of CRH and ACTH. In many depressed patients, this negative feedback system appears to be dysregulated leading to chronically high levels of CRH, ACTH, and cortisol. CRH interacts with other neurotransmitter systems, including the catecholamines (Berton and Nestler, 2006). It is postulated that in patients vulnerable to developing depression, the glucocorticoid receptors (GRs) lose their inhibitory effect leading to loss of the negative feedback loop but the exact mechanism remains unclear. The HPA axis response to stress is also clearly involved in the pathophysiology of anxiety disorders.

Some of the first evidence linking the immune system to mood and anxiety disorders came from the observation that patients with immune system disorders had increased rates of psychiatric disorders and many of the key symptoms of depression are also seen in immune-related illnesses. In later years, patients undergoing interferon (a form of cytokine) therapy for cancers or viral infections were observed to have high rates of depression (Hoyo-Becerra et al., 2014). Elevated levels of some cytokines and other inflammatory markers have since been found in depressed patients (Dowlati et al., 2010). However, it is still not known how these cytokines and other inflammatory markers lead to psychiatric symptomatology.

In recent years, focus has shifted on to the neurotrophic model which suggests that drug treatments act by inducing neuroplastic changes, for example, hippocampal neurogenesis, synaptogenesis, increased dendritic spines, and dendrites. The importance of neuroplasticity is emphasized by evidence of loss of neurons and glia in the hippocampus and prefrontal cortex in depression. Brain-derived neurotrophic factor (BDNF) is widely expressed throughout the brain, promoting neuronal survival and maturation, synaptic plasticity, and synaptic function. Low levels of BDNF have been found in postmortem brains of depressed patients, and BDNF can exert antidepressant activity. Other potential therapeutic targets include vascular endothelial growth factor (VEGF) and the transcription factor CREB (cAMP response element–binding protein; Duman and Aghajanian, 2012).

There have been few truly new developments in the pharmacological treatment of mood and anxiety disorders and there is still a lack of understanding of the molecular basis of these disorders. Genetic and environmental factors both contribute to the development of these disorders, but how they do so, and how they may interact, remains unsolved. Recent findings implicate microRNAs (miRNAs) and other epigenetic changes as potential "micromanagers" of these changes (Dalton et al., 2014). MiRNAs may also address the problematic issue of numerous genes of small effect, as one miRNA can potentially regulate hundreds of genes (Kolshus et al., 2014). With diagnosis of psychiatric disorders still relying on clinical judgment, the search for a blood biomarker to aid diagnosis, prognosis, and response to treatment is very much on going. MiRNAs may ultimately offer one such possibility.

## 2.4  ROLE OF MicroRNAs IN THE BRAIN

About half of all known miRNAs are expressed in the brain (Landgraf et al., 2007; Shao et al., 2010), where they play a role in a variety of processes including cell proliferation (Delaloy et al., 2010; Liu et al., 2010; Niu et al., 2013), neurogenesis (Morgado et al., 2015; Rago et al., 2014), cell specification (Smirnova et al., 2005), and synaptic plasticity (Schratt et al., 2006; Gao et al., 2010), among other functions. MiRNAs show cell and tissue-specific expression (Landgraf et al., 2007) and play an important role in neural cell-type specification. The various cell types in the brain, that is, neurons, astrocytes, oligodendrocytes, and microglia have distinct miRNA profiles (Jovicic et al., 2013) and various neuronal subpopulation, for example, glutamatergic and GABAergic neurons, also show distinct miRNA profiles (He et al., 2012a). Moreover, miRNAs are found to be localized in many different subcellular compartments such as axons and synapses (Lugli et al., 2008; Natera-Naranjo et al., 2010; Sasaki et al., 2014). MiRNA expression can occur in a temporal fashion in both the developing brain (Miska et al., 2004; Mineno et al., 2006) and after the induction of neuronal activity (van Spronsen et al., 2013). Thus, miRNAs can specifically alter local gene expression profiles indicating their potentially unique roles in the brain. Since miRNAs participate in such a variety of cellular processes, changes in miRNA levels can have profound and wide-ranging effects. Several lines of evidence exist to suggest that changes in miRNA levels are involved in the development and treatment of neuropsychiatric conditions such as mood- and anxiety-related disorders which will be discussed in the following sections.

## 2.5  SYSTEMATIC REVIEW OF MicroRNAs IN MOOD AND ANXIETY DISORDERS

### 2.5.1  INTRODUCTION

A number of reviews of miRNAs in psychiatric disorders have been published but this is a rapidly changing field with ongoing developments, and few of the existing reviews have been systematic. Here we offer an up-to-date systematic review of both preclinical and clinical evidences for the role of miRNAs in mood and anxiety disorders.

### 2.5.2  METHODS

Relevant preclinical and clinical studies were identified using searches of PubMed/Medline and Web of Science up to July 2014 with the following terms cross-referenced with "miRNA": "depression," "antidepressant," "bipolar," "anxiety," "panic," "OCD," "PTSD," "phobia," "psychiatr*." No language limit was used. The references from included articles were also reviewed. The results are presented in Tables 2.1 through 2.6 and discussed in the following sections.

## 2.6  MicroRNAs IN MOOD DISORDERS

### 2.6.1  PRECLINICAL STUDIES

A review of the literature identified 24 articles which were relevant for inclusion here. Preclinical studies assessing the role of miRNAs in mood disorders have so far focused on four major areas, that is, the role of miRNAs in the stress response, in depressive-like behaviors, and in the mechanism of action of antidepressants and mood stabilizers. The findings from these studies are outlined in Table 2.1 and summarized below.

**TABLE 2.1**

## Preclinical Studies of MiRNAs in Stress

| Author | Species | Tissue | Analysis | Main Findings |
|---|---|---|---|---|
| Uchida et al. (2008) | Fischer 344 rats; chronic RS; SH-SY5Y cells | PVN | RT-qPCR; northern blotting | • MiR-18a ↓ GR protein *in vitro*<br>• ↑ pre-miR-18a and ↑ mature miR-18a in F344 rat PVN under nonstressed and chronic stress conditions<br>• ↓ GR protein in F344 rat PVN |
| Vreugdenhil et al. (2009) | Long-Evans rats; NS1, A549, COS-1 cells | Frontal cortex, hippocampus | RT-qPCR | • MiR-18 and miR-124a ↓ GR-mediated events and ↓ GR protein levels<br>• GC induced GILZ activity impaired by overexpression of miR-124a and miR-18<br>• ↑ miR-18 and miR-124a in frontal cortex and hippocampus during postnatal development |
| Kawashima et al. (2010) | Rat primary cortical cultures | — | RT-qPCR | • BDNF ↑ miR-132 *in vitro*<br>• Exogenous ds-miR-132 ↑ postsynaptic proteins (NR2A, NR2B, GluR1)<br>• Dexamethasone ↓ BDNF-induced ERK1/2 activation, miR-132 expression, and postsynaptic proteins |
| Meerson et al. (2010) | Male rats; acute/ chronic RS | Amygdala, hippocampal CA1 | Microarray; RT-qPCR | • Acute RS amygdala: ↑ miR-106b, miR-134, miR-183, miR-382; ↓ let-7a-1, miR-202, miR-361, miR-376b, miR-381, miR-9-1<br>• Acute RS hippocampus: ↑ miR-1-2, miR-376b, miR-182*, miR-424, miR-190, miR-19a, miR-208, miR-216, miR-32; ↓ let-7f-2, miR-124a-1, miR-138-1. miR-15b, miR-202, miR-422a, miR-9-1<br>• Chronic RS amygdala: ↑ miR-1-2, miR-15a, miR-190, miR-193, miR-208, miR-22, miR-322, miR-361, miR-369, miR-376b, miR-381; ↓ let-7a-1, let-7c, let-7f-1, let-7f-2, miR-103-1, miR-134, miR-138-1, miR-182, miR-216, miR-222, miR-298, miR-323, miR-34a, miR-368, miR-9-1, miR-96<br>• Chronic RS hippocampus: ↑ miR-132, miR-17-5p, miR-208, miR-23a, miR-369, miR-376b, miR-410; ↓ let-7c, let-7f-1, miR-100, miR-134, miR-148a, miR-16-1, miR-182*, miR-219-1, miR-22, miR-221, miR-30a-3p, miR-330, miR-376a, miR-9-1, miR-96<br>• MiR-134 and miR-183 altered by RS, target SC35 mRNA |

(Continued)

**TABLE 2.1 (Continued)**
**Preclinical Studies of MiRNAs in Stress**

| Author | Species | Tissue | Analysis | Main Findings |
|---|---|---|---|---|
| Rinaldi et al. (2010) | Male CD1 mice; acute/chronic RS | Frontal cortex | Microarray; northern blotting | • Acute RS: ↑ miR-9, miR-9*, miR-26b, miR-29b, miR-29c, miR-30b, miR-30c, miR-30e, miR-125a, miR-126-3p, miR-129-3p, miR-207, miR-212, miR-351, miR-487b, miR-690, miR-691, miR-709, miR-711, let7a-e; ↓ miR-423, and miR-494<br>• Repeated RS: ↑ miR-29b, miR-29c, miR-129-3p, miR-207, miR-212, miR-351, miR-423, miR-487b, miR-494, miR-690, miR-691, miR-709, miR-711; ↓ miR-9, miR-9*, miR-26b, miR-30b, miR-30c, miR-30e, miR-125a, miR-126-3p, let7 a-e |
| Uchida et al. (2010) | Sprague-Dawley rats; MD | Medial prefrontal cortex | RT-qPCR; northern blotting | • MD ↑ pre-miR-132 ,-124-1, 9-1, -9-3, -212, -29a<br>• MD ↑ RE-1-containing genes—Glur2, Nr1, CamKIIα, L1, Adcy5, Kcnc1<br>• MD ↑ mature miR-132, -124, -9, and -29a (all have RE-1 site)<br>• REST4 overexpression in neonatal mice ↑ pre-miR-132, -212, and -9-3 and ↑ CamKIIα, Glur2, Adcy5 |
| Bai et al. (2012) | Male Sprague-Dawley rats; MD | Hippocampus | RT-qPCR | • MD ↓ BDNF mRNA and protein<br>• MD ↑ miR-16<br>• BDNF positively correlates with depressive-like behaviors in FST and OFT<br>• MiR-16 negatively correlates with depressive-like behaviors in FST and OFT |
| Rodgers et al. (2013) | C57BL/6:129 F1 hybrid mice; CUPS of sires during puberty or adulthood | Sperm | Microarray | • Paternal stress ↑ miR-29c, miR-30a, miR-30c, miR-32, miR-193-5p, miR-204, miR-375, miR-532-3p, miR-698<br>• mRNA targets: DNMT3a, Tnrc6b, Mtdh<br>• Offspring have ↓ HPA axis stress responsivity |
| Zhang et al. (2013a) | Sprague-Dawley rats; MD, CUPS | Nucleus accumbens | RT-qPCR | • MD + CUPS ↑ miR-504<br>• ↑ miR-504 negatively correlates with DRD2 expression<br>• DRD2 expression negatively correlates with immobility in FST |

(Continued)

**TABLE 2.1 (Continued)**
**Preclinical Studies of MiRNAs in Stress**

| Author | Species | Tissue | Analysis | Main Findings |
|---|---|---|---|---|
| Zucchi et al. (2013) | Long-Evans rats; MS | Dams: frontal cortex; offspring: whole brain | Microarray; RT-qPCR | • Dams: ↑ 147 miRNAs, ↓147 miRNAs; ↓miR-329, miR-380, miR-20a, miR-500, let-7c, miR-23b, miR-181, miR-186; ↑ miR-24-1 <br>• Offspring: ↑ 205 miRNAs, ↓ 131 miRNAs; ↓ miR-361, miR-17-5p, miR-425, miR-345-5p, miR-505, miR-103, miR-151, miR-145; ↑ miR-23a, miR-129-2, let-7f, miR-98, miR-9, miR-216-5p, miR-667, miR-219-2-3p, miR-323 |
| Issler et al. (2014) | 5HT neuronal cultures; male C57Bl/6 mice | — | Microarray; RT-qPCR | • 5HT neurons: ↑ miR-375, miR-376c, miR-7a, miR-137, mghv-miR-M1-2, miR-709, miR-291b-5p, miR-1224, miR-1892, miR-702, miR-139-3p, miR-762, miR-671-5p, miR-483*; ↓ miR-691, miR-4661, miR-17, miR-376b, miR-124, miR-218, miR-128, miR-140*, miR-148a, miR-340-5p, miR-181c, miR-210, miR-135a, miR-27a, miR-452, miR-370, miR-300, miR-376a, miR-127, miR-15b, miR-101a, miR-16, miR-324-5p, miR-434-5p, miR-92a, miR-669i; <br>• miR-135a ↑ by SSRI antidepressants <br>• miR-135a overexpression in 5HT neurons ↓ anxiety- and depression-like behaviors in mice <br>• Knockdown of miR-135a ↑ anxiety- and depression-like behaviors in mice |

*Note:* PVN = paraventricular nucleus; GR = glucocorticoid receptor; RT-qPCR = real-time quantitative polymerase chain reaction; GILZ = glucocorticoid-induced leucine zipper; BDNF = brain derived neurotrophic factor; ERK1/2 = extracellular signal-regulated kinase 1/2; RS = restraint stress; MD = maternal deprivation; LH = learned helplessness; FST = forced swim test; OFT = open field test; HPA = hypothalamic–pituitary–adrenal axis; MS = maternal separation; CUPS = chronic unpredictable stress.

### 2.6.1.1 MicroRNAs and Stress

Stress is considered to be a precipitating factor for the development of many psychiatric illnesses. Studies examining the role of miRNAs in the stress response have focused primarily on the role of miRNAs in the regulation of the GR, the effects of early-life stress on miRNAs, and the impact of various psychological stressors on miRNAs.

Studies assessing the role of miRNAs in GR regulation indicate that miR-18 plays an important role in this process. Uchida et al. (2008), investigating the role of miRNAs in the stress response and vulnerability to repeated stress, were the first to demonstrate that GR mRNA translation is inhibited by miR-18a *in vitro*. MiR-18a was also found to be increased in the paraventricular nucleus of F344 rats, a strain hyperresponsive to stress, following repeated restraint stress. A subsequent *in vitro* study by Vreugdenhil et al. (2009) confirmed the role of miR-18 in GR regulation by showing that overexpression of miR-18 reduces GR protein levels, attenuates GR-mediated transactivation, and reduces the induction of glucocorticoid-induced leucine zipper, a GR target gene (Vreugdenhil et al., 2009). These findings suggest that miR-18-mediated downregulation of the GR may be important in susceptibility to stress-related disorders.

As previously mentioned, increased levels of glucocorticoids and decreased levels of BDNF are common features of depressive disorders. Interestingly, a study by Kawashima et al. (2010) demonstrated that treatment of primary cortical neuronal cultures with the synthetic glucocorticoid dexamethasone attenuates BDNF and the BDNF-induced upregulation of miR-132 and postsynaptic glutamate receptors (NR2A, NR2B, GluR1). Increased miR-132 is critical for BDNF-induced dendritic outgrowth (Vo et al., 2005). Chronic stress and exogenous glucocorticoids are known to induce dendritic atrophy (Woolley et al., 1990; Watanabe et al., 1992) and this study is the first to suggest a possible mechanism by which this may occur, although further studies are required to fully evaluate this.

The adverse effects of early-life stress are suggested to contribute to the development of depressive-like behavior but little is known about the molecular mechanisms underlying the vulnerability to stress. Epigenetic mechanisms, such as those induced by miRNAs, appear to play a role with evidence coming from a number of studies that have investigated changes in miRNA levels in prenatal stress models. The majority of studies have so far focused on gestational stress or maternal deprivation (MD) with only one study examining the effects of paternal stress on miRNAs. The various forms of early-life stress used in these studies all result in increased depressive- and anxiety-like behaviors in the offspring (Uchida et al., 2010; Bai et al., 2012; Rodgers et al., 2013; Zhang et al., 2013a; Zucchi et al., 2013) but no clear role for miRNAs in the development of these behaviors has yet emerged.

Using an MD model, where young rats were separated from the mother during the early postnatal period, Bai et al. (2012) demonstrated increased levels of miR-16 in the hippocampus of offspring, corresponding with a decrease in BDNF. Interestingly, miR-16 levels negatively correlated with depressive-like behaviors. This finding is noteworthy because, as discussed in detail later, miR-16 has been shown to be a critical mediator of antidepressant action (Launay et al., 2011). Uchida et al. (2010) also investigated miRNA changes in the MD rat model and found increased levels of a number of brain-enriched precursor miRNAs and mature miRNA species in the medial prefrontal cortex (PFC; see Table 2.1). There was no overlap in the findings of this study and the one conducted by Bai et al. However, one issue with these studies is that they each examined miRNA levels in different brain regions and tissue types. As discussed earlier, different brain regions and cell types display different miRNA profiles and so this may account for the lack of consistency in the findings to date.

Two studies have examined the effects of parental stress on offspring, one a study of gestational stress in the mother (Zucchi et al., 2013) and the other a study of paternal stress (Rodgers et al., 2013). Both studies found that parental stress has a profound impact on offspring. Using a model of maternal stress (MS) in rats where dams were subjected to stress during late gestation, Zucchi et al. (2013) showed alterations in miRNA levels in the brains of both the mother and the offspring. MS disrupted antepartum maternal behavior and induced alterations in miRNA levels in the frontal cortex of dams (Table 2.1), a region of the brain involved in maternal care. MS also impacted on miRNA levels in the brains of offspring from stressed dams (Table 2.1) and these miRNAs were shown to theoretically target mRNA species implicated in apoptosis, brain pathologies, neurotransmission, neurodevelopment, angiogenesis, cell signaling, and the stress response. The perinatal period represents a period of particular vulnerability for the developing brain. Thus, the findings from this study suggest that stress during gestation can modify the epigenetic signature of both the mother and offspring during critical periods of fetal brain development, which may result in life-long consequences in the offspring. Examining the role of paternal stress on miRNAs, Rodgers et al. (2013) demonstrated that paternal stress induces robust changes in mouse sperm miRNAs as outlined in Table 2.1. Interestingly, four of these miRNAs (miR-29c, miR-30a, miR-30c, and miR-204) are found to target DNMT3a (DNA methyltransferase 3a), a critical regulator of *de novo* DNA methylation. Offspring from these mice, which were bred following exposure of sires to pubertal or adulthood stress, had a blunted HPA axis response to acute restraint stress. Importantly, this study is the first to suggest that paternal stress exposure may be transmitted to future generations by inducing changes in miRNA levels in sperm. Overall, these studies suggest that alterations in miRNAs during early life can have life-long consequences for the organism.

One of the predominant findings from the preclinical studies conducted to date is that different stressors act by different molecular mechanisms and have varying effects on miRNAs. For instance, Bai et al. compared the effects of MD on miRNAs with that of chronic unpredictable stress (CUPS) and found that while CUPS induced more pronounced depressive-like behaviors, the underlying molecular mechanisms were different, that is, MD-induced depression was associated with changes in miR-16 and BDNF while CUPS was not. Combining MD, an early-life stressor, with later exposure to CUPS results in more pronounced depressive-like behaviors in rats than either MD or CUPS alone (Zhang et al., 2013a), implying that early-life stress enhances the vulnerability to stress in later life. This was associated with an increase in miR-504 and a decrease in levels of the dopamine receptor D2 (DRD2) in the nucleus accumbens of rats.

Two studies by Meerson et al. (2010) and Rinaldi et al. (2010) found that restraint stress modulates miRNA levels quickly but the effects are not long lasting. There was overlap in a number of miRNAs identified by these studies (let-7a, miR-9) but the direction of change was in opposition. However, it must be noted that these studies used different species, rats (Meerson et al., 2010) and mice (Rinaldi et al., 2010), and analyzed different brain regions, amygdala and hippocampus versus frontal cortex. The findings from these studies suggest that the stress-induced effects on miRNAs are both temporally and regionally specific.

In summary, the preclinical studies conducted to date indicate that miRNAs play a role in modulating the response to stress. Various psychological stressors seem to act through different molecular mechanisms and induce differing changes in miRNA levels. In addition, gestational or early-life stress can significantly impact on miRNA levels which may result in life-long consequences for the organism.

## 2.6.1.2   Depressive-Like Behavior and Antidepressant Therapies

Preclinical miRNA studies on depression have so far focused on the role of miRNAs in the development of depressive-like behaviors and in the mechanism of action of antidepressants and mood stabilizers. These are outlined in Table 2.2 and summarized below.

### 2.6.1.2.1   *MicroRNAs and Depressive-Like Behavior*

Only three studies have so far directly examined the role of miRNAs in the development of depressive-like behaviors. Using the MD model, Bai et al. (2014) demonstrated that anhedonia is associated with upregulation of miRNA let-7a in the hippocampus of rats corresponding with downregulation of serotonin receptor 4 (Htr4). CUPS was again shown to induce similar depressive-like behaviors to MD but did not induce alterations in Htr4 or let-7a levels, once again suggesting that different psychological stressors have different neurobiological mechanisms. Notably, anhedonia is thought to act as a predictor of poor response to SSRI treatment in depressed patients. Thus, based on previous observations, it is postulated that upregulated let-7a and downregulated Htr4 may be linked to decreased hippocampal neurogenesis which may in turn contribute to SSRI resistance in depression although the evidence to support this is lacking at present.

The BDNF and other neurotrophins have been closely linked to depression (Duman and Aghajanian, 2012). Social defeat stress–induced depressive-like behavior was shown by Bahi et al. (2014) to be associated with decreased levels of BDNF and increased levels of hippocampal miR-124a, an miRNA known to target BDNF. Interestingly, the authors found that overexpression of miR-124a in the hippocampus exacerbated depressive-like behavior, whereas silencing hippocampal miR-124a reduced depressive-like behavior supporting a role for miR-124a in the development of depressive-like behaviors.

An individual's ability to cope with stress is critical in the development of major depressive disorder (MDD). Using the learned helplessness (LH) model, an animal model of stress-induced behavioral depression, Smalheiser et al. (2011) found that nonlearned helpless (NLH) rats show a normal response to inescapable shock and significant alterations in miRNA levels in the frontal cortex (Table 2.2). Half of

## TABLE 2.2
## Preclinical Studies of MiRNAs in Depression

| Author | Species | Tissue | Analysis | Main Findings |
|---|---|---|---|---|
| | | | | **Depressive-Like Behavior** |
| Smalheiser et al. (2011) | Male Holtzman rats; LH | Frontal cortex | Microarray | • LH ↑ miR-200b, miR-300, miR-miR-184, miR-106b*, miR-297a*, miR-136*, miR-496, miR-211, miR-214*, miR-369-3p, miR-18a*, miR-466d-3p, miR-467a*, miR-376a*, miR-142-3p, SNORD65, miR-22, miR-181a-1*, miR-29c*, miR-376a; ↓ miR-384-5p, miR-350 |
| Bahi et al. (2014) | Male Wistar rats; social defeat stress | Frontal cortex, hippocampus | RT-qPCR | • Social defeat stress ↓ BDNF and ↑ miR-124a in hippocampus but not frontal cortex<br>• Overexpression of hippocampal miR-124a ↑ depressive-like behavior<br>• Knockdown of miR-124a ↓ depressive-like behavior |
| Bai et al. (2014) | Male Sprague Dawley rats; MD vs. CUPS | Hippocampus | RT-qPCR; western blotting | • MD but not CUPS ↑ let-7a<br>• Let-7a negatively correlates with sucrose preference rate<br>• Let-7a correlates with hippocampal Htr4 mRNA and protein |
| | | | | **Antidepressants** |
| Baudry et al. (2010) | 1C11 neuroectodermal cell line; Male Swiss-Kumming mice; Flx treatment | — | RT-qPCR | • ↑ miR-16 in NA neurons but not 5HT neurons<br>• ↓ miR-16 in NA neurons ↑ SERT<br>• Flx ↑ miR-16 and ↓ SERT in raphe nuclei *in vivo*<br>• S100β ↓ miR-16 and ↑ SERT in NA neurons and *in vivo*<br>• Overexpression of miR-16 *in vivo* ↓ depressive-like behaviors following CUPS |
| Launay et al. (2011) | Male Swiss-Kumming mice; Flx treatment | Hippocampus | RT-qPCR | • Flx ↓ depressive-like behaviors following CUPS, ↓ hippocampal miR-16, ↑ SERT, ↑ neurogenesis<br>• Knockdown of hippocampal miR-16 ↓ depressive-like behaviors<br>• BDNF, Wnt2 and PGJ2 act synergistically to ↓ miR-16 and ↑ SERT in hippocampus |

*(Continued)*

**TABLE 2.2 (Continued)**
**Preclinical Studies of MiRNAs in Depression**

| Author | Species | Tissue | Analysis | Main Findings |
|---|---|---|---|---|
| Huang et al. (2012) | Male C57BL/6J, B6.129-Kdr^tm1lrn/J mice; primary hippocampal neurons; EE or CMS | Hippocampus | RT-qPCR | • EE ↑ antidepressant-like effects in TST in CMS mice<br>• MiR-107 inhibits HIF-1α/VEGF/Flk-1 signaling and negatively regulates dendritic spine formation<br>• MiR-107 overexpression blocks EE-induced antidepressant-like effects in the TST and HIF-1α |
| O'Connor et al. (2013) | Male Sprague-Dawley rats; MD; Flx, Ket or ECS treatment | Hippocampus | Microarray | • Flx, ECS and Ket ↑ miR-598-5p using microarray in nonstress animals<br>• Flx and Ket ↑ miR-598 using RT-qPCR in nonstress animals<br>• MD ↓ miR-451 using microarray and RT-qPCR<br>• Flx, ECS and Ket reverse MD-induced ↓ miR-451 using microarray<br>• Flx reverse MD-induced ↓ miR-451 using RT-qPCR<br>• MiR-451 mRNA targets: CREB5, GABAa receptor associated protein, muscarinic cholinergic receptor 5 |
| Ryan et al. (2013) | Male Sprague-Dawley rats; acute and chronic ECS | DG, hippocampus, frontal cortex, cerebellum, whole blood | RT-qPCR | • ↑ miR-212 in DG following both acute and chronic ECS<br>• Positive correlation between miR-212 in DG and whole blood following chronic ECS |
| Smalheiser et al. (2014) | Male Sprague-Dawley rats; LH model; enoxacin treatment | Frontal cortex | RT-qPCR | • Enoxacin ↓ LH behavior following inescapable shock<br>• Enoxacin ↑ let-7a, miR-124, miR-125a-5p and miR-132 in frontal cortex |

(Continued)

**TABLE 2.2 (Continued)**
**Preclinical Studies of MiRNAs in Depression**

| Author | Species | Tissue | Analysis | Main Findings |
|---|---|---|---|---|
| Yang et al. (2014) | Male Sprague-Dawley rats; primary hippocampal neuronal cultures; Ket treatment | Hippocampus | Microarray | • Ket ↑ miR-30e-5p, miR-218a-5p, miR-181a-5p, miR-181c-5p, miR-136-5p, miR-487b-3p, miR-132-3p, miR-345-5p, miR-598-3p, miR-98-5p, miR-221-3p, miR-138-5p, miR-219a-5p, miR-495, miR-497-5p, miR-99a-5p, miR-29c-3p, miR-124-5p, let-7c-5p, miR-29a-3p, miR-488-3p, miR-365-3p; ↓ miR-150-5p, miR-344b-1-3p, miR-299a-5p, miR-206, miR-103-1-5p, miR-344b-2-3p, miR-935, miR-132-5p, miR-340-3p, miR-465-3p, miR-3557-5p, miR-22-5p, miR-485-3p, miR-1839-3p, miR-3568, miR-221-5p, miR-214-3p, miR-3596c<br>• miR-206 ↓ following Ket<br>• miR-206 overexpression attenuates Ket-induced ↑ BDNF |
| | | | | **Mood Stabilizers** |
| Zhou et al. (2009) | Male Wistar-Kyoto rats; primary hippocampal neurons; Li or VPA treatment | Hippocampus | Microarray | • Li and VPA ↑ miR-144, miR-136 and ↓ Let-7b, let-7c, miR-105, miR-128a, miR-24, miR-30c, miR-34a, miR-221<br>• Predicted mRNA targets of these miRNAs: CAPN6, DPP10, ESRRG, FAM126A, GRM7, THRB<br>• miR-34a regulates GRM7 protein *in vitro* |
| Creson et al. (2011) | Rat primary cortical neuronal cultures; Li treatment | — | Western blotting | • Let-7b negatively regulates M1 receptor protein<br>• Lithium ↑ M1 receptor protein |
| Zhang et al. (2013b) | HEK293 cells; VPA treatment | — | Microarray | • VPA ↓ DICER mRNA and protein<br>• VPA ↑ 49 miRNAs, ↓ 68 miRNAs |

*Abbreviations:* Flx = fluoxetine; NA = noradrenergic; 5HT = serotonergic; SERT = serotonin transporter; Li = lithium; VPA = valproate; Ket = ketamine; RT-qPCR = real-time quantitative polymerase chain reaction; TST = tail suspension test; CMS = chronic mild stress; VEGF = vascular endothelial growth factor; BDNF = brain-derived neurotrophic factor; ECS = electroconvulsive stimulation; MD = maternal deprivation; EE = environmental enrichment; DG = dentate gyrus; CUPS = chronic unpredictable stress; Htr4 = serotonin receptor 4; LH = learned helplessness.

these miRNAs were found to target Creb1. On the other hand, LH rats had a blunted response to inescapable shock compared to NLH rats and show aberrant miRNA levels. MiRNA alterations in NLH rats may be interpreted as a homeostatic response to minimize the effects of stress, in particular on Creb1, while the inability of LH rats to mount a homeostatic response to stress might be accounted for by their abnormal miRNA response.

No clear role for miRNAs in the development of depressive-like behaviors has emerged from the studies conducted to date. As mentioned previously, different psychological stressors appear to act by different neurobiological mechanisms. Thus, these studies highlight the difficulties in using various animal models to investigate the underlying molecular mechanisms of depression.

### 2.6.1.2.2  MicroRNAs and Antidepressants

The majority of the preclinical work conducted to date examining the role of miRNAs in depression has focused on miRNAs in the mechanism of action of antidepressant therapies. Initial evidence for the interaction between antidepressants and miRNAs came from two important studies of the SSRI fluoxetine (Baudry et al., 2010; Launay et al., 2011). Treatment with SSRIs typically takes weeks before symptomatic relief is achieved, suggesting that changes to serotonin signaling and downstream cascades are necessary for antidepressant action. MiR-16 was identified as a regulator of the serotonin transporter (SERT) through computer analysis. Using human neuroectodermal cell lines and *in vivo* study of the raphe nucleus (RN) of mice, the authors showed that fluoxetine increased levels of miR-16, which previously had been blocked by Wnt signaling pathways (Baudry et al., 2010). Increased miR-16 levels in turn led to decreased SERT levels, which would result in increased serotonin signaling at the synapse. In addition, miR-16 induced an adaptational change in locus coeruleus neurons, from noradrenergic to serotonergic type. MiR-16 could also alter behavior in depression models in mice. Following on from this study, the same group examined the role of miR-16 in hippocampal neurogenesis (Launay et al., 2011). Although fluoxetine increases miR-16 maturation in the RN, it decreased miR-16 levels in the locus coeruleus and hippocampus. These changes were mediated by BDNF, Wnt2, and the prostaglandin 15d-PGJ2.

Regulation of miRNAs by other antidepressant therapies has since been investigated by others. These studies have primarily focused on the role of miRNAs in the response to electroconvulsive stimulation (ECS) (O'Connor et al., 2013; Ryan et al., 2013), the animal model equivalent of ECT, and ketamine (O'Connor et al., 2013; Yang et al., 2014).

The first study to examine miRNAs in the mechanism of action of ECT showed that treatment of rats with ECS increases levels of the BDNF-associated miRNA miR-212 in the rat dentate gyrus (Ryan et al., 2013). A positive association was found between miR-212 levels in the dentate gyrus and in whole blood indicating that miRNA changes in the periphery can reflect changes occurring in the brain following antidepressant treatment. Subsequently O'Connor et al. (2013) investigated changes in miRNA levels in the MD model of depression following treatment with ECS, ketamine, or fluoxetine. There was no overlap in the miRNAs altered by ECS in this study or the study by Ryan et al., but the differences in time points used post-ECS in these studies might account for this. Treatment of MD rats with ECS, ketamine, or fluoxetine reversed MD-induced changes in miRNA levels (Table 2.2). MiR-451 was identified as a common target of all three antidepressants in MD rats. Interestingly, bioinformatic analysis revealed that miR-451 theoretically regulates genes involved in the CREB pathway and in GABAergic and cholinergic neuro-transmission. The role of miRNAs in the mechanism of action of ketamine was further examined by Yang et al. (2014). Following treatment of rats with ketamine, an miRNA screen of the hippocampus identified changes in the levels of 40 miR-NAs (Table 2.2); however, again there was very little overlap between the miRNAs identified in this study and that of O'Connor and colleagues. The authors went on to further investigate the role of one miRNA, miR-206. Ketamine downregulated levels of miR-206 both *in vivo* and *in vitro* leading to an upregulation of BDNF. This study suggests that miR-206 may underlie the antidepressive effects of ketamine, although a role for other miRNAs cannot be ruled out.

### 2.6.1.2.3 *MicroRNAs and Mood Stabilizers*

Only three studies have so far examined the role of miRNAs in the mechanism of action of mood stabilizers (Zhou et al., 2009; Creson et al., 2011; Zhang et al., 2013b). These studies have focused on two drugs in particular, namely lithium and valproate. The first study in this area was conducted by Zhou et al. (2009) who showed that lithium and valproate alter the levels of a number of shared miRNAs (Table 2.2). Interestingly, bioinformatic analysis revealed that these miRNAs target genes implicated in BPAD such as those involved in neurite outgrowth, neurogenesis, and signaling pathways. Notably, treatment of primary neuronal cultures with either lithium or valproate lowered levels of miR-34a and elevated levels of its target gene GRM7, a glutamate receptor-encoding gene which has previously been identified as a candidate gene for BPAD.

Perturbations of the cholinergic system and alterations in muscarinic acetylcholine receptor levels have been implicated in mood disorders. Lithium is known to impact on the cholinergic system. Based on the findings of Zhou et al. (2009) which showed lithium-induced downregulation of let-7b, Creson et al. (2011) went on to further investigate the effects of chronic lithium treatment on let-7b and its target the presynaptic M1 muscarinic receptor (M1). They found that let-7b negatively regulates levels of M1 *in vitro* and that lithium significantly increases M1 levels *in vivo* in the frontal cortex. However, while there was an increase in M1 in the frontal cortex following lithium let-7b was not measured. Thus, while these results are interesting, there is a lack of evidence to support the idea that lithium-induced effects on let-7b underlies its therapeutic actions *in vivo* and further investigations are needed.

Zhang et al. (2013b) also investigated the effects of valproate *in vitro* and found that it induces the proteasomal degradation of Dicer and alters the levels of numerous miRNAs (Table 2.2) of which there was some overlap with the findings of Zhou and colleagues. Unexpectedly, the authors report an upregulation of 49 miRNAs, despite valproate-induced Dicer depletion, suggesting that these miRNAs may be activated via their hosting genes or may be generated by a Dicer-independent mechanism.

While interesting, the results from these studies have primarily been generated from *in vitro* work. A lot of details have yet to be determined about the effects of mood stabilizers on miRNAs. Moreover, whether such changes have any functional or behavioral consequences need to be examined in order to fully establish a role for miRNAs in the mechanism of action of mood stabilizers and to evaluate their potential as future therapeutic targets.

## 2.6.2 CLINICAL STUDIES

### 2.6.2.1 Clinical Studies in BPAD

Our systematic review identified 10 articles that were relevant for inclusion here. There is considerable evidence that BPAD and schizophrenia have at least a partially shared molecular basis and genetic risk (International Schizophrenia Consortium et al., 2009). It is therefore not surprising that the majority of clinical studies of miRNAs in BPAD have also included samples of schizophrenia patients. A series of postmortem brain expression studies (Zhu et al., 2009; Kim et al., 2010; Moreau et al., 2011; Miller et al., 2012; Smalheiser et al., 2014) as well as several studies (Whalley et al., 2012; Cummings et al., 2013; Guella et al., 2013) investigating a single nucleotide polymorphism (SNP) identified in a large genome-wide association study (GWAS; Schizophrenia Psychiatric Genome-Wide Association Study, 2011) make up the bulk of clinical studies in BPAD patients. One study has looked at the plasma levels of miRNAs following treatment for mania (Rong et al., 2011), and another study profiled miRNA expression in patients with postpartum psychosis (PP), which is closely related to BPAD (Weigelt et al., 2013). These studies are summarized in Table 2.3 and briefly outlined below.

#### 2.6.2.1.1 Postmortem Studies

Six studies to date have investigated the expression of miRNAs in postmortem brain samples. They all included samples from the Stanley Medical Research Institute (SMRI) brain bank (Zhu et al., 2009; Kim et al., 2010; Moreau et al., 2011; Miller et al., 2012) or a subset of these found in the Stanley Neuropathology Consortium (Smalheiser et al., 2014). One study also utilized brain samples from the University of California, Irvine Brain Bank (UCI; Guella et al., 2013) alongside samples from the SMRI. The brain banks contain samples from patients with schizophrenia, BPAD, and healthy controls.

There is considerable variation in the results from these studies, although they differ in many methodological ways. Reflecting the technological advances in this field in a relatively short time frame, the first study in 2009 only looked at a single miRNA (Zhu et al., 2009). Although there were differences in patients with schizophrenia, these changes did not meet statistically significant criteria in the BPAD group. By 2010, researchers were able to investigate the expression of 667 different miRNAs, and successfully validated four upregulated miRNAs (Kim et al., 2010). In contrast, Moreau et al. (2011) investigating the expression of 435 miRNAs found 5.5% of these to be downregulated and Miller et al. (2012) failed to find any change in levels in 800 miRNAs in BPAD patients. The most recent miRNA profiling study found 9 miRNAs out of 377 to have altered levels (Smalheiser et al., 2014). A final study, based on the hypothesis that miR-137 is linked to BPAD failed to find any difference in miR-137 expression between BPAD or schizophrenia patients and controls (Guella et al., 2013).

## TABLE 2.3
### Clinical Studies of MiRNAs in Bipolar Affective Disorder (BPAD)

| Author | Patients | Tissue | Analysis | Main Findings |
|---|---|---|---|---|
| | | | | **Postmortem Case–Control Brain Studies** |
| Zhu et al. (2009) | Scz = 35 BPAD = 35 Controls = 34 | DLPFC (BA 46) | RT-qPCR | • Nonsignificant reduction in miR-346 levels in BPAD patients • Target genes/gene pathways: CSF2RA, GRID1; Cytokine, glutamate transmission |
| Kim et al. (2010) | Scz = 35 BPAD = 35 Controls = 35 | DLPFC (BA 46) | Microarray; RT-qPCR | • ↑ miR-145*, miR-133b, miR-154*, miR-889 • Targeted genes/gene pathways: GRIN, DRD1, DLG3/4, ITPR1, CEP290, HTT, SHANK3 |
| Moreau et al. (2011) | Scz = 35 BPAD = 35 Controls = 35 | PFC (BA 9) | RT-qPCR | • ↓ miR-330, miR-33, miR-193b, miR-545, miR-138, miR-151, miR-210, miR-324-3p, miR-22, miR-425, miR-181a, miR-106b, miR-193a, miR-192, miR-301, miR-27b, miR-148b, miR-338, miR-639, miR-15a, miR-186, miR-99a, miR-190, miR-339 |
| Miller et al. (2012) | Scz = 35 BPAD = 31 Controls = 34 | DLPFC (BA 46) | Microarray; RT-qPCR | • No significant differences in BPAD group |
| Guella et al. (2013) | Scz = 42 BPAD = 40 Control = 43 | DLPFC | RT-qPCR | • No significant difference in groups |
| Smalheiser et al. (2014) | Scz = 15 BPAD = 15 Controls = 15 | DLPFC (BA 10) | RT-qPCR | • ↑ miR-17-5p, miR-579, miR-106b-5p, miR-29c-3p; ↓ miR-145-5p, miR-485-5p, miR-370, miR-500a-5p, miR-34a-5p |

*(Continued)*

## TABLE 2.3 (Continued)
## Clinical Studies of MiRNAs in Bipolar Affective Disorder (BPAD)

| Author | Patients | Tissue | Analysis | Main Findings |
|---|---|---|---|---|
| | | | | **Genotyping Studies** |
| Whalley et al. (2012) | High risk groups of: Scz = 44 BPAD = 90 Controls = 81 | DNA (venous blood) | PCR; genotyping assay | • No significant difference between BPAD subjects and controls |
| Cummings et al. (2012) | Scz = 573 SA = 123 BPAD = 125 | DNA | PCR; genotyping assay | • No association between SNP and diagnosis<br>• Risk allele associated with fewer and mood-congruent symptoms |
| Guella et al. (2013) | Scz = 42 BPAD = 40 Control = 43 | DNA (brain) | PCR; genotyping assay | • No association between SNP and diagnosis<br>• Homozygous controls had lower miR-137 levels |
| | | | | **Other Studies** |
| Rong et al. (2011) | BPAD = 25 Controls = 21 | Whole blood | RT-qPCR | • ↓ plasma miR-134 in mania patients vs. controls<br>• Treatment with lithium led to increase in miR-134 levels<br>• Targeted genes/gene pathways: Limk1, dendritic spine size regulation |
| Weigelt et al. (2013) | PP = 20 Controls = 40 | Monocytes | microarray; RT-qPCR | • MiR-146a ↓<br>• ↓ miR-212 in those with history of BPAD<br>• Targeted genes/gene pathways: Inflammatory markers; IL-6, ADM, CD4+ T cell subsets |

*Note:* ↑/↓ in all studies refer to cases (BPAD/PP) compared to controls.

*Abbreviations:* BA = Brodmann's area; DLPFC = dorsolateral prefrontal cortex; PFC = prefrontal cortex; qRT-PCR = quantitative reverse transcriptase polymerase chain reaction; Scz = schizophrenia; SA = schizoaffective disorder; PP = postpartum psychosis.

Some of these studies found an overlap between BPAD and schizophrenia patients (Kim et al., 2010; Moreau et al., 2011; Smalheiser et al., 2014), whereas others did not (Zhu et al., 2009; Miller et al., 2012; Guella et al., 2013). Only two miRNAs were identified in more than one study (miR-106b and miR-145-5p), although in opposite directions.

What does one make of this apparent lack of consistency in these results to date? Although the majority of brain samples are from the same source different brain regions were examined, and different brain regions may express different levels of miRNAs. There is still a lack of consensus as to the best statistical significance testing approach, correction for multiple testing, normalization strategy, and what endogenous controls are best suited to this type of analysis (Liu et al., 2014). The postmortem intervals differ considerably among the samples in the SMRI which may have influenced miRNA expression levels (Smalheiser et al., 2014). These findings emphasize the importance of basic scientific principles of confirmation and validation of findings using different experimental techniques and cohorts.

### 2.6.2.1.2   Genotyping and Rare Variants

One of the largest GWAS to date in schizophrenia, with over 40,000 individuals, found that the strongest association of any SNP with schizophrenia lies within the intron of miR-137 (Schizophrenia Psychiatric Genome-Wide Association Study, 2011). There were also SNPs found in a number of miR-137 targets such as *TCF4*. Those with the TT risk allele in rs1625579 are presumed to be at higher risk of schizophrenia and given the overlap between schizophrenia and BPAD it may also be a risk allele for BPAD (Guella et al., 2013). Three studies to date have examined this SNP in BPAD (Whalley et al., 2012; Cummings et al., 2013; Guella et al., 2013), but none of these genotyping studies supported an association between rs1625570 SNP and BPAD. This is in line with previous expression studies.

### 2.6.2.1.3   Other Studies

MiR-134 has been suggested as a useful blood marker of clinical status in BPAD. In an initial study of drug-free BPAD individuals with mania and controls, miR-134 levels were significantly decreased in manic subjects (Rong et al., 2011). Although numbers were low and treatment was open label, this study supports miR-134 as a potential peripheral biomarker in BPAD.

Postpartum psychosis is a rare, but very severe postpartum disorder that can present with manic and psychotic symptoms. There are strong links between BPAD and PP and underlying molecular mechanisms may well be shared between these disorders (Weigelt et al., 2013). An initial profiling experiment in eight PP patients subsequently led to the validation of miR-146a being significantly decreased in PP patients compared to controls. In further analysis, miR-212 was also associated with a past diagnosis of BPAD in PP patients. The levels of these miRNAs correlated with inflammatory genes such as IL-6 (miR-212) and subsets of CD4 T cells. However, these results should be treated with caution as multiple comparisons were carried out with no correction.

### 2.6.2.2   Clinical Studies in Depression

Our systematic review identified 15 articles that were relevant for inclusion. Research investigating the role of miRNA involvement in depressive disorder is rapidly gathering pace with a number of clinical studies in recent years. A number of studies have investigated miRNA expression in the brain (Smalheiser et al., 2012, 2014; Issler et al., 2014; Lopez et al., 2014a,b). Others have focused on peripheral sources of miRNA, typically in the search of a biomarker for depression. These include studies investigating miRNA expression in cerebrospinal fluid (CSF; Launay et al., 2011), blood (Belzeaux et al., 2012; Bocchio-Chiavetto et al., 2013; Li et al., 2013; Issler et al., 2014; Lopez et al., 2014b), and dermal fibroblasts (Garbett et al., 2015). Searching for rare variants that may be associated with depression has also yielded interesting results (Saus et al., 2010; Xu et al., 2010b; He et al., 2012b; Guintivano et al., 2014; Jensen et al., 2014). These results are summarized in Table 2.4 and expanded upon below.

### 2.6.2.3  Postmortem Studies

The first study to examine miRNA expression levels in human brain was in a sample of 18 antidepressant-free suicide and 17 matched nonpsychiatric controls (Smalheiser et al., 2012). Using multiplex PCR, the authors found a downregulation of 21 miRNAs. Validated predicted targets of these miRNAs included VEGFA, BCL-2, and DMNT3B, but when their respective protein levels were measured in the same cohorts, only DMNT3B was significantly upregulated.

The same group went on to study the expression of miRNAs in a subset of SMRI brain samples, along with patients with schizophrenia, BPAD, and healthy controls (Smalheiser et al., 2014). MiR-508-3p and miR-152-3p were both significantly downregulated but no correction for multiple testing was carried out.

Building on previous work, identifying the polyamine genes SAT1 and SMOX as playing a role in suicidal behavior, Lopez et al. (2014a) investigated PFC levels of miRNAs predicted to target these genes. Four miRNAs were upregulated in suicide completers. Two of these, miR-34c-5p and miR-320c, had a significant negative correlation with mRNA levels of SAT1, and miR-139-5p and miR-320c had a significant negative correlation with SMOX mRNA levels. The protein products of these mRNA transcripts were not measured, and multiple comparison correction was omitted.

The same group went on to examine miRNA expression in a larger sample of brains from the Douglas-Bell Canada Brain Bank in Quebec (Lopez et al., 2014b). A microarray-based approach tested for 866 miRNAs and found miR-1202 to be significantly decreased in the PFC of 14 depressed subjects compared to 11 controls. This miRNA was brain enriched and is only present in humans and primates. These findings were further validated with qRT-PCR in a sample of depressed subjects with a history of antidepressant use and controls. Interestingly, miR-1202 expression in those who had a history of antidepressant use was significantly different from those not exposed to antidepressants, with expression levels more similar to healthy controls. *GRM4*, a glutamate receptor-encoding gene, was predicted in silico to be targeted by miR-1202. *GRM4* levels were increased in these brain samples and negatively correlated with miR-1202 levels.

A final study, which investigated the role of miR-135 in mouse models went on to investigate levels of miR-135 and miR-16 in various subnuclei of the raphe in the brains of depressed suicide victims and controls (Issler et al., 2014). Significantly

## TABLE 2.4
## Clinical Studies of miRNAs in Depression

| Author | Patients | Tissue | Analysis | Main Findings |
|---|---|---|---|---|
| | | | | **Postmortem Case–Control Brain Studies** |
| Smalheiser et al. (2012) | Suicides = 18 Controls = 17 | PFC (BA 9) | RT-qPCR | • ↓ miR-142-5p, miR-137, miR-489, miR-148b, miR-101, miR-324-5p, miR-301a, miR-146a, miR-335, miR-494, miR-20a/b, miR-376a, miR-190, miR-155, miR-660, miR-130a, miR-27a, miR-497, miR-10a, miR-142-3p<br>• Targeted genes/gene pathways: CDK6, ELF1/6, NCOA2, DNMTB3, EZH2, MYCN, ICOS, SOX4, PTPRN2, MERTK, VEGFA, SLC16A1, SFRS11, TTK, AGTR1, BACH1, LDOC1, MATR3, TM6SF1, TAC1, CSF1, MAFB, MEOX2, HOXA1/5, SP1/3/4, RUNX1 |
| Smalheiser et al. (2014) | Scz = 15 BPAD = 15 Dep = 15 Controls = 15 | PFC (BA 10) | RT-qPCR | • ↓ miR-508-3p and miR-152-3p |
| Lopez et al. (2014a) | Dep = 15 Controls = 16 | PFC (BA44) | RT-qPCR | • ↑ miR-34c-5p, miR-139-5p, miR-195, miR-320c<br>• Targeted genes/gene pathways: SAT1, SMOX |
| Lopez et al. (2014b) | Dep = 64 Controls = 40 | PFC (BA44) | Microarray; RT-qPCR | • ↓ miR-1202<br>• Targeted genes/gene pathways: GRM4 |
| Issler et al. (2014) | Dep = 6 Control = 11 | Raphe nuclei | RT-qPCR | • ↓ miR-135a, miR-16 |
| | | | | **Peripheral Tissue Studies** |
| Launay et al. (2011) | MDD = 9 | CSF | RT-qPCR | • Following fluoxetine administration, miR-16 targeting molecules BDNF, Wnt2, 15d-PGJ2 levels ↑ in CSF<br>• Targeted genes/gene pathways: miR-16: SERT, Bcl-2 |

*(Continued)*

**TABLE 2.4 (Continued)**
## Clinical Studies of miRNAs in Depression

| Author | Patients | Tissue | Analysis | Main Findings |
|---|---|---|---|---|
| Bocchio-Chiavetto et al. (2012) | MDD = 10 | Whole blood | RT-qPCR | • ↑ miR-130b*, miR-505*, miR-29-b-2*, miR-26a/b, miR-22*, miR-664, miR-494, let7d/e/f/g, miR-629, miR-106b*, miR-103, miR-191, miR-128, miR-502-3p, miR-374b, miR-132, miR-30d, miR-500, miR-589, miR-183, miR-574-3p, miR-140-3p, miR-335, miR-361-5p<br>• ↓ miR-34c-5p, miR-770-5p<br>• Targeted genes/gene pathways: BDNF, NR3C1, NOS1, IGF1, FGF1, FGFR1, VEGFa, GDNF, CACn41C, CACNB4, SLC6A12, SLC8A3, GABRA4, 5HT-4. Neuroactive ligand–receptor interaction, axon guidance, LTP, signaling pathways |
| Belzeaux et al. (2012) | MDD = 9<br>Controls = 9 | PBMCs | Microarray; RT-qPCR | • ↑ miR-941, miR-589 |
| Li et al. (2013) | MDD = 40<br>Controls = 40 | Serum | RT-qPCR | • ↑ miR-132, miR-182<br>• Targeted genes: BNDF |
| Lopez et al. (2014b) | MDD = 32<br>Controls = 18 | Whole blood | RT-qPCR | • ↓ miR-1202 |
| Issler et al. (2014) | MDD = 11<br>Controls = 12 | Whole blood | RT-qPCR | • ↓ miR-135a |
| Garbett et al. (2014) | MDD = 16<br>Controls = 16 | Dermal fibroblasts | PCR array | • ↑ miR-132, miR-421, miR-542, miR-450a, miR-16-2*, miR-424, miR-628-3p, miR-629-5p, miR-4293, miR-661, miR-3909, miR-33a*, miR-135b, miR-7, miR-4267, miR-548a-3p, miR-548d-3p, miR-613, miR-3714, miR-1294, miR-429<br>• ↓ miR-122, miR-32, miR-196b*, miR-377, miR-193a-3-, miR-337-5p, miR-675*, miR-3176, miR-21*, miR-22, miR-425*, miR-185, miR-296-5p, miR-103a, miR-107, miR-186, miR-887 |

*(Continued)*

## TABLE 2.4 (Continued)
## Clinical Studies of miRNAs in Depression

| Author | Patients | Tissue | Analysis | Main Findings |
|---|---|---|---|---|
| | | | | **Genotyping Studies** |
| Xu et al. (2010) | MDD = 1088<br>Controls = 1102 | DNA | MiR-SNP | • Positive association between SNP in miR-30e precursor and MDD |
| He et al. (2012) | MDD = 314<br>Controls = 252 | DNA | MiR-SNP | • SNP in miRNA processing gene DGCR8 increased frequency<br>• SNP in miRNA processing gene AGO1 decreased frequency<br>• Associated with suicide risk and treatment response |
| Jensen et al. (2014) | 6725 subjects | DNA | Microarray | • Association between SNP in target site of miR-330-3p in MDD |
| Guintivano et al. (2014) | PM brains<br>MDD = 29<br>BPAD = 40<br>Scz = 29<br>Controls = 70<br>Blood<br>MDD = 75<br>BPAD = 15<br>Controls = 308 | DNA (PM brain, blood) | Microarray, pyrose-quencing | • Association between SNP in SKA2 and suicide, possibly mediated by miR-301a |

*Note:* ↑/↓ in all studies refer to cases (depressed) compared to controls.

*Abbreviations:* BA = Brodmann's area; BPAD = bipolar affective disorder; DLPFC = dorsolateral prefrontal cortex; LTP = long-term potentiation; MDD = major depressive disorder; PBMCs = peripheral blood mononuclear cells; PFC = prefrontal cortex; PM = postmortem; RT-qPCR = real-time quantitative polymerase chain reaction; Scz = schizophrenia.

lower levels of miR-135 and miR-16 were found in the dorsal raphe and raphe magnus compared to controls.

As discussed in the BPAD section, postmortem studies have significant limitations and to date the results in depressed cohorts suffer from a similar lack of correlation with other postmortem studies.

### 2.6.2.4   Peripheral Tissue Studies

Correlating miRNA level changes before and after treatment and matching them to clinical outcomes offers an exciting potential for biomarkers as well as the molecular basis of depression. For biomarkers to be clinically relevant, they need to be easily accessible from sources such as peripheral blood. Initial work using CSF had highlighted that targets of miR-16, previously implicated in the action of the antidepressant fluoxetine (Baudry et al., 2010), were increased following fluoxetine treatment (Launay et al., 2011). This highlighted the translational potential of miRNAs.

The first study investigating the effect of blood miRNA changes following treatment involved 10 treatment-naive depressed patients who showed a good response to 12 weeks of the antidepressant escitalopram (Bocchio-Chiavetto et al., 2013). Thirty miRNAs were significantly altered, and many of these are important gene expression regulators in the brain and have been implicated in other psychiatric disorders.

A further study investigating transcriptional signatures at different time points in depression identified two miRNAs that were upregulated at remission compared to healthy controls (Belzeaux et al., 2012). A panel of eight miRNAs was also identified that matched clinical response.

Li et al. (2013) investigated levels of two miRNAs that are thought to regulate BDNF, miR-132, and miR-182. In depressed patients, compared to controls, BDNF serum levels were lower, and miR-132 and miR-182 were upregulated. However, only miR-132 was significantly negatively correlated with BDNF levels. Of note, there was a significant positive correlation between the levels of these miRNAs and scores on a self-rating scale of depression, but no correction for multiple testing was performed.

Previously, we presented the evidence for the involvement of miR-1202 in depression based on postmortem findings (Lopez et al., 2014b). In a further experiment, miR-1202 blood levels were measured before and after treatment with the antidepressant citalopram. Compared to controls, miR-1202 levels were downregulated, matching the results seen in postmortem brain samples. Furthermore, after classifying the patients into those who remitted, and those who did not, the lower miR-1202 levels were specific to remitters, indicating the potential of miR-1202 to identify those who will respond to citalopram treatment.

In a similar vein, Issler et al. (2014) built on work investigating the role of miR-135 and miR-16 in mouse models and measured their levels in brain samples and in human blood. Compared to controls, miR-135a levels were significantly decreased. Of note, after 3 months of cognitive behavioral therapy, there was a significant increase in blood miR-135a levels compared to patients receiving the antidepressant escitalopram.

Finally, both mRNA and miRNA levels in dermal fibroblasts were investigated using a PCR array examining 1008 miRNAs (Garbett et al., 2015). Thirty-eight miRNAs were differentially expressed. Using miRNA-targeting software, 89% of the 38 miRNAs targeted at least one mRNA that was differentially expressed, indicating a close relationship between miRNA and mRNA networks. There was no correction for multiple testing, and culturing of fibroblasts represents a more time-consuming and delayed source of biomarkers in comparison to blood.

These studies in general need to be validated, and many may have a high false discovery rate. Some studies have investigated the role of specific miRNAs based on preclinical, in silico, or other tissues, such as miR-1202, miR-135a, and miR-132. MiR-132 was also found to be upregulated in dermal fibroblasts in depressed patients, whereas the others have not been identified in general profiling studies. Questions also arise over whether changes in miRNA levels in the periphery give any helpful information about what is happening in the brain (Kolshus et al., 2014). However, as any clinically useful biomarker for depression is likely to come from these peripheral sources, continued efforts in this field would be welcomed.

### 2.6.2.5   Genotyping and Rare Variants

A number of studies have searched the human genome for SNPs that may be associated with depression. One such large study of 1088 depressed patients and 1102 controls found a positive association between miR-30 and MDD (Xu et al., 2010b). This was carried out in an ethnically homogenous Han Chinese population (Xu et al., 2010a).

A key symptom in affective disorders is disturbed sleep, and disruption of circadian rhythms has been associated with depression (Germain and Kupfer, 2008). An SNP in miR-182 was found to be associated with late insomnia in 359 patients (341 controls) with MDD (Saus et al., 2010). Patients with this SNP had downregulated expression of genes previously associated with affective disorders and circadian rhythm such as *CLOCK* (Serretti et al., 2003).

SNPs in the genes involved in the miRNA processing machinery, like *DGCR8* and *AGO1*, were associated with increased risk of suicidal tendency and antidepressant treatment response in a sample of 314 patients and 252 controls (He et al., 2012b).

Using previous GWAS data and miRNA target prediction software, a recent publication identified a link between an miR-330-3p target site SNP (rs41305272) in mitogen-activated protein kinase 5 (MAP2K5) and restless leg syndrome (Jensen et al., 2014). This disorder is frequently comorbid with anxiety and depression, and this SNP was therefore investigated in a separate GWAS data set. This data set consisted of 6725 unrelated drug-dependent subjects and nondrug-dependent subjects recruited for a genetic study of dependence. There was an association between rs41305272 and MDD (OR = 2.64, $p = 0.01$) in subjects of African-American descent. However, these results should be treated with caution, as comorbidity with other disorders was very common in this sample.

Finally, a study investigating epigenetic and genetic markers of suicide has identified a potential role of miR-301a in suicide etiology (Guintivano et al., 2014). Incorporating postmortem and blood samples in various diagnostic groups and patient cohorts, the authors found an association between rs7208505, an SNP in a CpG site in the SKA2 gene, and suicide. In suicide cases, there was a higher level of DNA methylation of SKA2, as well as lower levels of SKA2 gene expression, associated with the SNP. Further analysis of related regions identified a possible role for miR-301a in SKA2 gene expression, although not in the suicide phenotype itself. MiR-301a has previously been found to be associated with suicide (Smalheiser et al., 2012).

### 2.6.3  SUMMARY

In summary, there has been a rapid growth in studies of miRNA involvement in depression as evidenced by expanding reviews in this field (Dwivedi, 2011; Mouillet-Richard et al., 2012; Kolshus et al., 2014; Maffioletti et al., 2014). Findings from preclinical studies are promising in terms of teasing out the contribution of miRNAs to depressive-like behaviors and the therapeutic antidepressant response. Moreover, the translational aspect of findings from preclinical studies is promising. However, despite a growth in miRNA studies in this field, there has been little progress in the way of validation and replication, and there are large variations in methodological approaches including normalization strategies (Liu et al., 2014). It is encouraging to see studies beginning to explore correlations between the brain and the periphery (Guintivano et al., 2014; Lopez et al., 2014b). However, a lot more work is needed in this area in order to fully evaluate the role of miRNAs in depression and their potential to act as biomarkers for diagnosis and treatment.

## 2.7  REVIEW OF MicroRNAs AND ANXIETY DISORDERS

### 2.7.1  PRECLINICAL STUDIES

Anxiety disorders are a heterogeneous collection of disorders with varying prevalence and presentation. Investigations into the role miRNAs play in these disorders are at an early stage with only a small number of preclinical studies published to date. A review of the literature identified 8 studies that were relevant for inclusion here. These are outlined in Table 2.5 and summarized below.

The first preclinical indication of a role for miRNAs in anxiety came from a study by Parsons et al. (2008). The authors examined basal miRNA levels in the hippocampus of four inbred mouse strains and showed 11 miRNA species significantly altered between strains. Of these, miR-34c, miR-323, and miR-212 correlated with behavioral measures for anxiety. MiR-34c has emerged as a key miRNA in anxiety-like behavior with two subsequent studies reinforcing these early findings (Haramati et al., 2011; Parsons et al., 2012). Parsons et al. (2012) demonstrated that basal hippocampal miR-34c levels positively correlate with anxiety-like behaviors in C57BL/6 J X DBA/2 J (BXD) recombinant inbred mice, reinforcing the significance of their early findings. Haramati et al. (2011) also showed that miR-34c is significantly upregulated in the amygdala of C57BL/6 mice following exposure to restraint or social defeat stress. Overexpression of miR-34c in the central amygdala of normal mice induces anxiolytic effects that were more pronounced following a stressful stimulus, suggesting a role for miR-34c in stress-induced anxiety. The stress-related corticotrophin-releasing factor receptor type I (CRFR1) was identified as an mRNA target of miR-34c along with other stress-related proteins such as the

## TABLE 2.5
## Preclinical Studies of MiRNAs in Anxiety

| Author | Species/Model | Tissue | Analysis | Main Findings |
|---|---|---|---|---|
| Parsons et al. (2008) | Male and female A/J, BALB/cJ, C57BL/6J, DBA/2J mice | Hippocampus | RT-qPCR | • Differential expression of miR-203, miR-451, miR-378, miR-195, miR-34a, miR-34c, miR-15b, miR-323, miR-301a, miR-212, miR-31 across strains<br>• MiR-34c and miR-323 correlate with anxiety-like behavior in EPM<br>• MiR-212 correlates with grooming duration in EPM, light-dark box, OFT |
| Haramati et al. (2011) | Dicer$^{flox}$/Dicer$^{flox}$ mice and C57BL/6 or C57BL/6X129SvJ mice | amygdala | microarray; RT-qPCR | • Dicer ablation ↑ anxiety-like behavior<br>• Acute restraint stress ↑ miR-34c, miR-100, miR-15b, miR-34a, miR-92a, miR-15a<br>• Overexpression of miR-34c ↓ anxiety-like behavior in dark-light box test and EPM |
| Lin et al. (2011) | Male C57/Bl6 mice; auditory cued fear conditioning | ILPFC | RT-qPCR | • Fear extinction learning ↑ miR-128b<br>• ↓ miR-128b impairs fear extinction memory, ↑ miR-128b enhances fear extinction memory<br>• MiR-128b negatively regulates Rcs mRNA |
| Parsons et al. (2012) | Male C57BL/6J X DBA/2 J recombinant inbred mice (BXD) | Hippocampus | RT-qPCR | • MiR-31 correlates with anxiety-like behavior in light-dark box and OFT<br>• MiR-34c correlates with anxiolytic behaviors in EPM |
| Ragu Varman et al. (2013) | Male Indian field mice; EE | Amygdala | RT-qPCR; semi-quantitative RT-PCR | • EE ↓ anxiety-like behavior in EPM and OFT<br>• EE ↑ miR-183<br>• EE ↓ SC35 mRNA, ↓ AChE levels, ↑ AChE-S mRNA |
| Yoon et al. (2013) | Rat C6 astroglioma cultures | — | RT-qPCR; miR-16 binding assay | • hnRNPK regulates SERT expression by antagonizing miR-16 binding |
| Shaltiel et al. (2013) | Male C57Bl/6J mice; predator scent stress | Hippocampus | RT-qPCR | • Predator scent stress ↑ anxiety in EPM<br>• Predator scent stress ↑ miR-132 ~220%<br>• ↑ miR-132 correlates with ↓ AChE activity |
| Durairaj and Koilmani (2014) | Male Indian field mice; EE | Amygdala | RT-qPCR; semi-quantitative RT-PCR; western blotting | • EE ↓ anxiety-like behavior in light-dark box<br>• EE ↑ Dicer, Ago-2 mRNA and protein, ↑ miR-124a and ↓ GR mRNA and protein |

*Abbreviations:* EPM = elevated plus maze; OFT = open field test; ILPFC = infralimbic prefrontal cortex; EE = environmental enrichment; AChE = acetylcholinesterase; SERT = serotonin transporter; GR = glucocorticoid receptor; RT-qPCR = real-time quantitative polymerase chain reaction.

5-hydroxytryptamine (serotonin) receptor 2C, $GABA_A$ receptor $\alpha 1$, and BDNF. It is postulated that miR-34c assists in stress recovery by downregulating stress-related proteins.

Environmental enrichment (EE) has been found in two studies to induce an anxiolytic phenotype in the Indian field mouse (Ragu Varman et al., 2013; Durairaj and Koilmani, 2014). However, no clear role for miRNAs in the anxiolytic effects of EE has yet emerged. Ragu Varman et al. (2013) showed that the anxiolytic behavioral effects of EE were accompanied by an increase in miR-183 levels and an miR-183-induced suppression of SC35, a serine–arginine protein implicated in the stress response, in the amygdala. Suppression of SC35 causes a shift in the splicing of AChE toward that of the AChE-S (synaptic) form and decreases stress and anxiety-like behaviors. Notably, anxiety has been linked to reduced AChE activity by another group (Shaltiel et al., 2013) who found that predator stress-induced anxiety is associated with reduced hippocampal AChE activity ($\sim 25\%$) and a long-lasting increase in hippocampal miR-132 levels ($\sim 220\%$). Transgenic mice overexpressing both miR-132 and AChE were shown to have an anxiogenic phenotype. Durairaj and Koilmani (2014) also went on to show that EE induces a reduction in anxiety-like behaviors. This anxiogenic effect was accompanied by increased levels of Dicer and Ago-2, members of the miRNA biogenesis pathway, and an upregulation of pre-miR-124a in the amygdala. These findings are notable as depletion of Dicer in the amygdala was previously shown by Haramati et al. (2011) to induce anxiety-like behaviors in mice. Of note, *in vitro* analysis revealed that miR-124a regulates levels of the GR, and the GR was downregulated in the amygdala of EE mice suggesting that these changes may contribute to the anxiolytic condition (Durairaj and Koilmani, 2014).

Another important finding regarding the role of miRNAs in anxiety comes from Yoon et al. (2013). In line with previous studies (Baudry et al., 2010; Launay et al., 2011), they showed that miR-16 regulates SERT levels. This was linked to heterogeneous nuclear ribonucleoprotein K (hnRNPK), a SERT distal polyadenylation element binding protein, which can regulate the expression of SERT by antagonizing miR-16 binding and depressing SERT translation. Serotonergic signaling plays a major role in anxiety-related behaviors and pharmacological therapies targeting the serotonergic system are often used to treat anxiety. Importantly, expression of SERT mRNA containing the distal polyadenylation element is known to be anxiolytic in humans and mice.

A number of other studies have also examined miRNAs in anxiety. Details of these are outlined in Table 2.5. Overall, no clear role for miRNAs has emerged from the preclinical studies conducted so far. MiR-34c holds promise as an anxiety-related miRNA; however, despite a number of studies having been conducted to date, further work is required in order to fully establish a role for it in anxiety.

## 2.7.2  CLINICAL STUDIES IN ANXIETY DISORDERS

Our systematic review identified eight articles that were relevant for inclusion here. Compared to the affective disorders, there is a paucity of clinical studies of miRNAs in anxiety disorders, despite them being common with a high burden of disease. As in other disorders, genotyping rare variants (Donner et al., 2008; Muinos-Gimeno et al., 2009, 2011; Hanin et al., 2014; Jensen et al., 2014) has been a productive approach to understand the genetic basis of anxiety disorders. Other studies have focused on establishing a blood biomarker in anxiety disorders (Katsuura et al., 2012; Honda et al., 2013; Zhou et al., 2014). These are summarized in Table 2.6 and expanded on below.

### 2.7.2.1 Genotyping and Rare Variants

To date, there is still a lack of solid evidence identifying genes predisposing to anxiety disorders, and heritability rates are variable. Identifying SNPs can therefore be a challenge, and one of the first studies in this regard investigated SNPs based on 17 anxiety genes previously identified in mice (Donner et al., 2008). In a sample of 974 Finnish subjects and controls with varying anxiety disorders, 208 SNPs, including some in miRNA target-binding sites, were examined for association with diagnostic status. One SNP, rs817782 in the aminolevulinate dehydratase (*ALAD*) gene, is a putative target site for miR-211 and miR-204, and was associated with social phobia.

Other genotyping studies have identified an association between an SNP in the target site of miR-485-3p and the hoarding phenotype in OCD. There was also an association between two newly identified SNPs and panic disorder (Muinos-Gimeno et al., 2009). The targets for these SNPs included neurotrophic tyrosine kinase receptor 3 gene (*NTRK3*), which has been implicated in animal models of anxiety (Dierssen et al., 2006). In a further study from three European countries, the authors found several SNPs in miRNAs associated with panic disorder, but these did not stand up to correction for multiple testing (Muinos-Gimeno et al., 2011).

In a previously discussed study by Jensen et al. (2014), a target site for miR-330-3p was associated with MDD. This group also examined the association between this SNP and anxiety disorders, and found rs41305272 was associated with agoraphobia. However, these findings are complicated by the high rate of comorbidity between substance dependence, affective, and anxiety disorders.

A final genotyping study examined the effect of an SNP in the target site of miR-608. This SNP (rs17228616) was in the acetylcholinesterase enzyme gene (AChE), previously implicated in anxiety (Shaltiel et al., 2013). The authors showed that rs17228616 bound less strongly to miR-608, and in postmortem brains, was associated with higher levels of AChE. Patients with this SNP had higher blood pressure and lower cortisol levels than those with the normal allele, both risk factors for anxiety-related disorders (Hanin et al., 2014).

Most of the current rare variant studies have examined miRNA target sites rather than miRNAs per se. These have supported the potential involvement of the *NTRK3* gene in anxiety disorders but, in general, sample sizes have been low and the diagnostic purity of patients has been low.

### 2.7.2.2 Expression Studies in Peripheral Blood

Although not a study of clinical anxiety, a study of 10 Japanese students before and after national examinations measured levels of miRNAs using a microarray (Katsuura et al., 2012). MiR-144/144* and miR-16 were significantly increased immediately after examinations and returned to normal after a week. Advancing this paradigm, the same group subsequently followed 25 medical students in a period

## TABLE 2.6
## Clinical Studies of MiRNAs in Anxiety Disorders

| Author | Patients | Tissue | Analysis | Main Findings |
|---|---|---|---|---|
| **Genotyping Studies** | | | | |
| Donner et al. (2008) | PD = 108<br>GAD = 73<br>SP = 58<br>AGO = 31<br>Other = 58<br>Controls = 653 | DNA | MiR-SNP | • An SNP in anxiety-related gene is a putative target for miR-211, miR-204 |
| Muinos-Gimeno et al. (2011) | PD = 626 | DNA | MiR-SNP | • No SNPs associated with PD following correction for multiple testing |
| Muinos-Gimeno et al. (2009) | OCD = 153<br>PD = 212<br>Controls = 324 | DNA | MiR-SNP | • Target site of miR-485-3p associated with hoarding phenotype in OCD<br>• Targeted genes/gene pathways: NTRK3 |
| Jensen et al. (2014) | 6725 subjects | DNA | Microarray | • Association between SNP in target site of miR-330-3p in agoraphobia |
| Hanin et al. (2014) | 461 healthy volunteers | DNA | Various | • Target site of miR-608 associated with SNP in AChE. SNP associated with altered gene expression of AChE and anxiety-related changes in peripheral blood |
| **Expression Studies in Peripheral Blood** | | | | |
| Katsuura et al. (2012) | 10 healthy volunteers | Whole blood | Microarray; RT-qPCR | • ↑ miR-144, miR-144*, miR-16 following exposure to exam situation |
| Honda et al. (2013) | 25 healthy volunteers | Whole blood | Microarray; RT-qPCR | • ↑ miR-16, miR-20b, miR-26b, miR-29a, miR-126, miR-144, miR-144* in pre-examination period<br>• Gene targets: WNT4, CCM2, MAK, FGFR1 |
| Zhou et al. (2014) | PTSD = 30<br>Controls = 42 | PBMC; lymphocytes | Microarray; RT-qPCR | • ↓ miR-125a, miR-181c downregulated<br>• Gene targets: IFN-gamma, Il-23, IL-8 |

*Abbreviations:* AGO = agoraphobia; GAD = generalized anxiety disorder; OCD = obsessive-compulsive disorder; PBMC; peripheral blood mononuclear cells; PD = panic disorder; PTSD = post-traumatic stress disorder; SP = social phobia; RT-qPCR = real-time quantitative polymerase chain reaction.

before and after national examinations (Honda et al., 2013). They found seven miR-NAs elevated in the pre-examination period accompanied by a downregulation of their target mRNAs. Levels of these miRNAs were significantly reduced 1 month after exams. In contrast to their previous study, the authors also found a significant correlation between state anxiety levels and miR-16.

Finally, a study of 30 PTSD patients and 42 controls examined miRNA expression in peripheral blood (Zhou et al., 2014). The results highlight the role that miRNAs may play in altered immune function in these patients. MiR-125a and miR-181c were significantly downregulated in PTSD subjects. MiR-125a was subsequently found to downregulate interferon-gamma in normal subjects. This suggests that alterations in miRNA levels may contribute to a raised inflammatory tone, which has been linked to the pathophysiology of PTSD (Pace and Heim, 2011).

### 2.7.3 SUMMARY

Despite some new studies in the area, there is still much to be done to clarify both what miRNAs might be involved in the molecular basis of anxiety and which might function as biomarkers. There is little overlap in miRNAs identified to date, but many studies have included either naturalistic models of anxiety or groups with multiple psychiatric disorders. Given the amount of "noise" inherent in epigenetic mechanisms like miRNAs that can target hundreds of genes, it would seem sensible to start with well-defined patient groups with little comorbidity, although these can be hard to find. Few of the traditional brain banks have samples of anxiety disorders, which limit this area of study. Studying the effect of anxiolytic treatment on miRNA levels could be a productive avenue.

## 2.8  CONCLUSIONS

In summary, we have carried out a systematic review of the evidence from preclinical and clinical studies of miRNA involvement in mood and anxiety disorders. The review has identified a number of studies in BPAD, MDD, and to a lesser extent anxiety disorders. Although the number of preclinical and clinical studies has rapidly increased in recent times, the psychiatric field is still behind other areas of medicine. Therefore, it is hoped that further understanding of miRNAs may help to elucidate the molecular mechanisms of psychiatric illnesses such as mood and anxiety disorders. MiRNAs play a key role in virtually all functions in the central nervous system and represent a truly novel therapeutic target for the treatment of psychiatric illnesses. In other medical fields, miRNAs have already been identified as biomarkers (Cho, 2010) and therapeutic targets (Janssen et al., 2013). The ability of miRNAs to target hundreds of genes is part of its attractiveness in polygenic disorders. Bioinformatic approaches can help to identify miRNA targets much more easily than experimental validation, but can potentially identify thousands of targets and it is inevitable that some of these targets might be of interest in psychiatric disorders. This puts an emphasis on validation and replication in well-defined samples, as the risk of false negatives is high. Future studies should measure not just miRNA levels but also downstream products (mRNA, protein) to better understand the impact of miRNAs. MiRNAs have great potential to inform us of pathophysiology, act as biomarkers and even be therapeutic agents or targets. It is expected that miRNAs will contribute to a leap forward in our understanding and treatment of psychiatric disorders. This will therefore be a fascinating field to follow over the coming years.

## ACKNOWLEDGMENTS

The authors thank the Health Research Board (TRA/2007/5; TRA/2007/5/R) and the Brain and Behaviour Research Foundation (22516) for their support.

## REFERENCES

American Psychiatric Association. 2013. *Diagnostic and Statistical Manual of Mental Disorders*, 5th Ed.. Arlington, VA: American Psychiatric Publishing.

Bahi, A., Chandrasekar, V., and Dreyer, J.L. 2014. Selective lentiviral-mediated suppression of microRNA124a in the hippocampus evokes antidepressants-like effects in rats. *Psychoneuroendocrinology* 46: 78–87.

Bai, M., Zhu, X., Zhang, Y. et al. 2012. Abnormal hippocampal BDNF and miR-16 expression is associated with depression-like behaviors induced by stress during early life. *PLoS One* 7, no. 10: e46921.

Bai, M., Zhu, X.Z., Zhang, Y. et al. 2014. Anhedonia was associated with the dysregulation of hippocampal Htr4 and microRNA Let-7a in rats. *Physiol Behav* 129: 135–141.

Baudry, A., Mouillet-Richard, S., Schneider, B. et al. 2010. MiR-16 targets the serotonin transporter: A new facet for adaptive responses to antidepressants. *Science* 329, no. 5998: 1537–1541.

Belzeaux, R., Bergon, A., Jeanjean, V. et al. 2012. Responder and nonresponder patients exhibit different peripheral transcriptional signatures during major depressive episode. *Transl Psychiatry* 2: e185.

Berton, O., and Nestler, E.J. 2006. New approaches to antidepressant drug discovery: Beyond monoamines. *Nat Rev Neurosci* 7, no. 2: 137–151.

Bocchio-Chiavetto, L., Maffioletti, E., Bettinsoli, P. et al. 2013. Blood microRNA changes in depressed patients during antidepressant treatment. *Eur Neuropsychopharmacol* 23, no. 7: 602–611.

Bostwick, J.M., and Pankratz, V.S. 2000. Affective disorders and suicide risk: A reexamination. *Am J Psychiatry* 157, no. 12: 1925–1932.

Cho, W.C. 2010. MicroRNAs: Potential biomarkers for cancer diagnosis, prognosis and targets for therapy. *Int J Biochem Cell Biol* 42, no. 8: 1273–1281.

Craddock, N., and Sklar, P. 2013. Genetics of bipolar disorder. *Lancet* 381, no. 9878: 1654–1662.

Creson, T.K., Austin, D.R., Shaltiel, G. et al. 2011. Lithium treatment attenuates muscarinic M(1) receptor dysfunction. *Bipolar Disord* 13, no. 3: 238–249.

Cummings, E., Donohoe, G., Hargreaves, A. et al. 2013. Mood congruent psychotic symptoms and specific cognitive deficits in carriers of the novel schizophrenia risk variant at miR-137. *Neurosci Lett* 532: 33–38.

Dalton, V.S., Kolshus, E., and McLoughlin, D.M. 2014. Epigenetics and depression: Return of the repressed. *J Affect Disord* 155: 1–12.

Delaloy, C., Liu, L., Lee, J.A. et al. 2010. MicroRNA-9 coordinates proliferation and migration of human embryonic stem cell-derived neural progenitors. *Cell Stem Cell* 6, no. 4: 323–335.

Dierssen, M., Gratacos, M., Sahun, I. et al. 2006. Transgenic mice overexpressing the full-length neurotrophin receptor TrkC exhibit increased catecholaminergic neuron density in specific brain areas and increased anxiety-like behavior and panic reaction. *Neurobiol Dis* 24, no. 2: 403–418.

Domschke, K., and Deckert, J. 2012. Genetics of anxiety disorders—Status Quo and Quo Vadis. *Curr Pharm Des* 18, no. 35: 5691–5698.

Donner, J., Pirkola, S., Silander, K. et al. 2008. An association analysis of murine anxiety genes in humans implicates novel candidate genes for anxiety disorders. *Biol Psychiatry* 64, no. 8: 672–680.

Dowlati, Y., Herrmann, N., Swardfager, W. et al. 2010. A meta-analysis of cytokines in major depression. *Biol Psychiatry* 67, no. 5: 446–457.

Duman, R.S., and Aghajanian, G.K. 2012. Synaptic dysfunction in depression: Potential therapeutic targets. *Science* 338, no. 6103: 68–72.

Durairaj, R.V., and Koilmani, E.R. 2014. Environmental enrichment modulates glucocorticoid receptor expression and reduces anxiety in Indian field male mouse *Mus booduga* through up-regulation of MicroRNA-124a. *Gen Comp Endocrinol* 199: 26–32.

Dwivedi, Y. 2011. Evidence demonstrating role of microRNAs in the etiopathology of major depression. *J Chem Neuroanat* 42, no. 2: 142–156.

Flint, J., and Kendler, K.S. 2014. The genetics of major depression. *Neuron* 81, no. 3: 484–503.

Gao, J., Wang, W.Y., Mao, Y.W. et al. 2010. A novel pathway regulates memory and plasticity via Sirt1 and miR-134. *Nature* 466, no. 7310: 1105–1109.

Garbett, K.A., Vereczkei, A., Kalman, S. et al. 2015. Coordinated messenger RNA/microRNA changes in fibroblasts of patients with major depression. *Biol Psychiatry* 77, no. 3: 256–265.

Germain, A., and Kupfer, D.J. 2008. Circadian rhythm disturbances in depression. *Hum Psychopharmacol* 23, no. 7: 571–585.

Group, U.E.R. 2003. Efficacy and safety of electroconvulsive therapy in depressive disorders: A systematic review and meta-analysis. *Lancet* 361, no. 9360: 799–808.

Guella, I., Sequeira, A., Rollins, B. et al. 2013. Analysis of miR-137 expression and rs1625579 in dorsolateral prefrontal cortex. *J Psychiatr Res* 47, no. 9: 1215–1221.

Guintivano, J., Brown, T., Newcomer, A. et al. 2014. Identification and replication of a combined epigenetic and genetic biomarker predicting suicide and suicidal behaviors. *Am J Psychiatry* 171, no. 12: 1287–1296.

Hanin, G., Shenhar-Tsarfaty, S., Yayon, N. et al. 2014. Competing targets of microRNA-608 affect anxiety and hypertension. *Hum Mol Genet* 23, no. 17: 4569–4580.

Haramati, S., Navon, I., Issler, O. et al. 2011. MicroRNA as repressors of stress-induced anxiety: The case of Amygdalar miR-34. *J Neurosci* 31, no. 40: 14191–14203.

He, M., Liu, Y., Wang, X. et al. 2012a. Cell-type-based analysis of microRNA profiles in the mouse brain. *Neuron* 73, no. 1: 35–48.

He, Y., Zhou, Y., Xi, Q. et al. 2012b. Genetic variations in microRNA processing genes are associated with susceptibility in depression. *DNA Cell Biol* 31, no. 9: 1499–1506.

Honda, M., Kuwano, Y., Katsuura-Kamano, S. et al. 2013. Chronic academic stress increases a group of microRNAs in peripheral blood. *PLoS One* 8, no. 10: e75960.

Hoyo-Becerra, C., Schlaak, J.F., and Hermann, D.M. 2014. Insights from interferon-alpha-related depression for the pathogenesis of depression associated with inflammation. *Brain Behav Immun* 42C: 222–231.

Huang, Y.F., Yang, C.H., Huang, C.C. et al. 2012. Vascular endothelial growth factor-dependent spinogenesis underlies antidepressant-like effects of enriched environment. *J Biol Chem* 287, no. 49: 40938–40955.

International Schizophrenia Consortium, Purcell, S.M., Wray, N.R. et al. 2009. Common polygenic variation contributes to risk of schizophrenia and bipolar disorder. *Nature* 460, no. 7256: 748–752.

Issler, O., Haramati, S., Paul, E.D. et al. 2014. MicroRNA 135 is essential for chronic stress resiliency, antidepressant efficacy, and intact serotonergic activity. *Neuron* 83, no. 2: 344–360.

Janssen, H.L., Reesink, H.W., Lawitz, E.J. et al. 2013. Treatment of HCV infection by targeting microRNA. *N Engl J Med* 368, no. 18: 1685–1694.

Jensen, K.P., Kranzler, H.R., Stein, M.B. et al. 2014. The effects of a MAP2K5 microRNA target site SNP on risk for anxiety and depressive disorders. *Am J Med Genet B Neuropsychiatr Genet* 165B, no. 2: 175–183.

Jovicic, A., Roshan, R., Moisoi, N. et al. 2013. Comprehensive expression analyses of neural cell-type-specific miRNAs identify new determinants of the specification and maintenance of neuronal phenotypes. *J Neurosci* 33, no. 12: 5127–5137.

Katsuura, S., Kuwano, Y., Yamagishi, N. et al. 2012. MicroRNAs miR-144/144* and miR-16 in peripheral blood are potential biomarkers for naturalistic stress in healthy Japanese medical students. *Neurosci Lett* 516, no. 1: 79–84.

Kawashima, H., Numakawa, T., Kumamaru, E. et al. 2010. Glucocorticoid attenuates brain-derived neurotrophic factor-dependent upregulation of glutamate receptors via the suppression of microRNA-132 expression. *Neuroscience* 165, no. 4: 1301–1311.

Kessler, R.C., and Bromet, E.J. 2013. The epidemiology of depression across cultures. *Annu Rev Public Health* 34: 119–138.

Kessler, R.C., Petukhova, M., Sampson, N.A. et al. 2012. Twelve-month and lifetime prevalence and lifetime morbid risk of anxiety and mood disorders in the United States. *Int J Methods Psychiatr Res* 21, no. 3: 169–184.

Kim, A.H., Reimers, M., Maher, B. et al. 2010. MicroRNA expression profiling in the prefrontal cortex of individuals affected with schizophrenia and bipolar disorders. *Schizophr Res* 124, no. 1–3: 183–191.

Kolshus, E., Dalton, V.S., Ryan, K.M. et al. 2014. When less is more—MicroRNAs and psychiatric disorders. *Acta Psychiatr Scand* 129, no. 4: 241–256.

Landgraf, P., Rusu, M., Sheridan, R. et al. 2007. A mammalian microRNA expression atlas based on small RNA library sequencing. *Cell* 129, no. 7: 1401–1414.

Launay, J.M., Mouillet-Richard, S., Baudry, A. et al. 2011. Raphe-mediated signals control the hippocampal response to SRI antidepressants via miR-16. *Transl Psychiatry* 1: e56.

Li, Y.J., Xu, M., Gao, Z.H. et al. 2013. Alterations of serum levels of BDNF-related miRNAs in patients with depression. *PLoS One* 8, no. 5: e63648.

Lin, Q., Wei, W., Coelho, C.M. et al. 2011. The brain-specific microRNA miR-128b regulates the formation of fear-extinction memory. *Nat Neurosci* 14, no. 9: 1115–1117.

Liu, C., Teng, Z.Q., Santistevan, N.J. et al. 2010. Epigenetic regulation of miR-184 by MBD1 governs neural stem cell proliferation and differentiation. *Cell Stem Cell* 6, no. 5: 433–444.

Liu, X., Zhang, L., Cheng, K. et al. 2014. Identification of suitable plasma-based reference genes for miRNAome analysis of major depressive disorder. *J Affect Disord* 163: 133–139.

Lopez, J.P., Fiori, L.M., Gross, J.A. et al. 2014a. Regulatory role of miRNAs in polyamine gene expression in the prefrontal cortex of depressed suicide completers. *Int J Neuropsychopharmacol* 17, no. 1: 23–32.

Lopez, J.P., Lim, R., Cruceanu, C. et al. 2014b. MiR-1202 is a primate-specific and brain-enriched microRNA involved in major depression and antidepressant treatment. *Nat Med* 20, no. 7: 764–768.

Lugli, G., Torvik, V.I., Larson, J. et al. 2008. Expression of microRNAs and their precursors in synaptic fractions of adult mouse forebrain. *J Neurochem* 106, no. 2: 650–661.

Maffioletti, E., Tardito, D., Gennarelli, M. et al. 2014. Micro spies from the brain to the periphery: New clues from studies on microRNAs in neuropsychiatric disorders. *Front Cell Neurosci* 8: 75.

Manji, H.K., Quiroz, J.A., Payne, J.L. et al. 2003. The underlying neurobiology of bipolar disorder. *World Psychiatry* 2, no. 3: 136–146.

McCarthy, M.J., Nievergelt, C.M., Kelsoe, J.R. et al. 2012. A survey of genomic studies supports association of circadian clock genes with bipolar disorder spectrum illnesses and lithium response. *PLoS One* 7, no. 2: e32091.

Meerson, A., Cacheaux, L., Goosens, K.A. et al. 2010. Changes in brain microRNAs contribute to cholinergic stress reactions. *J Mol Neurosci* 40, no. 1–2: 47–55.

Miller, B.H., Zeier, Z., Xi, L. et al. 2012. MicroRNA-132 dysregulation in schizophrenia has implications for both neurodevelopment and adult brain function. *Proc Natl Acad Sci USA* 109, no. 8: 3125–3130.

Mineno, J., Okamoto, S., Ando, T. et al. 2006. The expression profile of microRNAs in mouse embryos. *Nucleic Acids Res* 34, no. 6: 1765–1771.

Miska, E.A., Alvarez-Saavedra, E., Townsend, M. et al. 2004. Microarray analysis of microRNA expression in the developing mammalian brain. *Genome Biol* 5, no. 9: R68.

Moreau, M.P., Bruse, S.E., David-Rus, R. et al. 2011. Altered microRNA expression profiles in postmortem brain samples from individuals with schizophrenia and bipolar disorder. *Biol Psychiatry* 69, no. 2: 188–193.

Morgado, A.L., Xavier, J.M., Dionisio, P.A. et al. 2015. MicroRNA-34a modulates neural stem cell differentiation by regulating expression of synaptic and autophagic proteins. *Mol Neurobiol* 51, no. 3: 1168–1183.

Mouillet-Richard, S., Baudry, A., Launay, J.M. et al. 2012. MicroRNAs and depression. *Neurobiol Dis* 46, no. 2: 272–278.

Muinos-Gimeno, M., Guidi, M., Kagerbauer, B. et al. 2009. Allele variants in functional microRNA target sites of the neurotrophin-3 receptor gene (Ntrk3) as susceptibility factors for anxiety disorders. *Hum Mutat* 30, no. 7: 1062–1071.

Muinos-Gimeno, M., Espinosa-Parrilla, Y., Guidi, M. et al. 2011. Human microRNAs miR-22, miR-138-2, miR-148a, and miR-488 are associated with panic disorder and regulate several anxiety candidate genes and related pathways. *Biol Psychiatry* 69, no. 6: 526–533.

Murray, C.J., Vos, T., Lozano, R. et al. 2012. Disability-adjusted life years (DALYs) for 291 diseases and injuries in 21 regions, 1990–2010: A systematic analysis for the global burden of disease study 2010. *Lancet* 380, no. 9859: 2197–2223.

Natera-Naranjo, O., Aschrafi, A., Gioio, A.E. et al. 2010. Identification and quantitative analyses of microRNAs located in the distal axons of sympathetic neurons. *RNA* 16, no. 8: 1516–1529.

Niu, C.S., Yang, Y., and Cheng, C.D. 2013. MiR-134 regulates the proliferation and invasion of glioblastoma cells by reducing Nanog expression. *Int J Oncol* 42, no. 5: 1533–1540.

O'Connor, R.M., Grenham, S., Dinan, T.G. et al. 2013. MicroRNAs as novel antidepressant targets: Converging effects of ketamine and electroconvulsive shock therapy in the rat hippocampus. *Int J Neuropsychopharmacol* 16, no. 8: 1885–1892.

Pace, T.W., and Heim, C.M. 2011. A short review on the psychoneuroimmunology of posttraumatic stress disorder: From risk factors to medical comorbidities. *Brain Behav Immun* 25, no. 1: 6–13.

Pariante, C.M., and Lightman, S.L. 2008. The HPA axis in major depression: Classical theories and new developments. *Trends Neurosci* 31, no. 9: 464–468.

Parsons, M.J., Grimm, C.H., Paya-Cano, J.L. et al. 2008. Using hippocampal microRNA expression differences between mouse inbred strains to characterise miRNA function. *Mamm Genome* 19, no. 7–8: 552–560.

Parsons, M.J., Grimm, C., Paya-Cano, J.L. et al. 2012. Genetic variation in hippocampal microRNA expression differences in C57BL/6 J X DBA/2 J (BXD) recombinant inbred mouse strains. *BMC Genomics* 13: 476.

Rago, L., Beattie, R., Taylor, V. et al. 2014. Mir379-410 Cluster miRNAs regulate neurogenesis and neuronal migration by fine-tuning N-cadherin. *EMBO J* 33, no. 8: 906–920.

Ragu Varman, D., Marimuthu, G., and Rajan, K.E. 2013. Environmental enrichment upregulates micro-RNA-183 and alters acetylcholinesterase splice variants to reduce anxiety-like behavior in the little Indian field mouse (*Mus booduga*). *J Neurosci Res* 91, no. 3: 426–435.

Rinaldi, A., Vincenti, S., De Vito, F. et al. 2010. Stress induces region specific alterations in microRNAs expression in mice. *Behav Brain Res* 208, no. 1: 265–269.

Rodgers, A.B., Morgan, C.P., Bronson, S.L. et al. 2013. Paternal stress exposure alters sperm microRNA content and reprograms offspring HPA stress axis regulation. *J Neurosci* 33, no. 21: 9003–9012.

Rong, H., Liu, T.B., Yang, K.J. et al. 2011. MicroRNA-134 plasma levels before and after treatment for bipolar mania. *J Psychiatr Res* 45, no. 1: 92–95.

Ryan, K.M., O'Donovan, S.M., and McLoughlin, D.M. 2013. Electroconvulsive stimulation alters levels of BDNF-associated microRNAs. *Neurosci Lett* 549: 125–129.

Sasaki, Y., Gross, C., Xing, L. et al. 2014. Identification of axon-enriched microRNAs localized to growth cones of cortical neurons. *Dev Neurobiol* 74, no. 3: 397–406.

Saus, E., Soria, V., Escaramis, G. et al. 2010. Genetic variants and abnormal processing of pre-miR-182, a circadian clock modulator, in major depression patients with late insomnia. *Hum Mol Genet* 19, no. 20: 4017–4025.

Schizophrenia Psychiatric Genome-Wide Association Study, C. 2011. Genome-wide association study identifies five new schizophrenia loci. *Nat Genet* 43, no. 10: 969–976.

Schratt, G.M., Tuebing, F., Nigh, E.A. et al. 2006. A brain-specific microRNA regulates dendritic spine development. *Nature* 439, no. 7074: 283–289.

Serretti, A., Benedetti, F., Mandelli, L. et al. 2003. Genetic dissection of psychopathological symptoms: Insomnia in mood disorders and clock gene polymorphism. *Am J Med Genet B Neuropsychiatr Genet* 121B, no. 1: 35–38.

Shaltiel, G., Hanan, M., Wolf, Y. et al. 2013. Hippocampal microRNA-132 mediates stress-inducible cognitive deficits through its acetylcholinesterase target. *Brain Struct Funct* 218, no. 1: 59–72.

Shao, N.Y., Hu, H.Y., Yan, Z. et al. 2010. Comprehensive survey of human brain microRNA by deep sequencing. *BMC Genomics* 11: 409.

Smalheiser, N.R., Lugli, G., Rizavi, H.S. et al. 2011. MicroRNA expression in rat brain exposed to repeated inescapable shock: Differential alterations in learned helplessness vs. non-learned helplessness. *Int J Neuropsychopharmacol* 14, no. 10: 1315–1325.

Smalheiser, N.R., Lugli, G., Rizavi, H.S. et al. 2012. MicroRNA expression is down-regulated and reorganized in prefrontal cortex of depressed suicide subjects. *PLoS One* 7, no. 3: e33201.

Smalheiser, N.R., Lugli, G., Zhang, H. et al. 2014. Expression of microRNAs and other small RNAs in prefrontal cortex in schizophrenia, bipolar disorder and depressed subjects. *PLoS One* 9, no. 1: e86469.

Smirnova, L., Grafe, A., Seiler, A. et al. 2005. Regulation of miRNA expression during neural cell specification. *Eur J Neurosci* 21, no. 6: 1469–1477.

Spedding, M., Jay, T., Costa e Silva, J. et al. 2005. A pathophysiological paradigm for the therapy of psychiatric disease. *Nat Rev Drug Discov* 4, no. 6: 467–476.

Uchida, S., Nishida, A., Hara, K. et al. 2008. Characterization of the vulnerability to repeated stress in fischer 344 rats: Possible involvement of microRNA-mediated down-regulation of the glucocorticoid receptor. *Eur J Neurosci* 27, no. 9: 2250–2261.

Uchida, S., Hara, K., Kobayashi, A. et al. 2010. Early life stress enhances behavioral vulnerability to stress through the activation of REST4-mediated gene transcription in the medial prefrontal cortex of rodents. *J Neurosci* 30, no. 45: 15007–15018.

van Spronsen, M., van Battum, E.Y., Kuijpers, M. et al. 2013. Developmental and activity-dependent miRNA expression profiling in primary hippocampal neuron cultures. *PLoS One* 8, no. 10: e74907.

Vo, N., Klein, M.E., Varlamova, O. et al. 2005. A camp-response element binding protein-induced microRNA regulates neuronal morphogenesis. *Proc Natl Acad Sci USA* 102, no. 45: 16426–16431.

Vreugdenhil, E., Verissimo, C.S., Mariman, R. et al. 2009. MicroRNA 18 and 124a down-regulate the glucocorticoid receptor: Implications for glucocorticoid responsiveness in the brain. *Endocrinology* 150, no. 5: 2220–2228.

Watanabe, Y., Gould, E., and McEwen, B.S. 1992. Stress induces atrophy of apical dendrites of hippocampal Ca3 pyramidal neurons. *Brain Res* 588, no. 2: 341–345.

Weigelt, K., Bergink, V., Burgerhout, K.M. et al. 2013. Down-regulation of inflammation-protective microRNAs 146a and 212 in monocytes of patients with postpartum psychosis. *Brain Behav Immun* 29: 147–155.

Whalley, H.C., Papmeyer, M., Romaniuk, L. et al. 2012. Impact of a microRNA miR137 susceptibility variant on brain function in people at high genetic risk of schizophrenia or bipolar disorder. *Neuropsychopharmacology* 37, no. 12: 2720–2729.

Wittchen, H.U., Jacobi, F., Rehm, J. et al. 2011. The size and burden of mental disorders and other disorders of the brain in Europe 2010. *Eur Neuropsychopharmacol* 21, no. 9: 655–679.

Woolley, C.S., Gould, E., and McEwen, B.S. 1990. Exposure to excess glucocorticoids alters dendritic morphology of adult hippocampal pyramidal neurons. *Brain Res* 531, no. 1–2: 225–231.

Xu, Y., Li, F., Zhang, B. et al. 2010a. MicroRNAs and target site screening reveals a pre-microRNA-30e variant associated with schizophrenia. *Schizophr Res* 119, no. 1–3: 219–227.

Xu, Y., Liu, H., Li, F. et al. 2010b. A polymorphism in the microRNA-30e precursor associated with major depressive disorder risk and P300 waveform. *J Affect Disord* 127, no. 1–3: 332–336.

Yang, X., Yang, Q., Wang, X. et al. 2014. MicroRNA expression profile and functional analysis reveal that miR-206 is a critical novel gene for the expression of BDNF induced by ketamine. *Neuromolecular Med* 16, no. 3: 594–605.

Yoon, Y., McKenna, M.C., Rollins, D.A. et al. 2013. Anxiety-associated alternative polyadenylation of the serotonin transporter mRNA confers translational regulation by hnRNPK. *Proc Natl Acad Sci USA* 110, no. 28: 11624–11629.

Zhang, Y., Zhu, X., Bai, M. et al. 2013a. Maternal deprivation enhances behavioral vulnerability to stress associated with miR-504 expression in nucleus accumbens of rats. *PLoS One* 8, no. 7: e69934.

Zhang, Z., Convertini, P., Shen, M. et al. 2013b. Valproic acid causes proteasomal degradation of dicer and influences miRNA expression. *PLoS One* 8, no. 12: e82895.

Zhou, J., Nagarkatti, P., Zhong, Y. et al. 2014. Dysregulation in microRNA expression is associated with alterations in immune functions in combat veterans with post-traumatic stress disorder. *PLoS One* 9, no. 4: e94075.

Zhou, R., Yuan, P., Wang, Y. et al. 2009. Evidence for selective microRNAs and their effectors as common long-term targets for the actions of mood stabilizers. *Neuropsychopharmacology* 34, no. 6: 1395–1405.

Zhu, Y., Kalbfleisch, T., Brennan, M.D. et al. 2009. A microRNA gene is hosted in an intron of a schizophrenia-susceptibility gene. *Schizophr Res* 109, no. 1–3: 86–89.

Zucchi, F.C., Yao, Y., Ward, I.D. et al. 2013. Maternal stress induces epigenetic signatures of psychiatric and neurological diseases in the offspring. *PLoS One* 8, no. 2: e56967.

# 3 MicroRNA Dysregulation in Schizophrenia and Functional Consequences

*Liliana Laskaris, Ting Ting Lee,*
*Piers Gillett, and Gursharan Chana*

## CONTENTS

Abstract ..................................................................................................................60
3.1   Introduction ..................................................................................................60
3.2   Evidence for MicroRNA Dysregulation in SCZ...........................................64
    3.2.1   Evidence for Mature MicroRNA Dysregulation
        in Schizophrenia: Postmortem Studies....................................64
    3.2.2   Peripheral MicroRNA Biomarkers for SCZ....................................68
3.3   Genetic MicroRNA Dysregulation in SCZ ..................................................70
    3.3.1   Introduction ......................................................................................70
    3.3.2   Genome-Wide Association Studies ..................................................70
    3.3.3   GWAS: Replication Successes and Failures....................................70
    3.3.4   Exploring the Functional Implications of GWAS Data....................71
    3.3.5   Delving Deeper into miR-137: An Illustration of
        Functional Studies.............................................................................73
    3.3.6   Is It Global: The Investigation of MicroRNA Biogenesis
        Dysregulation in SCZ ......................................................................75
    3.3.7   A Model of Dysregulated MicroRNA Biogenesis: The Dgcr8
        Haploinsufficiency Mouse................................................................77
3.4   Effect of Antipsychotic Medications on MicroRNA Expression ..................80
3.5   What Does It All Mean: In Search of a Functional Narrative
    of MicroRNA Dysregulation ........................................................................81
    3.5.1   Introduction ......................................................................................81
    3.5.2   Are a Disproportionate Number of SCZ Risk Genes
        also Targets of MicroRNAs? ............................................................82
    3.5.3   What Effect Do Dysregulated MicroRNAs Have on Their
        Target Molecules?.............................................................................85
        3.5.3.1   MiR-219-NMDAR ...............................................................85
        3.5.3.2   MiR-132-NMDAR ...............................................................86
        3.5.3.3   MiR-132, miR-212-PGD, Tyrosine Hydroxylase ................87

        3.5.3.4  MiR-107-CHRM1 ................................................................. 87

        3.5.3.5  MiR-195-BDNF ................................................................. 88

    3.5.4  Pathways Section .............................................................. 89

3.6   Where to Now: Future Directions in MicroRNA Research ........................... 91

3.7   Conclusions ..................................................................... 92

References ........................................................................... 93

## ABSTRACT

This chapter describes the role of microRNA dysregulation in the development of schizophrenia (SCZ) with particular reference to functional changes arising from this dysregulation. The chapter begins with evidence from both postmortem and peripheral blood studies that microRNAs are indeed dysregulated in SCZ. This dysregulation is believed to arise in part from genetic mutations in the microRNA transcript; therefore, a thorough review of the studies that identify microRNA single-nucleotide polymorphisms (SNPs) in SCZ population is subsequently provided. Particular emphasis is placed on the mechanism by which these microRNA SNPs may give rise to functional microRNA and downstream gene expression changes. In addition to these local SNPs, a second subset of studies indicates that microRNA dysregulation in SCZ may arise in part from global mutations in the microRNA biogenesis pathway. These studies are discussed with particular reference to the links that have been discovered between dysregulation of genes involved in the microRNA biogenesis pathway and copy number variations that are established candidate risk factors for SCZ, such as the 22q11.2 microdeletion. This evidence is then followed by an interrogation of the pharmacological studies, some of which have shown that antipsychotic medication used in the treatment of SCZ has a measurable effect on microRNA expression. Finally, as the information related above is highly heterogeneous, an attempt is made to collate the evidence around the mechanisms through which these diverse microRNA candidate SNPs affect mRNA expression and how this in turn has downstream effects on the expression of genes and proteins that have already been associated with SCZ. To this end, there is a detailed examination of studies that have examined the functional implications of microRNA dysregulation on SCZ-related targets. Evidence is presented from groups that have employed pathway analysis to determine the downstream effects of microRNA dysregulation, which have consistently implicated pathways involved in nervous system development and functioning. In conclusion, this chapter provides a thorough review of the body of evidence surrounding microRNA dysregulation in SCZ with the aim of determining the functional significance of this dysregulation. This investigation may indicate potential convergence pathways for a genetically heterogeneous disorder while simultaneously shedding light on the method by which differential microRNA expression produces dysregulation in these pathways.

## 3.1   INTRODUCTION

This chapter will be focused on the dysregulation of microRNAs (miRNAs) in schizophrenia (SCZ), with particular emphasis on the functional consequences of these perturbations. MiRNAs are small single-stranded RNA molecules 19–23 nucleotides in length that are capable of binding to multiple mRNA transcripts and altering their expression (Filipowicz et al. 2008). The downstream effect of miRNA binding is either the degradation of the mRNA target molecule or the inhibition of its translation, both resulting in the silencing of protein expression. Nonetheless, cases of increased protein expression as a result of miRNA binding have also been recorded (Huntzinger and Izaurralde 2011). MiRNAs are transcribed from independent genes or intronic regions of the protein-coding genome into pri-miRNA molecules which undergo a biogenesis process to produce a mature miRNA molecule capable of binding to its target mRNA (Krol et al. 2010b; Figure 3.1).

The target of the mature miRNA is located in the 3′ untranslated region (3′-UTR) end of the mRNA molecule and the binding is relatively nonspecific, requiring complete complementarity only between nucleotides 2–8 termed the miRNA seed region

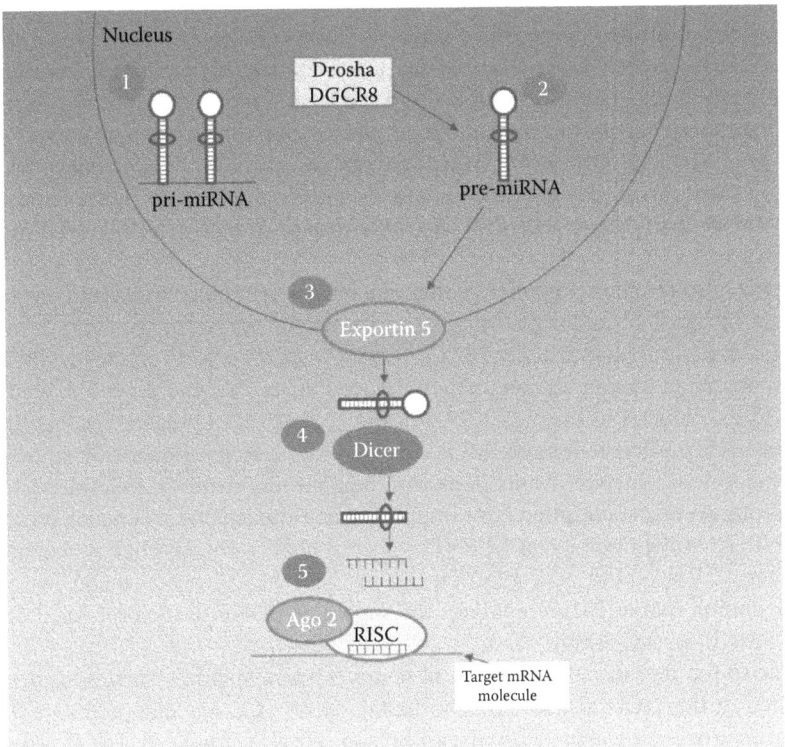

**FIGURE 3.1**  Graphical depiction of miRNAs biogenesis. (1) Intronic regions of protein-coding genome or independent genes are transcribed in the nucleus by RNA polymerase II to produce pri-miRNA transcript (100–1000 nts). (2) Pri-miRNAs fold into a hairpin structure and form a substrate for the Drosha and DGCR8 microprocessor complex, which cleaves them into pre-miRNAs (60–100 nts). (3) Pre-miRNAs are exported into the cytoplasm by Exportin 5. (4) In the cytoplasm, the enzyme Dicer processes pre-miRNAs to a ~20-bp miRNA/miRNA* duplex. (5) One strand of the duplex (mature miRNA) is then loaded into the miRNA-induced silencing complex (miRISC), which together with proteins such as Argonaute (AGO) target complementary sequences in mRNA molecules.

(Kim et al. 2010). This incomplete complementarity allows one miRNA molecule to potentially affect the expression of multiple mRNA molecules and protein (Krol et al. 2010b). This binding variability is particularly interesting within the context of a heterogeneous disorder such as SCZ, whose etiology appears to be increasingly multigenic, arising not from a single rare mutation but rather a combination of several common mutations and their interaction with environmental factors (Beveridge and Cairns 2012).

Schizophrenia is a neuropsychiatric disorder characterized by at least one of the following symptoms: delusions, hallucinations, disorganized speech, and behavior that persist for more than 6 months (American Psychiatric Association 2013). A significant difference between this recent edition of the DSM V and DSM IV is that the SCZ subtypes (i.e., paranoid, disorganized, catatonic, undifferentiated, and residual types) have been abolished. This change will increase the validity of SCZ diagnosis, as these subtypes have previously been criticized for their low reliability and limited diagnostic stability. In terms of clinical diagnosis, there has been increasing emphasis placed on the identification of potential biomarkers to aid not only in diagnosis but primarily in the identification of molecular subtypes of this heterogeneous condition and their progression throughout the patient's life span (Hickie et al. 2009). The investigation into the role of molecules such as miRNAs that might play key roles in affecting the expression of multiple proteins may provide useful biomarkers for assessing groups of molecules involved in SCZ etiology.

Schizophrenia as a disorder of the synapse is a theory supported by many experts in the field and which has the potential of revealing the links between genetic mutations associated with the disorder leading to synaptic dysfunction and subsequent disordered connectivity (Harrison and Weinberger 2005; Gauthier et al. 2010; Faludi and Mirnics 2011; Fornito et al. 2012). Evidence for this theory comes from postmortem studies that demonstrate decreased dendritic spine density and loss of synaptic connections with neighboring neurons, in addition to the numerous genetic studies that have implicated neuroplasticity genes such as BDNF and NRG1 in the etiology of SCZ (Davis et al. 2014). Interestingly, connections between miRNAs and neuroplasticity genes have already been made, with one study finding that miR-195 regulates the supply of BDNF to the prefrontal cortex (PFC; Mellios et al. 2009).

Due to the fact that the majority of neurons and synaptic connections are formulated in the prenatal and early postnatal years, SCZ has emerged as a disorder of neurodevelopment (Fatemi and Folsom 2009; Catts et al. 2013). Prenatal inflammation has also been shown to lead to aberrant neuronal development and synapse formation and subsequent SCZ phenotype in animal models of this disorder (Meyer et al. 2011). Indeed there is evidence from epidemiological studies that various prenatal factors such as infection arising in the mother in the second or third trimester of pregnancy are associated with increased risk of SCZ (Brown 2011). Synapse development does not end with birth but rather continues into early adulthood with the onset in adolescence of synaptic pruning that coincides with the age of onset for SCZ (Faludi and Mirnics 2011). This process of eliminating underutilized synapses has also been modeled in animals (Zhan et al. 2014) and when disrupted been shown to result in weak synaptic transmission and neuro-psychiatric behavioral deficits. In addition to these findings, many mutations that show consistent association with SCZ are in genes coding for proteins involved in synapse formation, stabilization, and function, throughout the process of neurode-velopment (Harrison and Weinberger 2005; Gauthier et al. 2010). In conclusion, there is evidence from multiple studies using a variety of methodologies that SCZ may arise from the interplay of genetic mutations and environmental factors dur-ing the neurodevelopmental period ranging from prenatal infection right through to urban living and stress in adolescence (Anderson and Maes 2013; Catts et al. 2013). This interaction of genetic predisposition and environment appears to play out at the level of the synapse thus affecting the formation of synaptic connec-tions, leading on the one hand to the aberrant connectivity observed in patients using functional imaging studies (Zhou et al. 2007; Whitfield-Gabrieli et al. 2009; Fornito et al. 2012) and on the other to the neuropil reduction and enlarged ven-tricles, which constitute one of the most replicated findings associated with SCZ (Steen et al. 2006; Vita et al. 2006).

In order to better understand the etiology of SCZ, it is necessary to move beyond associative evidence and examine the mechanisms involved at every step of the neu-rodevelopmental process from gene to protein expression, subsequent synapse for-mation, and adult connectivity patterns. As such miRNAs with their aforementioned ability to affect the expression of multiple proteins that are involved in a variety of synaptic networks constitute a unique target for etiological investigation (Beveridge and Cairns 2012; Maffioletti et al. 2014).

This chapter is focused on the exploration of miRNAs strongly associated with SCZ with an aim toward determining whether there are functional pathways that appear to be commonly perturbed and which may provide clues as to the etiology of this highly heterogeneous disorder. We will begin with a presentation of the evidence for miRNA gene expression dysregulation in SCZ, evidenced firstly from postmor-tem studies and subsequently peripheral miRNA studies. Following presentation of miRNA gene expression dysregulation in SCZ, we will examine single-nucleotide polymorphisms (SNPs) in miRNAs implicated in SCZ. This will be followed by an interrogation of the evidence for mutations arising in genes coding for proteins involved in the miRNA biogenesis pathway that have been associated with SCZ. This is due to the fact that the theory has been put forward that the dysregulation of miRNAs seen in SCZ may not be due to single mutations on individual miRNA genes but rather a general dysregulation of their biogenesis process (Beveridge et al. 2010) and any body of work examining miRNA dysregulation in SCZ needs to con-sider this possibility.

We will then elucidate known evidence relating to the effects of psychotropic medications used in the treatment of SCZ on miRNA expression from both animal and postmortem studies.

Finally we will attempt to collate and amalgamate the evidence around the mech-anisms through which these diverse miRNA candidate SNPs affect mRNA expres-sion and how this in turn has downstream effects on the expression of genes that have already been associated with SCZ. Evidence will be presented from groups that have employed both *in vitro* functional assays and pathway analysis to determine the downstream effects of miRNA dysregulation, which have consistently implicated pathways involved in nervous system development and functioning.

## 3.2   EVIDENCE FOR MicroRNA DYSREGULATION IN SCZ

### 3.2.1   EVIDENCE FOR MATURE MicroRNA DYSREGULATION IN SCHIZOPHRENIA: POSTMORTEM STUDIES

MiRNAs are not limited to a single target gene or chromosomal location but can regulate numerous loci across the genome. The large number of target sites can lead to a wide range of complications if the function or activity of any given miRNA is compromised or lost. The potential for such problems is particularly pertinent in the central nervous system (CNS), where the correct development of the CNS relies on a number of miRNA functioning correctly.

Human work has also taken place with some previously identified miRNAs in model organisms having their human homologue identified and implicating them in human CNS development. An example of these homologues is miR-9, previously identified in zebrafish studies. MiR-9 was found along with miR-124a, to be an important regulator in neural lineage differentiation in cultures of human embryonic stem cells (Krichevsky et al. 2006). Following in this line of work, Le et al. (2009) undertook neural differentiation of the human neuronal cell line SH-SY5Y and were able to show a significant increase in expression of six particular miRNAs. Of these six, miR-125b was shown to be particularly important for correct neurite outgrowth to occur. There are a number of different differentiation paths throughout neuronal development and many involve miRNAs.

Development does not always proceed correctly and there are a vast number of steps where development can falter. The dysregulation or production of dysfunctional miRNA is one such aspect of development that can go awry. Hence, it is not entirely unexpected that issues with miRNAs have been associated with disorders such as SCZ, which is believed by some to be of a neurodevelopmental nature (Marenco and Weinberger 2000; Piper et al. 2012). A summary of many of the studies that have associated miRNA with SCZ is summarized in Table 3.1.

The results of the different studies on the subject have identified a whole host of different miRNA as being implicated in SCZ. An investigation by Miller et al. (2012) focusing on in excess of 800 miRNAs that are known to be involved in human brain function identified miR-132 specifically as being at significantly lower levels in SCZ brains. Another individual miRNA, miR-195, was identified by Zhu et al. (2009) and was found at significantly lower levels in the SCZ brain. The discovery of a role for miR-195 in SCZ is not completely unexpected. MiR-195 regulates brain-derived neurotrophic factor which in turn regulates both neuropeptide Y (NPY) and somatostatin levels, both of which have been previously highlighted as important factors in SCZ and brain development, particularly as pertains their role in relation to GABAergic interneurons that have been shown to be implicated in the pathophysiology of SCZ (Moghaddam and Javitt 2012).

Not all associated miRNA are downregulated, in fact upregulation of miR-181b has been seen in SCZ brains (Beveridge et al. 2008). The implication arising

## TABLE 3.1
## The Key Findings from Studies of Postmortem Brain Tissue Demonstrating a Link between Schizophrenia and Dysfunctional or Dysregulated MicroRNAs (MiRNAs)

| | Author(s) | Year | Key Findings |
|---|---|---|---|
| MiRNA-132 dysregulation in schizophrenia has implications for both neurodevelopmental and adult brain function | Miller et al. | 2012 | Of >800 miRNA investigated, only miR-132 and miR-132* were at decreased levels of significance (did show a correlation with the age of the sample). The decreased levels of miR-132 caused upregulation of five CNS pathways including the glutamatergic system |
| MiRNA expression profiling in the prefrontal cortex of individuals affected with schizophrenia and bipolar disorders | Kim et al. | 2010 | Found seven miRNA to be significantly differentially expressed as compared to controls. MiR-34a showed the greatest difference from its normal levels. Targets of these miRNA are previously identified loci of interest in SCZ, including genes within the glutamatergic system |
| Dysregulation of miRNA 181b in the temporal cortex in schizophrenia | Beveridge et al. | 2008 | A significant upregulation of miR-181b. As a result, there is significant decrease in gene targets of miR-181b, including GRIA2 (part of the glutamatergic system) and VSNL1 |
| Molecular determinants of dysregulated GABAergic gene expression in the prefrontal cortex of subjects with schizophrenia | Mellios et al. | 2009 | Significantly decreased levels of miR-195 that was shown to in turn affect levels of BDNF and its targets |
| An miRNA gene is hosted in an intron of a schizophrenia-susceptibility gene | Zhu et al. | 2009 | Expression levels of miR-346 are at significantly lower levels in SCZ patients compared to controls |
| Altered miRNA expression profiles in postmortem brain samples from Individuals with schizophrenia and bipolar disorder | Moreau et al. | 2011 | 24 total miRNA were found to show a significant association with SCZ |
| Upregulation of dicer and MiRNA expression in the dorsolateral prefrontal cortex brodmann area 46 in schizophrenia | Santarelli et al. | 2011 | 25 miRNA were significantly upregulated and 3 downregulated in BA46 in addition to an increase in Dicer expression. Predicted miRNA targets based on sequence homology identified nine targets of these miRNA belonging to the glutamatergic system |

*(Continued)*

**TABLE 3.1** (*Continued*)
**The Key Findings from Studies of Postmortem Brain Tissue Demonstrating a Link between Schizophrenia and Dysfunctional or Dysregulated MicroRNAs (MiRNAs)**

| | Author(s) | Year | Key Findings |
|---|---|---|---|
| Schizophrenia is associated with an increase in cortical. miRNA biogenesis | Beveridge et al. | 2010 | 59 and 26 miRNAs were significantly upregulated in the STG and DLPFC, respectively, of SCZ individuals as compared to controls. Dicer was shown to be significantly upregulated in the DLPFC of SCZ individuals |
| MiRNA expression in the prefrontal cortex of individuals with schizophrenia and schizoaffective disorder | Perkins et al. | 2007 | 15 miRNAs identified to be upregulated and 1 downregulated in the PFC of SCZ sufferers |
| Chromosome 8p as a potential hub for developmental neuropsychiatric disorders: implications for schizophrenia, autism and cancer | Tabares-Seisdedos and Rubenstein | 2009 | The 8p region has been linked to SCZ a number of times this region contains seven miRNA that have been linked to SCZ |

from both down- and upregulated miRNA being found in SCZ brains is that global changes in miRNA regulation are potentially contributing to the disorder. Global changes in miRNA regulation are more likely in many cases than a single dysfunctional miRNA being the sole causative factor in SCZ.

A more widespread dysregulation of miRNA has been identified in a number of papers. One such study by Kim et al. (2010) identified seven miRNA as differentially expressed in the SCZ brain. The significance of this paper does not arise from the number of miRNA identified but the predicted targets of the identified miRNA. The targets of these miRNA have themselves been associated with SCZ or neurodevelopmental pathways, which include mRNA for tyrosine hydroxylase (TH) and phosphogluconate dehydrogenase (PGD). Just as particular genes have been previously associated with SCZ, whole genomic regions have also been associated. The p arm of chromosome 8 is one particular example of this. This region was found to harbor, in addition to protein-coding genes associated with SCZ, seven genes encoding miRNAs (Tabares-Seisdedos and Rubenstein 2009).

Perkins et al. (2007) increased the number of SCZ candidate miRNA by identifying a total of 16 miRNA, 15 downregulated and 1 upregulated in SCZ samples. Adding to the list was Moreau et al. (2011) who demonstrated an association linking SCZ to the dysregulation of 24 miRNA encoding genes. Santarelli et al. (2011) has further identified 28 miRNA of interest in SCZ: 25 of which are upregulated and 3 that are downregulated, seemingly conflicting with the work of Perkins et al. (2007)

who found a significant excess of miRNA being downregulated as opposed to upregulated. Finally, a total of 21% (59) of miRNA expressed in the superior temporal gyrus of SCZ cases as well as 9.5% (26) of miRNA in the DLPFC as compared to psychiatrically healthy samples. This is of particular interest, given that both these regions have demonstrated structural abnormalities in SCZ brains, as well as aberrant activation during cognitive tasks (Kyriakopoulos et al. 2012; Vita et al. 2012). The finding of miRNA dysregulation in brain regions known to be implicated in SCZ, as well as the sheer number of miRNA arising in this work makes a compelling case for their role in SCZ but these association studies have limitations. More in-depth analysis of each implicated miRNA is required before any kind of causative link can be established. Nonetheless, there remains value in these associations as part of the significance of the implicated miRNA arises from the types of genes which the miRNA target. The targets of miRNAs associated with SCZ may provide understanding of the etiology of SCZ.

The glutamatergic signaling system is key to the correct functioning of the CNS as it is the primary excitatory signaling system (Robinson and Coyle 1987). It is a system particularly pertinent to SCZ as the glutamatergic system has been strongly linked to SCZ (Moghaddam and Javitt 2012). This association extends to miRNA as many of the targets, bioinformatically predicted or experimentally tested, are for genes encoding the glutamatergic systems functional components. The targets range from metabotropic glutamate receptors 1, 3, 5, and 7 (Beveridge et al. 2008; Beveridge et al. 2010; Kim et al. 2010; Santarelli et al. 2011), subunits of NMDA receptors including *GRIN1, GRIN2A, GRIN2B*, and *GRIN2C* (Beveridge et al. 2010; Kim et al. 2010; Santarelli et al. 2011), and ionotropic glutamate receptor subunits *GRIA2, GRIK2*, and *GRID* (Beveridge et al. 2008; Santarelli et al. 2011). All these predicted targets of dysregulated miRNAs are key aspects in correct glutamatergic system functioning.

Finally, as many studies have identified a number of miRNAs that are dysregulated within the brain as opposed to single dysfunctional miRNA genes, an overarching regulatory problem may in fact be the issue that could explain the widespread miRNA dysregulation in the brain. This has spurred interest in genes that encode miRNA biogenesis elements as if their products are dysfunctional, this may go some way to explaining the pathophysiology seen in the SCZ brain. The role of dysfunctional biogenesis is supported by findings that precursor elements to mature miRNA, pri-miRNA, and pre-miRNA are found in abnormal ratios in schizophrenic brains, as compared to the ratios seen in healthy individuals (Beveridge et al. 2010; Moreau et al. 2011). Decreased levels of pre-miRNA could be explained by significantly higher levels of Dicer being found in SCZ brains (Beveridge et al. 2010), Dicer being the enzyme that cleaves pre-miRNA resulting in the final miRNA product. Another aspect of miRNA biogenesis is the DiGeorge syndrome critical region gene 8 (DGCR). The protein product of this gene is an important aspect of the pri-miRNA cleavage complex, from which pre-miRNA results, in which Drosha is also a subunit. The gene for DGCR8 is found in a region of the genome, 22q11, which has been highly associated with SCZ (Murphy et al. 1999). Thus, dysfunctional DGCR8 or even just aberrantly functioning DGCR8 could have significant downstream effects on miRNA. With their products having such pivotal roles in the biogenesis of miRNA, mutations in miRNA biogenesis genes certainly can go some way to explain the seemingly high numbers of miRNA being associated with SCZ. For a more detailed examination of miRNA biogenesis, see Section 3.6.

The postmortem studies discussed here have generated some interesting results that certainly warrant further attention. With predicted gene targets of some implicated miRNA being previously identified genes of interest in SCZ as well as the pathways in which they function being of interest, miRNA are providing further support for existing hypotheses surrounding SCZ and as a result making their role in SCZ pathogenesis appear far more credible. While more detailed research remains to be undertaken, the results thus far certainly present miRNA-based hypotheses as plausible and as both worthy and an interesting direction for future SCZ research to take.

### 3.2.2  PERIPHERAL MicroRNA BIOMARKERS FOR SCZ

Free circulating miRNAs have been stably detected in various peripheral samples including whole blood, serum, plasma, and cerebral spinal fluid (CSF), urine, and saliva (Jin et al. 2013; Maffioletti et al. 2014). However, it is not certain whether these miRNAs are secreted by healthy cells to regulate their extracellular environment or whether they constitute a by-product of apoptotic cells. As reviewed by Maffioletti et al., miRNAs are known to exist in three extracellular forms that are protective against RNase degradation: within lipid microvesicles, RNA-binding protein bound, as well as associated with high-density lipoproteins (Jin et al. 2013; Maffioletti et al. 2014).

A large body of evidence suggests that miRNAs may be potential disease biomarkers, primarily due to their stable expression level in plasma and serum. MiRNAs are known to be stable against RNAase in the blood, while they also remain intact in extreme temperature and pH environments (Chen et al. 2008). Moreover, miRNAs can be easily measured in a small amount of sample using quantitative real-time polymerase chain reaction (RT-PCR). Extracellular miRNAs and their expression changes are also found to be correlated in some cases with that of CNS miRNAs (Xu et al. 2010; Beveridge and Cairns 2012; Gardiner et al. 2012). Since their discovery, miRNAs have become favorable as disease biomarkers due to their easy and noninvasive sample collection process as opposed to biopsies (Jin et al. 2013).

To date, the number of studies that have investigated the expression level of miRNAs in peripheral samples of SCZ patients is limited. Xu et al. identified an upregulation of miR-30e in the blood of patients with SCZ. A subsequent genome-wide investigation of miRNA expression revealed a SCZ-associated miRNA profile when comparing the expression levels in peripheral blood mononuclear cells (PBMCs) between 112 SCZ patients and 76 healthy controls (Gardiner et al. 2012). Among the 33 downregulated miRNAs, a significant subgroup of 17 are transcribed from the imprinted DLK1-DIO3 region on chromosome 14q32, which indicates genetic involvement in the underlying miRNA pathophysiology of SCZ. Furthermore, miRNAs in this domain, including miR-134 and miR-329, are believed to play an important role in brain function and development, as they are involved in neuronal morphology regulation and dendritic development (Gardiner et al. 2012).

Lai et al. (2011) utilized a similar approach of miRNA expression investigation and found a seven-miRNA profile (miR-34a, miR-449a, miR-564, miR-432, miR-548d, miR-572, and miR-652) that differs between SCZ patients and healthy controls. The expression levels of these seven miRNAs were also correlated with clinical symptoms, neurocognitive performance, and neurophysiological function of SCZ patients, suggesting that these peripheral miRNAs may act as a marker of symptom severity (Lai et al. 2011). Among the miRNAs showing altered expression, miR-34a expression has previously been found to be altered in the prefrontal cortex of patients with SCZ, as well as the hippocampus of rat and *in vitro* in cultured cells treated with mood stabilizer (Shibata et al. 2009; Zhou et al. 2009; Kim et al. 2010).

Shi et al. (2012) also conducted a study investigating the expression of nine miRNA within the serum of 115 SCZ patients and 40 healthy controls using qRT-PCR. Four of the miRNAs (miR-181b, miR-219-2-3p, miR-1308, and let-7g) were significantly upregulated, while miR-195 was found to be downregulated, indicating another miRNA signatures associated with SCZ.

We can see therefore that while peripheral miRNA SCZ signatures appear to exist, they nonetheless show great diversity across studies. This may be due to variability in recruitment criteria, experimental methodologies, the diversity of biological samples measured across studies, or merely the heterogeneity innate in the pathophysiology of SCZ. Variation is also seen when peripheral studies are compared to those that examine miRNA expression in postmortem brain tissue. For example, while miR-30 and miR-181b have been found to be upregulated in both peripheral (Xu et al. 2010; Shi et al. 2012) and postmortem studies (Perkins et al. 2007; Beveridge and Cairns 2012), other miRNAs differ in their expression profile between postmortem and peripheral studies. MiR-107 for example was found to be upregulated in postmortem studies (Beveridge and Cairns 2012; Scarr et al. 2013) but downregulated in a peripheral study (Gardiner et al. 2012). This discrepancy may be due to the different expression levels of these miRNAs in the brain and blood, postulating that these molecules may play different roles in different tissues. Furthermore, the microenvironment in the brain is different to that of peripheral tissue, therefore dysregulation of miRNAs in the blood does not necessarily reflect the SCZ pathophysiology in the brain.

The literature suggests that the alteration in peripheral miRNA expression mentioned above is not limited to SCZ, but involved in several other neuropsychiatric and neurodevelopmental disorders. For instance, miR-134 was also found downregulated in plasma samples of patients with bipolar disorder compared to healthy controls (Rong et al. 2011). Among Alzheimer's disease patients, miR-34a was found altered in the blood, while miR-449 was implicated in the CSF (Schipper et al. 2007; Cogswell et al. 2008).

Taken together, this evidence suggests that while peripheral miRNA expression may be used as a biomarker for SCZ, the results have suffered from a lack of replication. Furthermore, their alteration signatures associated with other neuropsychiatric and neurodevelopmental disorders need to be investigated further in order to develop a high accuracy peripheral miRNA signature specific to SCZ. It is also crucial to identify if peripheral expression of miRNAs reflects that of the CNS by quantifying the blood and CSF levels in the same patient cohort.

## 3.3    GENETIC MicroRNA DYSREGULATION IN SCZ

### 3.3.1    INTRODUCTION

MiRNAs have many target genes, therefore a mutation in a given miRNA may have significant downstream effects (Beveridge and Cairns 2012; Maffioletti et al. 2014). Studies have shown that one variant such as a SNP can affect both the expression of the miRNA molecule and its target, as well as the binding of that miRNA to its target mRNA sites (Feng et al. 2009; Guella et al. 2013). A mutation on an miRNA target site can likewise affect the binding of the miRNA molecule and has significant downstream effects (Begemann et al. 2010; Liu et al. 2012). Given the multitargeted nature of miRNAs and the ability of miRNA variants to exert significant downstream effects, it is therefore imperative to examine these miRNA variants in relation to SCZ.

### 3.3.2    GENOME-WIDE ASSOCIATION STUDIES

Despite their promising beginnings, genome-wide association studies (GWASs) are prone to a collection of well-cited problems, which need to be kept in mind when examining any miRNA genetic association study. Among these problems are the issues surrounding multiple testing which may lead to a high rate of false positives, as well as the established fact that GWASs are unable to account for epistatic effects (Gershon et al. 2011; Wray et al. 2011). While it is true that currently the value of miRNA SNPs in diagnosis and treatment is limited, their utility may lie in the identification of functional pathways driving the SCZ phenotype (Cummings et al. 2013). We will begin with a detailed description of miRNA GWAS in the SCZ field focusing mainly on the miRNAs that were successfully replicated across studies, as well as those that were not replicated. Subsequently, studies that went beyond GWAS data and attempted to explore the functional implications of those associations are examined. As an illustration of the wide variety of functional investigations that can be performed, the subsequent section examines studies focusing on the miR-137 SNP, shown to have the greatest association with SCZ in a large scale GWAS (Schizophrenia Psychiatric Genome-Wide Association Study 2011). This is followed by a discussion of the evidence for the theory that the dysregulation takes place on a global scale rather than at the individual SNP level and that it is in fact miRNA biogenesis that is disrupted in SCZ. The section concludes with discussion of the Dgcr8 heterogeneous knockout (KO) mouse model, which is designed to investigate the consequences of miRNA dysregulation.

### 3.3.3    GWAS: REPLICATION SUCCESSES AND FAILURES

The first study examining miRNA SNPs and their association with SCZ was conducted on three Scandinavian population and identified SNPs in miR-206 and miR-198 that were significant in the Norwegian and Danish population, although not the Swedish (Hansen et al. 2007). A test for homogeneity indicated there was significant heterogeneity in the odds ratios of the three population, which is consistent with the lack of an overall significant association across the combined samples. This study therefore highlights the need for obtaining samples of a similar genetic makeup, even in a population that is largely homogeneous. Interestingly, miR-198, one of the SNPs significantly associated with SCZ in this study was subsequently replicated by Guo and colleagues (2010). The replication of one out of only two SNPs shown to be risk factors indicates the specificity obtained in some GWASs, especially considering that over 100 miRNA SNPs were examined by Hansen et al. (2007). Having said this, the association between miR-198 and SCZ demonstrated in these two studies could not be replicated when the gene was directly sequenced in the Han Chinese population (Xu et al. 2010). This outcome again highlights the variability in GWAS results obtained across different populations, as well as the need to exercise caution when interpreting such results.

The difficulty in replicating miRNA SNPs that constitute risk factors for SCZ across studies and population is crystalized in the Sun et al. (2009) and Feng et al. (2009) studies, that were performed in the same laboratory utilizing similar methods and on the same patient and control samples. Even under these favorable conditions, they were only able to identify one SNP that was replicated across both studies, which was a C/G transversion in miR-502. One of the reasons for the replication of this SNP may be due to its location in the miR-502 genome which is immediately adjacent to the Drosha/Dgcr8 cleavage site that has consistently been shown to lower protein expression. If this is indeed the reason for the replication of miR-502, these results implicate miRNA biogenesis in the pathogenesis of SCZ, as both Drosha and Dgcr8 are part of the microprocessor complex that is an essential component of miRNA biogenesis. The theory that miRNA biogenesis is globally disrupted in SCZ is discussed further in Section 3.6.

### 3.3.4   Exploring the Functional Implications of GWAS Data

Hansen et al. (2007) and Xu et al. (2010) went beyond the identification of miRNA SNPs and incorporated their results into a wider functional narrative. In the case of Hansen et al. (2007), they used the two miRNA risk alleles as a catalyst for identifying targets common to both miR-206 and miR-198 using online software provided for the identification of miRNA targets (TargetScan; Lewis et al. 2003). Having identified targets that were common to both their miRNAs, they constructed signaling networks that incorporated those targets and their first-degree interaction partners. Interrogation of this *in silico* interaction network enabled them to identify important proteins such as CREB and transcriptional factors expressed in the PFC, as well as two genes previously associated with SCZ. By building on their GWAS data in this manner, they therefore established an SCZ-related network that could potentially be examined from a biomolecular perspective with a view toward determining the molecular mechanisms driving these genetic associations.

Xu et al. (2010) collected miRNA candidates that were aberrantly expressed in postmortem studies of SCZ and focused their investigations on identifying SNPs located on these candidate miRNAs that might account for this aberrant expression (Perkins et al. 2007; Beveridge et al. 2010). They detected 39 SNPs inherent in 8 miRNAs, they examined and out of these variants only one SNP in miR-30e was strongly associated with SCZ (Xu et al. 2010). Interestingly, members of the miR-30 family were also implicated in an *in silico* miRNA regulatory network (Guo et al. 2010). When peripheral levels of mature miR-30e were examined in blood taken from SCZ patients, they were found to be increased but a correlation analysis between the identified risk variant of miR-30e and peripheral miR-30 levels could not be performed due to the low prevalence of the CT allele (Xu et al. 2010). This study illustrates the unfortunate fact that clinical samples cannot always be interrogated for the impact of a given SNP, due to low population prevalence. Nonetheless, there are at least two innovative ideas contained in this study that take it beyond simple genetic association data. Firstly, by narrowing their focus to a collection of miRNAs already shown to be dysregulated in SCZ postmortem studies, they were able to reduce the number of statistical tests and therefore false positives which are a common problem of GWAS. Of course there are problems with narrowing the search when the pool of studies to draw from is relatively small, but this innovative study nonetheless indicates an approach that could be applied in the future when the initial GWAS findings are more robust. Secondly, the reversal of this process, that is, examining the peripheral levels of an miRNA shown to be genetically associated with SCZ allows the assessment of potential miRNA biomarkers.

Potkin et al. (2010) conducted a GWAS with the aim of determining which of the identified SNPs were related to the BOLD fMRI signal from the DLPFC emitted during a working memory (WM) task. Begemann et al. (2010) also focused on functionality and attempted to determine whether one SCZ candidate gene, complexin 2 (CPLX2) was related to 3000 phenotypic data points they collected from their participants, which ranged from environmental factors such as prenatal infection through to tests of WM and executive function. Therefore, we have on the one hand, a study where an entire GWAS data set is interrogated for SNPs that relate to brain activity during particular cognitive task (Potkin et al. 2010), while on the other, a study that attempts to determine whether SNPs on a SCZ risk gene are associated with a range of SCZ cognitive phenotypes (Begemann et al. 2010). In both cases, although the emphasis is different, the attempt to link function to SNPs in the context of SCZ is clear.

Begemann et al. (2010) discovered one SNP on the 3'-UTR binding site of CPLX2 that was correlated with the cognitive phenotype. This SNP resulted in reduced binding of miR-498 *in vitro* and was linked to the SCZ phenotype when Cplx2-null mutant mice produced cognitive impairments analogous to those seen in SCZ patients. They were thus able to confirm that the SNP discovered in that gene disrupted this miRNA–target interaction. They were then able to relate the disrupted miRNA–target interaction to cognitive impairment in Cplx2-null mutant mice, which was similar to the impairment demonstrated by patients with this SNP. These results constitute a good example of how GWAS can firstly be enhanced by relating the data to a cognitive phenotype and secondly, once SNPs are identified how these may be related back to functional interactions. In this case, it is of course highly unlikely that the single SNP on CPLX2 and the reduced binding to miR-498 are solely responsible for the SCZ phenotype, but it provides evidence that miRNA and SCZ risk gene interactions are important in producing this phenotype.

Potkin et al. (2010) also investigated functional interactions, producing in their case, mixed results. They subjected each SCZ-related SNP identified in their GWAS, to genome enrichment analysis (GEA), in order to determine whether there were any predetermined functional gene sets enriched for their candidate genes. They found that these functional gene sets included multiple targets of miRNAs including most notably miR-137 that was later strongly associated with SCZ in a large scale GWAS (Schizophrenia Psychiatric Genome-Wide Association Study 2011). However, the *p* values of the GEA were not significant, the authors merely ranked the strongest nonsignificant associations. This indicates that GWAS candidate genes even when examined in relation to a cognitive phenotype, still suffer from an inability to be replicated and cannot always be categorized neatly into functional categories.

The functional investigation of SNPs has taken additional forms, which include *in vitro* assays designed to determine the effect of a particular miRNA SNP on the expression of the mature miRNA molecule and its proposed target. Guella et al. (2013) and Feng et al. (2009) with the exception of miR-510 showed that a risk-associated miRNA SNP when cloned into the relevant miRNA molecule had no effect on its mRNA target. In other studies, the particular miRNA mutation had no effect on the expression of the mature miRNA molecule (Sun et al. 2009; Guella et al. 2013). Taken together these studies show that a given risk SNP has no effect on either the expression of the miRNA molecule itself or its proposed target and call into question the clinical utility of the identification of individual SNPs in a polygenic condition such as SCZ.

### 3.3.5  DELVING DEEPER INTO miR-137: AN ILLUSTRATION OF FUNCTIONAL STUDIES

The miR-137 SNP has been examined from various angles in the SCZ literature, ranging from phenotypic association studies (Cummings et al. 2013; Green et al. 2013; Lett et al. 2013) to functional *in vitro* studies (Guella et al. 2013; Kwon et al. 2013) and therefore constitutes a good example of how these techniques can be used to determine the functional implications of a given SNP. The SNP in miR-137 refers to rs1625579 that was identified within an intron of the primary transcript for miR-137. This SNP was discovered in one of the largest SCZ GWAS to date and deemed the most significant locus with several other significant loci emerging from this study found to be targets of miR-137 (Schizophrenia Psychiatric Genome-Wide Association Study 2011).

Guella et al. (2013) found that the TT risk allele reduced the expression of miR-137 in the DLPFC but only in controls, the significance of which is unclear (Guella et al. 2013). A finding that was consistent across both controls and patients was that the TT risk allele increased the expression of TCF4, which is one of the targets of miR-137. However, the mechanism by which the TT risk allele resulted in increased expression of TCF4 is unclear, as we can rule out changes in miR-137, as miR-137 was only altered in controls. The authors suggested that this particular miR-137 SNP may not be the functional variant influencing miR-137 expression but is tagging another allele that has an effect on the functional expression of miR-137.

The existence of such genetic networks involving miR-137 was illustrated by the GWAS which discovered this risk allele (Schizophrenia Psychiatric Genome-Wide Association Study 2011). The results of this study indicated that targets of miR-137 namely TCF4, CACNA1C, CSMD1, and C10orf26 were more likely to be identified as SCZ risk genes, a finding which implicates miR-137 in a genetic network of SCZ pathogenesis. All these genes have been previously linked to SCZ (Shi et al. 2009; Stefansson et al. 2009; Green et al. 2010; Schizophrenia Psychiatric Genome-Wide Association Study 2011; Navarrete et al. 2013). In particular, TCF4 has been strongly associated with SCZ in several independent studies (Stefansson et al. 2009; Navarrete et al. 2013), while CACNA1C is an L-type voltage-dependent calcium channel which has been implicated in variety of neuropsychiatric disorders including SCZ (Shi et al. 2009; Green et al. 2010). Furthermore, these genes have been also validated as miR-137 targets by *in vitro* methods (Kim et al. 2012; Kwon et al. 2013), a finding which confirms their functional role within a genetic network potentially regulated by miR-137.

Another set of studies that illustrates a way in which the functional implications of an SNP can be scrutinized, are those devoted to examining the relationship between miR-137 and clinical presentation. Lett et al. (2013) found no association between the miR-137 risk allele and any of the clinical symptoms of SCZ (Lett et al. 2013). Instead they found that this risk allele was related to morphological characteristics of SCZ, such as larger left ventricles, smaller hippocampus volume, and lower white matter integrity, which are among some of the most consistent findings in SCZ (Steen et al. 2006; Vita et al. 2012). This finding, coupled with the fact that patients with this allele had a significantly earlier age of onset, serves to illustrate how studies can begin to draw links between miRNA risk alleles and well-established features of SCZ.

In contrast to the study by Lett et al. (2013), the following two studies found that the miR-137 risk allele was associated with at least one of the symptoms of SCZ. Cummings et al. (2013) found that the risk allele was associated with cognitive impairment in the form of lower performance on attention and episodic memory tasks (Cummings et al. 2013). Likewise, Green et al. (2013) found that the miR-137 risk allele predicted membership to the cognitive deficit group of SCZ patients when examined in conjunction with the negative symptom score of their patients. It is indicative that none of the three studies found any association between the miR-137 risk allele and the positive symptoms of SCZ (Cummings et al. 2013; Green et al. 2013; Lett et al. 2013). These results coupled with the fact that Potkin et al. (2010) linked this mutation to increased activity in the DLPFC during a WM task indicate that this risk allele may be specifically targeting the cognitive symptoms of SCZ (Potkin et al. 2010). Further support for this theory comes from the finding that patients with this risk allele had a smaller hippocampus volume (Lett et al. 2013), which is a structure heavily involved in memory and cognition (Shohamy and Turk-Browne 2013). The hippocampus is also implicated in neurogenesis, which is a process that has also implicated miR-137 (Silber et al. 2008). It is possible, therefore, that the miR-137 risk allele affects functional pathways involved in neurogenesis in the hippocampus leading to the cognitive deficits observed in carriers of this allele (Cummings et al. 2013; Green et al. 2013).

Following on from this, further research could focus on the neurobiological mechanisms underlying these cognitive deficits with an aim toward determining the role miR-137 may play in those interactions. Taken together, these studies demonstrate that the utility of an SNP, such as the miR-137 SNP is enhanced when examined in relation to clinical phenotypes and the potential molecular mechanisms generating these phenotypes.

Having said this, the fact still remains that the miR-137 SNP was responsible for less than 1% of the variance in neuropsychological scores between patients and controls (Cummings et al. 2013). This indicates that when examined on its own rather than in conjunction with target genes as part of a regulatory network (Schizophrenia Psychiatric Genome-Wide Association Study 2011) or in relation to a specific cognitive phenotype (Cummings et al. 2013; Green et al. 2013) the miR-137 SNP has limited diagnostic power.

Given the complex polygenic nature of SCZ it is unlikely that risk associated with miRNA dysregulation stems from a single miRNA and instead it has been postulated that it is the process of miRNA biogenesis that is globally dysregulated in SCZ (Beveridge et al. 2010; Santarelli et al. 2011). Accordingly, several studies have set out to explore whether miRNA biogenesis is indeed dysregulated in this disorder, employing a variety of methods that range from assaying for the expression of miRNA biogenesis genes (Beveridge et al. 2010; Santarelli et al. 2011; Zhang et al. 2012) to animal models of Dgcr8 haploinsufficiency (Stark et al. 2008; Fenelon et al. 2011; Schofield et al. 2011; Earls et al. 2012; Ouchi et al. 2013).

### 3.3.6   IS IT GLOBAL: THE INVESTIGATION OF MicroRNA BIOGENESIS DYSREGULATION IN SCZ

One of the strongest arguments of miRNA biogenesis involvement in the pathophysiology of SCZ is the fact that in most studies there is no difference between patients and controls in the expression levels of the initial pri-miRNA transcript (Beveridge and Cairns 2012; Maffioletti et al. 2014). The difference arises later at the pre-miRNA and mature miRNA stage, by which time miRNA biogenesis proteins have interacted significantly with the developing miRNA molecule (Krichevsky et al. 2006; Krol et al. 2010b). These proteins include Drosha and Dgcr8, which form the microprocessor complex as well as Exportin 5 that transports pre-miRNA into the cytoplasm and Dicer that performs the final cleavage to form the mature miRNA molecule (Krol et al. 2010b). As each of these proteins performs a crucial role in the miRNA biogenesis pathway, it is possible that mutations in any of these biogenesis genes may affect the formation and expression of mature miRNA molecules, resulting in aberrant mRNA regulation and the development of SCZ.

A particularly strong candidate for dysregulation in SCZ is Dgcr8. Dgcr8 is one of the genes implicated in Di George syndrome, which has a 30% conversion rate to SCZ (Drew et al. 2011), a fact which indicates it may be of primary importance in the miRNA biogenesis/SCZ narrative. In addition to this, mouse models of Dgcr8 haploinsufficiency examined in Section 3.3.7 display a genetic and behavioral phenotype analogous to SCZ (Stark et al. 2008; Earls et al. 2012; Ouchi et al. 2013). Dgcr8 is also the most consistently upregulated miRNA biogenesis gene in human postmortem studies, in both the DLPFC and superior temporal gyrus (STG) (Beveridge et al. 2010; Santarelli et al. 2011). The consistent upregulation of Dgcr8 in SCZ patients, contrasts that of nearly all the other main miRNA biogenesis proteins. Exportin 5 showed no difference in expression across two studies, while Dicer was upregulated in the DLPFC but not the STG of SCZ patients (Beveridge et al. 2010; Santarelli et al. 2011). Drosha on the other hand, was only upregulated in one study (Beveridge et al. 2010), a finding which was not subsequently replicated (Santarelli et al. 2011). The fact that Drosha is inconsistently upregulated in SCZ raises interesting questions as both Drosha and the consistently upregulated Dgcr8 are part of the microprocessor complex and are both deemed to be rate limiting (Krol et al. 2010a). It is expected therefore that both genes should show dysregulation in SCZ if miRNA biogenesis is indeed globally dysregulated in this disorder. In contrast to this, Beveridge et al. (2010) found that although both the miRNAs tested (miR-181b and miR-26b) showed overexpression at the pre-miRNA and mature transcript level which would indicate microprocessor involvement, Dgcr8 was the only one to show concomitant upregulation. This indicates that Dgcr8 was solely responsible for the overexpression of the two candidate miRNAs in patients with SCZ.

Although there is strong evidence for the involvement of Dgcr8 in miRNA biogenesis dysregulation, this does not explain how the consistent upregulation of an miRNA biogenesis gene leads to the downregulation of selected miRNA molecules in some postmortem studies of SCZ (Kim et al. 2010; Moreau et al. 2011; Maffioletti et al. 2014). One would expect that if widespread upregulation of the miRNA biogenesis machinery is a feature of SCZ, this would result in equally widespread upregulation of miRNAs. Further research is therefore needed to determine the mechanisms by which miRNA biogenesis genes target miRNAs which may serve to explain why some miRNAs are upregulated while others remain at normal levels or are even downregulated (Beveridge et al. 2008; Beveridge et al. 2010; Moreau et al. 2011; Maffioletti et al. 2014). Secondly, given that a study by Zhang et al. (2012) into 59 miRNA machinery genes yielded only one mutation significantly associated with SCZ, it would be interesting to determine whether specific miRNA biogenesis genes such as Dgcr8 are particularly sensitive to mutation or alternatively whether they interact with other SCZ risk genes (Zhang et al. 2012). Finally, given that Dgcr8 itself did not survive correction for multiple testing in Zhang et al. (2012), it will be instructive to continue research into biogenesis gene expression with expanded postmortem samples to resolve some of the discrepancies noted above (Beveridge et al. 2010; Santarelli et al. 2011).

A potential explanation for the discrepancies in miRNA biogenesis expression in SCZ is that the disorder is not attributable to the dysregulation of a particular miRNA biogenesis gene but rather that specific regions of the chromosome are particularly susceptible to genetic and environmental perturbations that lead to miRNA biogenesis gene dysregulation. This would appear to be the case in the 22q microdeletion syndrome, which is associated with both DiGeorge and velocardiofacial syndrome, both of which display a 30% increased risk of SCZ (Drew et al. 2011).

Just such a sensitive region was discovered by Gardiner et al. (2012) who found that 17 of the 33 downregulated miRNAs in PBMCs from patients with SCZ could be localized to a particular region of the genome (Gardiner et al. 2012). This was the DLK1-DIO3 region on chromosome 14q32 that consists of over 5% of all the known human miRNA genes (Cavaille 2007). The fact that the study by Gardiner et al. (2012) implicated a particular region of the chromosome rather than a diverse array of SNPs distributed throughout the genome indicates that GWAS into miRNA dysregulation might benefit from focusing on the identification of similarly critical regions. Gardiner et al. (2012) attributed their results to a possible common aberration in the transcriptional regulation process of this particular chromosomal cluster. In contrast to this speculation, however, a study by Moreau et al. (2011) into miRNA expression levels in the postmortem brain found that none of the miRNAs that were downregulated in their SCZ sample arose from genomically proximal hairpins, a finding which suggests that aberrant transcriptional regulation of those miRNAs is unlikely to be the cause behind their dysregulation (Moreau et al. 2011). These two studies demonstrate that there is still uncertainty about whether the aberrant expression of miRNAs seen in SCZ patients can be attributed to regional dysregulation of the process of transcription or to a collection of common mutations dispersed throughout the genome.

Another possibility put forward by Gardiner et al. (2012) to explain their regionally localized miRNA results was that the 14q chromosomal region had been shown to be particularly sensitive to miRNA biogenesis gene mutations, such as mutations to Dgcr8. In particular, Dgcr8 heterozygous mice were found to have significant downregulation of miRNAs in the syntenic region of chromosome 12 (Stark et al. 2008). The global dysregulation of miRNA expression localized to that particular chromosomal region could in that case be explained by a disruption of the miRNA biogenesis process (Gardiner et al. 2012). The miRNA biogenesis process is also implicated in the study by Moreau et al. (2011) who showed that eight of their aberrantly expressed miRNAs were also dysregulated in the Dgcr8 heterozygous mouse model (Stark et al. 2008). Taken together, these studies suggest that although there are still doubts about whether miRNA dysregulation is chromosomally localized, the process of miRNA biogenesis in the form of mutations in crucial proteins such as Dgcr8 may be the common cause behind the diverse findings of miRNA dysregulation in SCZ. In order to elucidate the role that Dgcr8 biogenesis protein plays in the development of SCZ, several studies have used the mouse model of Dgcr8 haploinsufficiency examined in the following section (Stark et al. 2008; Fenelon et al. 2011; Schofield et al. 2011; Earls et al. 2012; Ouchi et al. 2013).

### 3.3.7  A MODEL OF DYSREGULATED MicroRNA BIOGENESIS: THE DGCR8 HAPLOINSUFFICIENCY MOUSE

The Dgcr8 haploinsufficiency mouse model has one allele of the Dgcr8 gene deleted as a model of the 22q microdeletion syndrome that includes such disorders as DiGeorge and velocardiofacial syndrome, which are associated with a 30% risk of developing SCZ (Drew et al. 2011).

This mouse model displays one of the main phenotypes of SCZ in the form of a spatial WM deficit, which has been replicated across various studies and laboratories (Stark et al. 2008; Ouchi et al. 2013). In some cases, male mice are the only ones susceptible to this deficit (Ouchi et al. 2013) a fact which may confirm the estrogen protective effects on SCZ observed in the clinical population (Jackson et al. 2013). The Dgcr8 mouse model also presents with various affective traits such as increased fear and greater anhedonia (Stark et al. 2008; Ouchi et al. 2013), as well as some deficits in nesting and social interaction (Ouchi et al. 2013), which can be seen as correlates of the negative symptoms of SCZ. Finally, as regards the positive symptoms of SCZ, the Dgcr8 mouse has impaired performance on tests of prepulse inhibition (PPI; Stark et al. 2008), which is considered a sensorimotor correlate of psychosis. The Dgcr8 haploinsufficient mice present therefore with all three symptoms of SCZ (positive, cognitive, and negative), a finding which is particularly interesting when one considers that the only element that distinguishes them from wild-type (WT) mice is that they carry only one allele of Dgcr8 and clearly implicates miRNA biogenesis in the development of SCZ symptomatology.

Nonetheless, a mouse model can never capture the complexity of this human neuropsychiatric disorder and indeed the usefulness of the Dgcr8 mouse model of SCZ, lies more in its use as a method for exploring the molecular mechanisms by which Dgcr8 deletion may lead to the development of SCZ symptoms. In line with this aim, Ouchi et al. (2013) demonstrated that the Dgcr8 haploinsufficiency mouse model displayed not only cognitive impairment in line with SCZ but also reduced hippocampus neurogenesis and downregulation of a specific subset of SCZ risk genes found to be reduced in postmortem and peripheral tissue from clinical population. This study therefore highlights a potential molecular pathway by which Dgcr8 haploinsufficiency leads to decreased hippocampus neurogenesis and cognitive impairment. In addition, this study served to validate the Dgcr8 mouse model as it was shown to demonstrate a behavioral and genetic profile similar to that seen in SCZ patients.

One of the most replicated and hardly unexpected findings is that Dgcr8 haploinsufficiency reduces Dgcr8 expression by approximately 50% (Stark et al. 2008; Schofield et al. 2011). In accordance, with the reduction in Dgcr8 expression, the expression of many but by no means all, miRNA molecules is also reduced (Schofield et al. 2011). In particular, Earls et al. (2012) showed that Dgcr8 haploinsufficiency led to decreased expression of miR-25 and miR-185 in the hippocampus, whereas the same mutation in the same species produced decreased expression of another set of miRNAs, namely miR-134 and miR-491, in the medial PFC (mPFC) (Schofield et al. 2011). It is indicative that this collection of mouse studies fails to show uniform miRNA dysregulation, while showing strong replication of the behavioral phenotype. One might expect that the haploinsufficiency of the Dgcr8 protein, critical to miRNA biogenesis would produce widespread effects that would vary between individual miRNA molecules. While these differences could be attributed to brain region, these varied results highlight the need for stringent investigation and categorization of miRNA expression as a result of Dgcr8 haploinsufficiency across multiple regions of the mouse cortex. This exercise could then inform human postmortem examinations of miRNA expression, due to the fact that many of the seed regions of miRNAs are conserved across both species (Schofield et al. 2011).

Nonetheless, despite the variability in miRNA expression, there is general consensus in these studies that Dgcr8 haploinsufficiency leads to a reduction in dendritic complexity in both mPFC (Fenelon et al. 2011; Schofield et al. 2011) and the hippocampus (Stark et al. 2008). This would indicate that mutations in Dgcr8 likely affect the development of neurons and the subsequent formation of neuronal circuits. In light of this, it seems strange that while the reduction in dendritic complexity is a well-replicated result, the electrophysiological results from these studies remain varied. While there is consensus that Dgcr8 haploinsufficiency leads to a reduction in excitatory synaptic activity in the mPFC (Fenelon et al. 2011; Schofield et al. 2011), enhanced excitatory activity represented by increased long-term potentiation (LTP) was demonstrated in the hippocampus (Earls et al. 2012). It is possible, therefore, that neurons which share the characteristic of reduced dendritic complexity are acted upon by other factors downstream of the Dgcr8 mutation, to produce sometimes opposing electrophysiological results. These factors may include but are not restricted to, environmental cues that may vary between brain regions, which would explain the diversity in electrophysiological activity observed between hippocampus and mPFC. In order for these downstream factors to be identified, however, further experiments are required potentially utilizing viral tracers and conditional KO models, which would elucidate the functional mechanisms by which a mutation in Dgrc8 affects miRNA expression and subsequent neuronal development.

One of the ways of exploring the functional mechanism of this miRNA biogenesis gene may be by determining which miRNAs are consistently downregulated as a result of this mutation. While there is great diversity in miRNA expression in the Dgcr8 mouse (Schofield et al. 2011; Earls et al. 2012), two mi-RNAs, miR-491 and miR-134, have been found to be consistently downregulated in the hippocampus and mPFC (Stark et al. 2008; Schofield et al. 2011). Of the two, miR-134 is of particular interest, as it was also downregulated in PBMCs from a human clinical population and occupied a position of prime importance in the 14q32 cluster identified by Gardiner et al. (2012) as particularly susceptible to both miRNA biogenesis mutations and increased SCZ risk. These findings provide an association between the miRNA biogenesis gene Dgcr8 and miR-134 in the context of SCZ in both mouse and clinical studies. It is particularly interesting to note that miR-134 was shown to regulate dendritic morphology in the hippocampus (Fiore et al. 2009), a finding which indicates the widely replicated reduction in dendritic complexity in the Dgcr8 mouse may be due to the concomitant reduction of miR-134. It is associations of this kind between miRNA biogenesis mutations and their downstream effects on miRNA expression, neuronal morphology, and subsequent electrophysiological functionality that can and should be pursued in the elucidation of the pathophysiology of SCZ.

In conclusion, there is some evidence that the varied and sometimes inconsistent results arising from GWAS could be attributed to global miRNA biogenesis dysregulation in SCZ rather than the contribution of individual SNPs. However, this does not obscure the fact that investigations into miRNA biogenesis genes and their relationship to SCZ are themselves plagued by inconsistency (Beveridge et al. 2010; Santarelli et al. 2011). Furthermore, many of the genes regulating miRNA biogenesis and functioning are not associated with SCZ (Zhang et al. 2012).

Nonetheless, Dgcr8 is an miRNA biogenesis gene that appears to have strong links with SCZ. This is due to the fact that its expression is most consistently downregulated in SCZ patients especially when compared with other miRNA biogenesis genes (Beveridge et al. 2010; Santarelli et al. 2011). Furthermore, Dgcr8 is implicated in DiGeorge syndrome which has a 30% conversion rate to SCZ (Drew et al. 2011), while the mouse model of Dgcr8 haploinsufficiency displays both aberrant miRNA expression and SCZ impairments (Stark et al. 2008; Ouchi et al. 2013). Further research could focus on the molecular mechanisms by which aberrant miRNA expression in this mouse model gives rise to dysregulated neuronal development that would be of particular importance to clinical applications, which may target particular miRNAs known to be vulnerable in these population.

Taken together, these studies simultaneously caution us against the overreliance on algorithms to identify potential SNPs as well as putative genetic targets and also constitute a testament to the importance of functional assays in determining the accuracy of those predictions.

## 3.4    EFFECT OF ANTIPSYCHOTIC MEDICATIONS ON MicroRNA EXPRESSION

While the assessment of the effects of various antipsychotic medications on expression of miRNAs has been limited, nonetheless studies to date provide evidence indicating that these drugs can alter levels of miRNAs. One of the first studies to assess antipsychotic medication effects on miRNA expression was that by Perkins et al., in 2007. In this investigation, the authors assessed the expression of a panel of 264 miRNAs within the PFC of subjects with SCZ versus psychiatrically normal controls and found that 15 miRNAs were expressed at significantly lower levels and 1 expressed at higher levels in patients with SCZ versus controls. In order to control for medication effects, the authors exposed rats to haloperidol for 4 weeks and compared to nonexposed rats. Of interest, they found that haloperidol caused an increase in miR-199a, miR-128a, and miR-128b (Perkins et al. 2007), with none of these miRNAs being altered in the PFC of subjects with SCZ. This finding points to the potential for antipsychotic medications in having treatment-specific effects that are separate to correcting any dysregulation in miRNA species that may be due to the underlying etiology of the disorder. However, in relation to antipsychotics ability to restore normal miRNA levels, it has also been demonstrated that pretreatment with haloperidol or clozapine prevented a reduction in miR-219 in the PFC of mice that was elicited through exposure to the NMDA antagonist dizocilipine (Kocerha et al. 2009). This is of interest given that other NMDA antagonists such as phencyclidine can induce psychotomimetic effects that mimic that seen in SCZ and is in keeping with an NMDA hypofunctioning as an underlying cause. Another, more recent investigation found that miR-193 was reduced in expression in the whole brain of mice treated with olanzapine as well as miR-434-5p and miR-22 being reduced by haloperidol (Santarelli et al. 2013). While the results for haloperidol are different to the previous study by Perkins, it is important to note that the latter study was conducted in whole brain and therefore may have masked the ability of the authors to pick up PFC-specific miRNA changes as well as adding new dysregulation data from other regions. This is important in the context of regional as well as gender differences in miRNAs that have been recently reported within the human brain (Ziats and Rennert 2014) and that are likely to be reflected in rodent brains. In relation to the ability of antipsychotics in modulating miRNAs together with their known therapeutic mechanism as dopamine D2 receptor antagonists a recent paper published by Chun et al., demonstrates a series of elegant experiments in a *Df(16)1/+* mouse model of SCZ that mimics a 22q11DS, with 30% of individuals with this deletion developing SCZ. *Df(16)1/+* mice have aberrant elevation of DRD2 within the thalamus, with haploinsufficiency in *Dgcr8* with resulting synaptic defects leading to reductions in excitatory postsynaptic currents (EPSCs) in thalamocortical neurons. Importantly, exposure to either haloperidol or clozapine were able to reverse these synaptic defects and normalize EPSC amplitudes to WT levels (Chun et al. 2014). This correction of *Dgcr8*-mediated synaptic defects by antipsychotic medications both new and old potentially provides the strongest evidence to date of miRNAs being a potential therapeutic target for antipsychotic medications. A preliminary study administering antipsychotic medications (olanzapine, quetiapine, ziprasidone, and risperidone) to SCZ patients has found that miRNA-181b was significantly downregulated in patients with SCZ and that there was a correlation between this downregulation and symptomatology improvement (Song et al. 2014). Finally, it is noteworthy that while miRNAs may offer potentially novel therapeutic targets their dysregulation may also be associated with side effects that are known to occur with antipsychotic medications. This has been demonstrated in a recent paper by Gardiner et al., whereby the authors investigated peripheral gene expression and miRNA changes in T-lymphocytes exposed to chlorpromazine, haloperidol, and clozapine *in vitro*. Their findings demonstrated that there were interactions between dysregulated miRNAs (upregulation of miR-942, miR-362-5p, and miR-421 (chlorpromazine); downregulation of miR-17-3p (clozapine); downregulation of miR-200c-3p, miR-28-5p, and miR-624-5p (haloperidol)) that were associated with known metabolic pathways linked to side effects such as weight gain with antipsychotic medications (Gardiner et al. 2014). Therefore, while miRNAs may offer novel therapeutic targets, they may also offer the potential as molecules that can be monitored in relation to adverse effects for drugs used to treat psychosis and SCZ.

## 3.5    WHAT DOES IT ALL MEAN: IN SEARCH OF A FUNCTIONAL NARRATIVE OF MicroRNA DYSREGULATION

### 3.5.1    Introduction

In the investigation of the relationship between miRNA dysregulation and SCZ, it is important to consider the functional implications of any significant associations that may arise from these studies. As we have seen, the GWAS, postmortem, and peripheral studies in miRNA dysregulation in SCZ have produced varied results, which often suffer from a lack of replication. It is important therefore, to examine those associations from a mechanistic perspective in order to determine the precise nature of the interaction between miRNA and target molecule and the functional networks that are perturbed as a result. By exploring these aspects of miRNA dysregulation, we may start to unravel the narrative of miRNA dysregulation and begin to answer some of those pervasive questions that include:

1. Is aberrant miRNA expression in SCZ a local (SNP based) or a global phenomenon arising from the dysregulation of miRNA biogenesis?
2. Are a disproportionate number of SCZ risk genes also targets of miRNAs?
3. What effects do dysregulated miRNAs have on their target molecules?
4. Are these consistently replicated miRNA molecules participating in particular functional pathways (e.g., neurogenesis) implicated in SCZ and is that the basis of their effect?

The first question was explored in the previous section devoted entirely to the genetics of miRNAs and the relationship of individual SNPs to the pathogenesis of SCZ and the promising area of miRNA biogenesis investigation. The final questions will be investigated in this section, with specific focus on studies devoted to exploring the interaction between miRNA and their target molecules. Subsequently, the functional pathways frequently implicated in these investigations will be presented and discussed with a final section on future directions in miRNA research.

### 3.5.2 ARE A DISPROPORTIONATE NUMBER OF SCZ RISK GENES ALSO TARGETS OF MicroRNAs?

This is a crucial question as if answered in the affirmative it positions miRNA dysregulation in the forefront of SCZ pathogenesis. It is also difficult to answer due to the fact that an individual miRNA may have many targets. In this section, we examine the targets of miRNAs with particular emphasis on those independently established SCZ risk genes that are consistently replicated across studies. Each replicated SCZ risk gene is then discussed in terms of the neurotransmitter system that is implicated in its dysregulation. Over the course of the study of SCZ, several neurotransmitter systems have been put forward as the underlying cause of SCZ symptomatology (Howes and Kapur 2009; Moghaddam and Javitt 2012) and it is therefore important to determine if miRNA dysregulation is implicated in the functioning of those systems. Finally, this section concludes with a discussion of the premise that a disproportionate number of SCZ risk genes are targets of miRNAs.

Several studies have implicated a larger number of SCZ risk genes as putative targets of miRNAs. Table 3.2 lists postmortem and peripheral studies with selected miRNAs that were found to be dysregulated and their target SCZ risk genes.

This table can only be considered a summary of results, highlighting the SCZ risk genes most consistently implicated in miRNA dysregulation. This is due to the fact that each miRNA targets many genes and among them, a substantial portion of SCZ genes. An example of this is evident in the study by Zhu et al. (2009), who investigated all known miRNAs and compared the frequencies with which they targeted SCZ-associated genes (Zhu et al. 2009). The two miRNAs they discovered that most consistently targeted SCZ genes, miR-346 and miR-566 targeted 121 and 162 SCZ risk genes, respectively, which gives us an idea of the large number of SCZ-related targets an individual miRNA may have.

The highlighted results in the table demonstrate SCZ risk genes that have consistently been associated with dysregulated miRNAs. Early growth response protein (EGR3) is a molecule shown by Guo et al. (2010) to be the main regulator of their *in silico* SCZ-related network and in light of this, it is interesting that it is also validated by Santarelli et al. (2011). Other validated targets include well-known SCZ risk genes that are associated with the glutamatergic system such as AMPA and GRIA2 (Beveridge et al. 2008) as well as the subunits of the NMDA receptor GRIN1-3, which appear multiple times (Kim et al. 2010;

## TABLE 3.2
## MicroRNAs (MiRNAs) Found to be Dysregulated in Postmortem and Peripheral SCZ Studies and their Target SCZ Risk Genes

| Study | MicroRNA | Target SCZ Risk Gene |
|---|---|---|
| Feng et al. (2009) | Let7f, miR-18b, miR-510, miR-188, miR-502, miR-505, miR-325 | *DISC1, NRG1, RGS4*, GRM3 |
| Zhu et al. (2009) | MiR-346 various but the above miRNA targets SCZ genes with greater frequency than other genes | CSF2RA, *GRIN2C*, DGCR6L, and others |
| Beveridge et al. (2008) | MiR-181b | AMPA, GRIA2 |
| Beveridge et al. (2010) | MiR-15 family, miR-107, miR-181b, miR-16, and miR-20a | *BDNF, NRG1, RELN, DRD1*, HTR4, GABR1, *GRIN1*, GRM7, *CHRM1*, and ATXN2 |
| Guo et al. (2010) | Various including: miR-195 | *BDNF, EGR3* |
| Kim et al. (2010) | MiR-132 | *GRIN1, GRIN2A, GRIN2B* (NMDA receptor subunits), *BDNF*, and *DRD1* |
| Santarelli et al. (2011) | MiR-328, miR-17-5p, miR-134, miR-652, miR-382, and miR-107 (28 total: the above validated by RT-PCR) | 299 SCZ risk genes including: NMDA, *GRIN1–3*, five serotonin receptors, *NRG1* and *BDNF*, *RELN, EGR3* |
| Miller et al. (2012) | MiR-132 | DNMT3A, GATA2, DPYSL3 |
| Gong et al. (2013) | MiR-124 | *RGS4* |
| Kwon et al. (2013) | MiR-137 | CSMD1, C10orf26, CACNA1C, and *TCF4* |
| Navarette et al. (2013) | MiR-137 | *TCF4* |
| Scarr et al. (2013) | MiR-107 | *CHRM1* |
| Wong et al. (2013) | MiR-17 | NPAS3 |

*Note:* The genes in italics indicate genes that have been identified multiple times in miRNA studies.

Santarelli et al. 2011). This result implicates miRNA regulation in the pathophysiology of the NMDA hypofunctioning theory of SCZ (Moghaddam and Javitt 2012). In light of this, it is interesting that one of the studies examined below also implicates miR-219 in the functional regulation of the NMDA receptor (Kocerha et al. 2009).

Another enduring theory regarding the pathophysiology of SCZ is the dopaminergic hypothesis, arising from the clinical effectiveness of antipsychotic drugs that antagonized the D2 receptor (Howes and Kapur 2009). MiRNA dysregulation in SCZ also impacts the DA system, as replicated miRNA targets include the dopaminergic risk gene DRD1 (Beveridge et al. 2010; Kim et al. 2010).

Furthermore, there is also some evidence for the cholinergic system being involved in miRNA dysregulation, with the cholinergic muscarinic receptor 1 (CHRM1) also appearing as a target of dysregulated miRNAs (Beveridge et al. 2010; Scarr et al. 2013). This finding is important given the role the cholinergic system plays in learning and memory, both of which are cognitive traits dysregulated in SCZ (Volpicelli and Levey 2004).

Given the importance of synapse formation in neurodevelopment for the maturation of cognition (Lu et al. 2014), it is interesting that BDNF is one of the most widely replicated SCZ risk gene targets of aberrantly expressed miRNAs (Beveridge et al. 2010; Kim et al. 2010; Santarelli et al. 2011). Indeed, one study discussed in the following section showed that an interaction between BDNF and miR-195 affected the functioning of the GABAergic system (Mellios et al. 2009). The GABAergic system is further implicated when we take into account that another replicated SCZ target gene is reelin (RELN), which is expressed mainly on GABAergic interneurons (Beveridge et al. 2010; Santarelli et al. 2011). Interestingly, the dysfunctioning of GABAergic interneurons in the PFC is also one of the main tenants of the hypoglutamatergic hypothesis of SCZ (Moghaddam and Javitt 2012). This taken in conjunction with the fact that the miRNAs frequently implicated in SCZ target both RELN and the NMDA receptor, lend support to the theory that miRNA dysregulation may contribute to the hypoglutamatergic state shown in some cases to characterize SCZ. Nonetheless, we cannot disregard the involvement of either the DA or the noradrenergic systems, as both have been shown to be impacted by miRNA dysregulation (Howes and Kapur 2009; Scarr et al. 2013). Indeed it seems more likely that SCZ is a product of the interaction of multiple neurotransmitter systems, each of which has been associated with a variety of miRNAs.

Finally, regardless of the neurotransmitter systems involved, it would appear that the studies discussed above indicate that miRNAs disproportionally target SCZ risk genes. A study by Zhu et al. (2009) compared the frequency with which miRNAs targeted SCZ susceptibility genes versus a random cross section of genes. Their sample consisted of all miRNAs known at the time (>500) as well as 455 SCZ susceptibility genes selected using the Schizophrenia Gene database. They found that for all but two miRNAs examined, there was no difference in the frequency with which they targeted SCZ risk genes versus a random cross section of human genes. These results strike a cautionary note in the interpretation of results regarding the involvement of SCZ risk genes in miRNA dysregulation. In particular, they indicate that the main reason that a vast number of SCZ susceptibility genes emerge as targets of miRNAs is simply that miRNAs have many targets. For example, the average number of overall predicted targets for a given miRNA in this study was approximately 1500 genes, with some miRNAs having up to 3000 putative genetic targets. These results, therefore, indicate that the vast majority of miRNAs do not preferentially target SCZ susceptibility genes. This does not, however, exclude the possibility that interactions between miRNAs and SCZ susceptibility genes contribute to the pathophysiology of SCZ. Rather it indicates that a different approach is needed, moving away from interactions established using bioinformatic algorithms, to a mechanistic approach that is capable of assessing the strength and implications of those interactions. The following section is, therefore, devoted entirely to studies that investigate the effect an identified miRNA has on its target molecules from a mechanistic perspective and the downstream implications of that interaction.

### 3.5.3 WHAT EFFECT DO DYSREGULATED MicroRNAs HAVE ON THEIR TARGET MOLECULES?

As seen in the previous section, miRNAs shown to be dysregulated in SCZ (Kim et al. 2010; Santarelli et al. 2011) target a large number of NMDA receptor subunit genes. This finding combined with the prevalence of the hypoglutamatergic theory of SCZ that attributes the dysfunctioning of multiple neurotransmitter systems to a hypoactive NMDA receptor (Garrido et al. 2009; Moghaddam and Javitt 2012), make the following study that investigated the interaction between NMDA receptor and a candidate miRNA particularly interesting.

#### 3.5.3.1 MiR-219-NMDAR

Kocerha et al. (2009) showed that antagonism of the NMDAR caused a marked decrease (50%) in miR-219 in the mouse PFC and behavioral deficits considered sensorimotor correlates of psychosis. These results indicate that there may be an interaction between the NMDAR and miR-219, which when perturbed by NMDAR antagonism, results in an SCZ phenotype.

Through a series of experiments, this study confirmed an interaction between the NMDA receptor and miR-219 but was not able to show that the decrease in miR-219 was necessarily responsible for the SCZ phenotype. The interaction between NMDA and miR-219 was confirmed by introducing a GRIN1 mutation into the NMDA receptor, which resulted in a sustained decrease in miR-219 in the mouse PFC. Furthermore, as mentioned above, when the NMDAR was antagonized pharmacologically miR-219 was significantly decreased and the mice displayed an SCZ-related phenotype. It is here, however, that the study demonstrates that the miR-219 decrease is not essential to the behavioral phenotype of SCZ. Firstly, the miR-219 decrease observed upon administration of an NMDAR antagonist lasted for a period of only 2 h, which calls into question the idea that the NMDAR–miR-219 interaction is linked to the pathogenesis of a chronic condition such as SCZ. Secondly, none of the SCZ impairments were observed when miR-219 was inhibited *in vivo* in the absence of NMDAR antagonism, which indicates that a decrease in miR-219 is not an essential component of the behavioral pathophysiology of SCZ.

While this is most likely true, this study elucidated a potential pathway which may be contributing to SCZ pathophysiology. In particular, they showed that *in vitro* stimulation of miR-219 led to a 40% reduction of one of its putative targets CAMKII, which is a key component of the NMDAR signaling pathway. Interestingly, antagonism of the NMDAR increased CAMKII protein levels. These findings indicate therefore that a hypo-functioning NMDA receptor leads to a reduction in miR-219 that then leads to an increase in CAMKII. While this may not be the case in all SCZ patients, future investigations to determine the downstream effects of an increase in CAMKII, as well as whether miR-219 is dysregulated in SCZ patients are warranted. This study demonstrates that while the possibilities that a single miRNA would be responsible for the SCZ phenotype are close to zero, the functional interactions between an miRNA and other molecules are a fruitful avenue of investigation as we attempt to unravel the mechanisms surrounding the complex disorder that is SCZ.

### 3.5.3.2  MiR-132-NMDAR

A similar association between the NMDAR signaling pathway and miR-132 was established in a study by Miller et al. (2012). Administration of an NMDAR antagonist in mice resulted in the downregulation of miR-132 but in contrast to the Kocerha study, both the downregulation and the SCZ behavioral phenotype persisted throughout adulthood (Miller et al. 2012). MiR-132 may therefore be more integral to the functioning of the NMDA receptor than miR-219, as its downregulation by an NMDAR antagonist lasts much longer than 2 h recorded by Kocerha et al. (2009). Alternatively, the sustained effect of the NMDAR antagonist could be due to the timing of its administration, which was during the early stages of postnatal development. Interestingly, this study showed that this early neurodevelopmental period coincides with synaptic pruning, which is one of the proposed causes of SCZ, leading to neuronal loss and enlarged ventricles (Haukvik et al. 2013). It is possible therefore that administrating an NMDAR antagonist at this crucial neurodevelopmental time point resulted in a decrease in miR-132 which affected synaptic pruning and lead to neuronal loss and a sustained rather than an acute behavioral phenotype.

In accordance with this SCZ phenotype, we would expect the clinical population to have low levels of miR-132 expression, which is precisely what Miller et al. (2012) found in the clinical postmortem component of their study. The functional narrative would therefore be that genetic and environmental factors have resulted in hypofunctioning NMDA receptors in SCZ patients, which have led to a reduction in miR-132, an increase in synaptic pruning, and the behavioral impairments of SCZ.

In contrast to this theory, however, the overexpression of miR-132 is also associated with impairments in novel object recognition (Hansen et al. 2010) and spatial WM (Hansen et al. 2013), both of which constitute cognitive deficits associated with SCZ. If, as may be concluded from the study by Miller et al. (2012), miR-132 underexpression is a contributor to the SCZ phenotype, we would expect its overexpression to lead to an improvement of cognitive symptoms, not the deficit seen in the mouse studies (Hansen et al. 2010, 2013). In fact, this deficit is more in keeping with the results obtained by Kim et al. (2010) who showed increased levels of miR-132 in the PFC of SCZ patients. It is most likely, therefore, given these contradictions that miR-132 expression needs to be kept at optimal levels and that both overexpression as in Hansen et al. (2013) and Kim et al. (2010) and underexpression as in Miller et al. (2012) can lead to cognitive deficits associated with SCZ.

This study highlights the importance of timing in miRNA expression as it showed that a decrease in miR-132 at a crucial neurodevelopmental time point resulted in behavioral impairments that persisted throughout adulthood. In light of this, it would be interesting to examine miR-132 expression in brains at different stages of development in order to determine whether the results of the mouse studies indicating a rise in miR-132 coinciding with synaptic pruning can be replicated. This would serve to support the proposed role for miR-132, which in interaction with the NMDAR may result in dysregulated synaptic pruning and SCZ-related behavioral deficits. Finally, this study shows the importance of miRNA expression being kept at optimal levels as both decreased (Miller et al. 2012) and increased miRNA expression (Kim et al. 2010) have been associated with SCZ.

### 3.5.3.3    MiR-132, miR-212-PGD, Tyrosine Hydroxylase

As mentioned earlier, Kim et al. (2010) also identified miR-132 as one of the seven miRNAs shown to be dysregulated in SCZ. The aim of the study was to examine all known miRNAs for potential dysregulation in SCZ patients using a postmortem design. Having discovered seven dysregulated miRNAs, they then selected three to perform further analyses. Two of three miRNAs (miR-132 and miR-212) were found to be negatively correlated with the expression of their predicted targets TH and PGD. Interestingly, PGD was not a predicted target of either miR-212 or miR-132 in the bioinformatic databases, which indicates the importance of functionally confirming bioinformatic predictions.

This study implicates the DA system, due to the fact that TH, the other target of miR-132 and miR-212, is the enzyme responsible for catalyzing the conversion of tyrosine to L-DOPA. The glutamatergic system is also implicated, as the predicted targets of miR-132 include GRIN1, GRIN2A, and GRIN2B, which code for different subunits of the NMDAR (Weickert et al. 2013). Considering that the NMDAR has been implicated in all the studies discussed above, we can conclude that there is evidence that NMDAR/miRNA interaction plays a significant role in SCZ (Kocerha et al. 2009; Kim et al. 2010; Miller et al. 2012).

The study by Kim et al. (2010) has therefore shown an association between a small number of miRNAs also validated by other studies (Perkins et al. 2007; Miller et al. 2012) and two of the major SCZ neurotransmitter candidates, the glutamatergic and dopaminergic systems (Bonoldi and Howes 2013; Weickert et al. 2013). In so doing, it highlights the importance of expanding miRNA expression studies with further investigation into potential interactions between these dysregulated miRNAs and previously identified SCZ risk factors.

### 3.5.3.4    MiR-107-CHRM1

A study conducted by Scarr et al. (2013) implicated miRNA interaction with another neurotransmitter system in the pathophysiology of SCZ. This was the cholinergic system that has been shown to be involved in cognitive functions such as learning and memory that are impaired in SCZ (Volpicelli and Levey 2004). In particular, a subset of SCZ patients had previously been discovered that displayed a 75% reduction in the CHRM1 (Scarr et al. 2009). This subset of patients termed Def-SZ constitutes approximately 25% of the total SCZ population. Interestingly, this Def-SZ cohort showed a significant increase in miR-107 expression in the DLPFC, which was not seen in either the controls or any of the other SCZ subgroups (Scarr et al. 2013). While the mRNA levels of *CHRM1* were significantly decreased in all SCZ patients, the protein levels of CHRM1 were only decreased in the Def-SZ group, which lends support to the theory that the increased miR-107 expression seen in this group is contributing to the posttranscriptional silencing of CHRM1. The direct effect of miR-107 on CHRM1 was also confirmed using a functional assay, which demonstrated the expected inverse relationship between the activation of miR-107 and the expression of CHRM1. This study therefore was able to attribute the downregulation of CHRM1 exhibited by a subset of SCZ patients to the upregulation of a particular miRNA, further illuminating the role these molecules may play in the pathophysiology of SCZ. It is also interesting to note that there is considerable collaborative evidence from both postmortem and peripheral studies that miR-107 is indeed dysregulated in SCZ (Beveridge et al. 2010; Gardiner et al. 2012), although while it is upregulated in the Beveridge et al. (2010) and Scarr et al. (2013) studies, it had been shown to be downregulated in PBMCs from patients with SCZ (Gardiner et al. 2012).

This discrepancy between miRNA expression in postmortem versus peripheral studies is also seen in the case of miR-181b, which like miR-107 was upregulated in postmortem and some peripheral studies (Beveridge et al. 2008; Shi et al. 2012) but downregulated in others (Gardiner et al. 2012). This finding indicates the variability in miRNA expression and calls into question the utility of putting undue emphasis on the direction of the dysregulation of miRNAs. While it is true that some studies have found a relatively consistent upregulation of miRNA (Beveridge et al. 2010), other studies have revealed significant miRNA downregulation in SCZ (Perkins et al. 2007). This lack of unidirectional expression may indicate that the focus of future investigations should be placed on miRNA dysregulation in SCZ regardless of direction.

### 3.5.3.5  MiR-195-BDNF

Mellios et al. (2009) discovered a strong inverse correlation between BDNF and miR-195 levels, indicating that miR-195 may be involved in the regulation of BDNF. Building on those results, they showed that BDNF was also correlated with levels of GABAergic transcript, NPY, which indicated that BDNF may be involved in the regulation of GABAergic transcripts in the PFC (Mellios et al. 2009). Having established these relationships they sought to determine whether they were dysregulated in SCZ.

They found significant upregulation of miR-195 in the PFC of patients with SCZ, accompanied by reduced levels of both NPY and BDNF in the PFC of these patients (Mellios et al. 2009). Their findings therefore support the theory that miR-195 negatively regulates BDNF expression and that when miR-195 is upregulated as in SCZ, it leads to decreased BDNF levels that act to reduce the expression of GABAergic genes in the PFC. Support for this theory is weakened, however, when one considers that the SCZ patients in their study did not in fact have decreased BDNF protein levels. This finding significantly weakens the case for BDNF being responsible for the observed differences in GABAergic transcript levels between patients and controls and makes the regulation of BDNF by miR-195 less relevant than it otherwise would have been.

Having said this, there is collaborative evidence from other studies for the involvement of miR-195, BDNF, and GABAergic transcripts in SCZ. Dysregulation of GABAergic interneurons in the PFC is a main tenant of the hypoglutamatergic theory of SCZ (Moghaddam and Javitt 2012), while a number of studies have shown reduced mRNA, protein, and serum levels of BDNF in patients with SCZ (Weickert et al. 2013; Lu et al. 2014). Furthermore, BDNF has repeatedly been found to be a target of miRNAs that show dysregulation in SCZ (Beveridge et al. 2010; Kim et al. 2010; Santarelli et al. 2011). Finally, miR-195 was found to be a core component of an SCZ-related network responsible for the regulation of six SCZ risk genes

(Guo et al. 2010). These studies indicate that there is strong evidence that the individual components of the pathway (miR-195, BDNF, GABAergic transcripts) proposed by Mellios et al. (2009) are associated with SCZ. Future studies could perhaps focus on the role miR-195 plays in the regulation of BDNF, employing larger sample sizes, which may show the SCZ-related decrease in BDNF that was not demonstrated in Mellios et al (2009). This study strongly demonstrates the importance of formulating a hypothesis and then employing appropriately mechanistic tools to determine its accuracy. Although BDNF was not conclusively linked to the SCZ sample, this study provides strong evidence for the involvement of miR-195 in this disorder and illustrates some of the mechanisms by which this involvement is realized.

Taken together, these studies demonstrate the importance of examining the effects miRNAs has on their target molecules. The figure below illustrates a number of these interactions between miRNAs found to be dysregulated in SCZ and the functional effect they have on their molecular targets, which include SCZ risk genes.

As we can see from Figure 3.2, a number of correlations have been found between miRNAs that are dysregulated in SCZ brains (e.g., miR195, miR-107, and miR-132) and molecules such as BDNF, CHRM1, and TH that have been implicated in SCZ. MiR-134 found to be dysregulated in SCZ brains is also decreased in the DGCR8 haploinsufficiency mouse. DGCR8 is an miRNA biogenesis gene that has been associated with Di George syndrome, which has a 30% conversion rate to SCZ (Drew et al. 2011). Accordingly, the DGCR8 mouse shows decreased hippocampus neurogenesis and cognitive deficits and decreased DRD2 and EPSCs that are reversed upon administration of antipsychotics (Ouchi et al. 2013; Chun et al. 2014).

The results of these studies can be used as a basis for formulating new theories about the way in which miRNA dysregulation may interact with molecules known to be associated with SCZ and the downstream consequences of this interaction.

### 3.5.4   Pathways Section

Following the identification of miRNA targets, many studies have conducted pathway analysis with the aid of online databases such as KEGG. In keeping with their involvement in DNA transcription and protein translation, many pathways fundamental to cellular functioning are implicated in the miRNAs found to be dysregulated in SCZ. These include such pathways as MAPK and CREB signaling (Beveridge et al. 2008; Santarelli et al. 2011; Gardiner et al. 2012; Miller et al. 2012). Most importantly, many pathways relating to neuronal functioning that have been found to be dysregulated in SCZ are also implicated by the dysregulated miRNAs. These include pathways such as LTP, Axon guidance, Wnt signaling, and neurogenesis (Beveridge et al. 2008; Kim et al. 2010; Lai et al. 2011; Santarelli et al. 2011; Gardiner et al. 2012; Miller et al. 2012). The most important point is that many of these pathways are related to neurodevelopment, which therefore implicates miRNA dysregulation in one of the strongest theories of SCZ as a disorder of neuronal development and connectivity (Marenco and Weinberger 2000; Fatemi and Folsom 2009). If for example a set of miRNAs involved in neurogenesis and axon guidance is dysregulated in SCZ patients, this could form part of an explanation for the altered neuronal connectivity identified in these patients (Gauthier et al. 2010; Faludi and Mirnics 2011; Fornito et al. 2012).

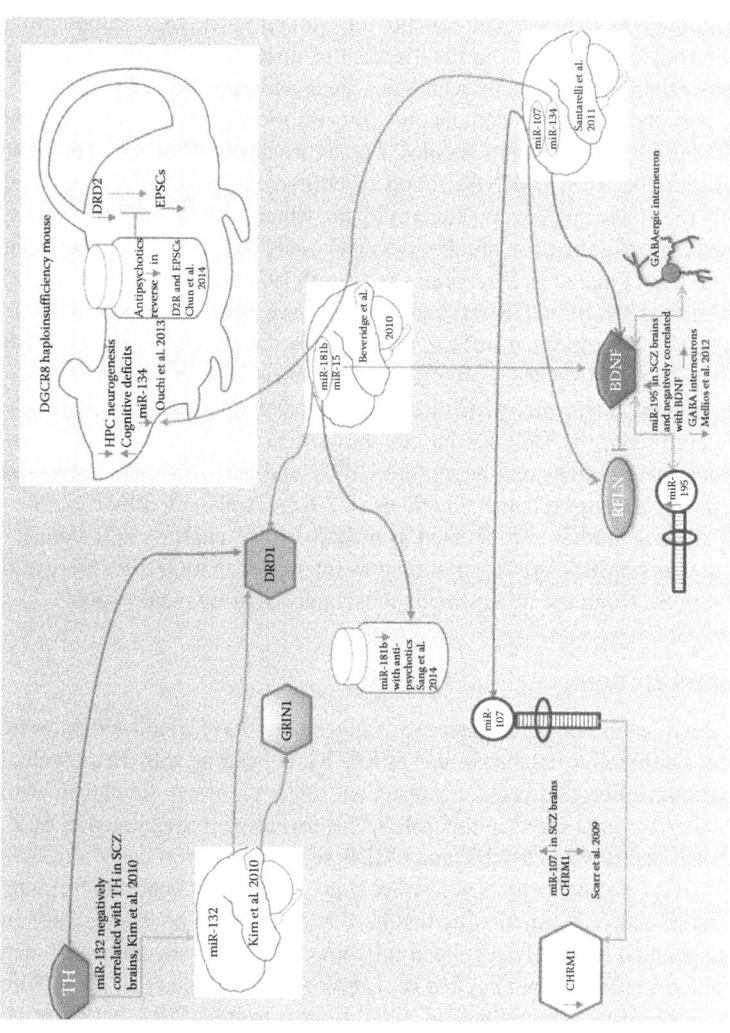

**FIGURE 3.2** Illustration of miRNA dysregulation identified in animal and postmortem brain findings related to SCZ. Arrows link miRNAs that are found to be dysregulated in SCZ to their target molecules and to animal studies that have shown that miRNA expression can be influenced by antipsychotics. Abbreviations: BNDF, brain-derived neurotrophic factor; CHRM1, cholinergic muscarinic receptor 1; DRD1, dopamine receptor 1; DRD2, dopamine receptor 2; GRIN1, glutamate receptor ionotropic N-methyl-D-aspartate subunit 1; HPC, hippocampus; RELN, reelin; TH, tyrosine hydroxylase.

## 3.6    WHERE TO NOW: FUTURE DIRECTIONS
## IN MicroRNA RESEARCH

Investigations into the role that miRNAs play in normal neurodevelopment may serve to establish a baseline against which any deviations from normal miRNA expression can be measured. Significant steps toward this goal were taken by a study that investigated miRNA expression across the neurodevelopmental trajectory, employing 97 samples from nonpsychiatric cohorts ranging from 2 months to 78 years (Beveridge et al. 2014). They found a global decrease in miRNA expression from neonates to adulthood, with significant decreases taking place in adolescence, which is frequently when the onset of SCZ symptoms occur. Interestingly, this study also demonstrated that a number of miRNAs (miR-137, miR-132, and miR-181b) that had previously been associated with SCZ (Beveridge et al. 2008, 2010; Kim et al. 2010; Schizophrenia Psychiatric Genome-Wide Association Study 2011; Gardiner et al. 2012; Shi et al. 2012) also showed a significant expression decline with age. This indicates that these risk miRNAs may be implicated in the pathogenesis of SCZ, primarily through their failure to show age-related decline, which then leads to the inhibition of their target genes (Beveridge et al. 2014). This extreme inhibition of the expression of target genes may then result in the cessation of various downstream maturational processes, which together with environmental factors may contribute to the pathogenesis of SCZ. This study therefore demonstrates the importance of continuing to study miRNA expression in the normal brain, which may then inform theories of SCZ pathogenesis.

One of the main problems currently facing miRNA studies is the lack of replication inherent in GWAS. More extensive GWAS with larger sample sizes and an increasing emphasis on the homogeneity of those samples will no doubt yield further miRNA SNPs that constitute risk factors for SCZ. However, these studies do not deal with the problem of epistatic interaction, which is an integral component of biological reality. In order to address this, current investigations are focusing on compiling algorithms that are capable of revealing complex interactions between two or more genes (Arkin et al. 2014). These improved GWAS will be able to reveal more about the relationship between particular miRNA risk SNPs, as well as between an miRNA SNP and independently discovered SCZ risk genes. This will enable us to discover potential SCZ networks consisting of multiple risk genes, which may explain the variability seen in current miRNA GWAS as different combinations of mutations in each individual may perturb the network, leading nonetheless to a common phenotype (McIntosh et al. 2014). With the discovery of these gene interaction networks, will come the potential of pharmacological targeting of these networks with therapies that are tailored to the individual genetic profile of the patient.

Future steps could also involve combining these genetic studies with studies aimed at determining the clinical phenotype of the patient, such as imaging studies. Steps toward combining genetic and imaging studies in the field of miRNAs were taken by Begemann et al. (2010), Cummings et al. (2013), and others and are listed in the genetics section of this chapter. The combination of large scale GWAS with gene set enrichment analysis and imaging have been shown to be potent tools in the field of cognition (Heck et al. 2014) and the development of standardized imaging banks such as ENIGMA (Thompson et al. 2014) will no doubt prove useful in the study of SCZ. Furthermore, the discovery of new miRNA SNPs associated with SCZ will be aided by the wide-spread application of next generation sequencing (NGS), which are demonstrably more accurate than the microarrays currently being used.

In conjunction with these genetic studies, further research is required into the downstream effects of each individual miRNA molecule. Steps have been taken using new gene targeting technology such as clustered regularly interspaced short palindromic repeats (CRISPR) and transcription activator-like effector nucleases (TALENS) to determine the effect of silencing individual miRNAs (Hu et al. 2013). One study built up a library of TALENS, that is, restriction enzymes capable of targeting 274 miRNAs (Kim et al. 2014). By generating KO cells for two particular miRNAs, they were able to show that despite their similarity, they had substantially different sets of targets. CRISPR is another genetic engineering technique that allows particular mutations to be inserted in target genes, the effect of which can then be examined *in vivo* in a mouse (Sung et al. 2014). By utilizing CRISPR and TALENS in the context of SCZ, we can potentially determine both the downstream consequences of a specific miRNA mutation discovered using GWAS, as well as the behavioral results of this mutation in an animal model of this disorder. The combined use of animal models and specific miRNA mutations can then allow us to test for the efficacy of various pharmacological and behavioral interventions, with a view to clinical treatment.

Finally, the problem regarding the absence of unidirectional miRNA dysregulation in SCZ and the contrasting results occasionally obtained from postmortem versus peripheral studies (Beveridge et al. 2010; Gardiner et al. 2012) can begin to be addressed by experimenting with different methods of sampling miRNAs. Emerging work at the Walter and Eliza Hall Research Institute (WEHI) in Melbourne has shown that an accurate miRNA expression profile can be obtained from exosomes (Hill et al. 2014). This highlights the importance of developing techniques that enable miRNA expression profiles to be obtained with ease and accuracy, which can then allow these profiles to be compared across different methods.

## 3.7  CONCLUSIONS

We can see from the studies discussed in this chapter that investigations into miRNA expression, miRNA–target interaction as well as GWAS involving miRNAs have produced some interesting results for SCZ but have nonetheless been plagued by inconsistencies. These inconsistencies could be due to a number of causes including sample variability and technical variation across laboratories. The study of miRNAs in relation to SCZ is a relatively new field and with increasing sample sizes, technical improvements, and streamlining across laboratories, it is possible that these occasionally disparate results will begin to converge on one primary theory. Alternatively, as has been suggested before (Fatemi and Folsom 2009; Brown 2011), these varied results could be due to the fact that SCZ is a multifaceted condition with multiple interacting causes, which cannot be discerned by the study of one type of molecule or one technique. Therefore, the study of miRNAs and their contribution to the etiology of SCZ needs a concerted effort that spans researchers from different backgrounds and utilizing different techniques and models.

## REFERENCES

American Psychiatric Association. 2013. *Diagnostic and Statistical Manual of Mental Disorders: DSM-5* (5th ed.). Arlington, VA: American Psychiatric Publishing.

Anderson, G., and M. Maes. 2013. Schizophrenia: Linking prenatal infection to cytokines, the tryptophan catabolite (TRYCAT) pathway, NMDA receptor hypofunction, neurodevelopment and neuroprogression. *Prog Neuropsychopharmacol Biol Psychiatry* 42:5–19. doi: 10.1016/j.pnpbp.2012.06.014.

Arkin, Y., E. Rahmani, M. E. Kleber et al. 2014. EPIQ-efficient detection of SNP-SNP epistatic interactions for quantitative traits. *Bioinformatics* 30 (12):i19–25. doi: 10.1093/bioinformatics/btu261.

Begemann, M., S. Grube, S. Papiol et al. 2010. Modification of cognitive performance in schizophrenia by complexin 2 gene polymorphisms. *Arch Gen Psychiatry* 67 (9):879–88. doi: 10.1001/archgenpsychiatry.2010.107.

Beveridge, N. J., and M. J. Cairns. 2012. MicroRNA dysregulation in schizophrenia. *Neurobiol Dis* 46 (2):263–71. doi: 10.1016/j.nbd.2011.12.029.

Beveridge, N. J., E. Gardiner, A. P. Carroll, P. A. Tooney, and M. J. Cairns. 2010. Schizophrenia is associated with an increase in cortical microRNA biogenesis. *Mol Psychiatry* 15 (12):1176–89. doi: 10.1038/mp.2009.84.

Beveridge, N. J., D. M. Santarelli, X. Wang et al. 2014. Maturation of the human dorsolateral prefrontal cortex coincides with a dynamic shift in microRNA expression. *Schizophr Bull* 40 (2):399–409. doi: 10.1093/schbul/sbs198.

Beveridge, N. J., P. A. Tooney, A. P. Carroll et al. 2008. Dysregulation of miRNA 181b in the temporal cortex in schizophrenia. *Hum Mol Genet* 17 (8):1156–68. doi: 10.1093/hmg/ddn005.

Bonoldi, I., and O. D. Howes. 2013. The enduring centrality of dopamine in the pathophysiology of schizophrenia: *In vivo* evidence from the prodrome to the first psychotic episode. *Adv Pharmacol* 68:199–220. doi: 10.1016/B978-0-12-411512-5.00010-5.

Brown, A. S. 2011. The environment and susceptibility to schizophrenia. *Prog Neurobiol* 93 (1):23–58. doi: 10.1016/j.pneurobio.2010.09.003.

Catts, V. S., S. J. Fung, L. E. Long et al. 2013. Rethinking schizophrenia in the context of normal neurodevelopment. *Front Cell Neurosci* 7:60. doi: 10.3389/fncel.2013.00060.

Cavaille, J. 2007. MicroRNAs: Biosynthesis: Mechanisms of action and biological functions. *Ann Pathol* 27 Spec No 1:1S31–2.

Chen, X., Y. Ba, L. Ma et al. 2008. Characterization of microRNAs in serum: A novel class of biomarkers for diagnosis of cancer and other diseases. *Cell Res* 18 (10):997–1006. doi: 10.1038/cr.2008.282.

Chun, S., J. J. Westmoreland, I. T. Bayazitov et al. 2014. Specific disruption of thalamic inputs to the auditory cortex in schizophrenia models. *Science* 344 (6188):1178–82. doi: 10.1126/science.1253895.

Cogswell, J. P., J. Ward, I. A. Taylor et al. 2008. Identification of miRNA changes in Alzheimer's disease brain and CSF yields putative biomarkers and insights into disease pathways. *J Alzheimers Dis* 14 (1):27–41.

Cummings, E., G. Donohoe, A. Hargreaves et al. 2013. Mood congruent psychotic symptoms and specific cognitive deficits in carriers of the novel schizophrenia risk variant at MIR-137. *Neurosci Lett* 532:33–8. doi: 10.1016/j.neulet.2012.08.065.

Davis, J., S. Moylan, B. H. Harvey, M. Maes, and M. Berk. 2014. Neuroprogression in schizo-phrenia: Pathways underpinning clinical staging and therapeutic corollaries. *Aust N Z J Psychiatry* 48 (6):512–29. doi: 10.1177/0004867414533012.

Drew, L. J., G. W. Crabtree, S. Markx et al. 2011. The 22q11.2 microdeletion: Fifteen years of insights into the genetic and neural complexity of psychiatric disorders. *Int J Dev Neurosci* 29 (3):259–81. doi: 10.1016/j.ijdevneu.2010.09.007.

Earls, L. R., R. G. Fricke, J. Yu et al. 2012. Age-dependent microRNA control of synaptic plasticity in 22q11 deletion syndrome and schizophrenia. *J Neurosci* 32 (41):14132–44. doi: 10.1523/JNEUROSCI.1312-12.2012.

Faludi, G., and K. Mirnics. 2011. Synaptic changes in the brain of subjects with schizophre-nia. *Int J Dev Neurosci* 29 (3):305–9. doi: 10.1016/j.ijdevneu.2011.02.013.

Fatemi, S. H., and T. D. Folsom. 2009. The neurodevelopmental hypothesis of schizophrenia, revisited. *Schizophr Bull* 35 (3):528–48. doi: 10.1093/schbul/sbn187.

Fenelon, K., J. Mukai, B. Xu et al. 2011. Deficiency of Dgcr8, a gene disrupted by the 22q11.2 microdeletion, results in altered short-term plasticity in the prefrontal cortex. *Proc Natl Acad Sci USA* 108 (11):4447–52. doi: 10.1073/pnas.1101219108.

Feng, J., G. Sun, J. Yan et al. 2009. Evidence for X-chromosomal schizophrenia associated with microRNA alterations. *PLoS One* 4 (7):e6121. doi: 10.1371/journal.pone.0006121.

Filipowicz, W., S. N. Bhattacharyya, and N. Sonenberg. 2008. Mechanisms of post-transcrip-tional regulation by microRNAs: Are the answers in sight? *Nat Rev Genet* 9 (2):102–14. doi: 10.1038/nrg2290.

Fiore, R., S. Khudayberdiev, M. Christensen et al. 2009. Mef2-mediated transcription of the miR379-410 cluster regulates activity-dependent dendritogenesis by fine-tuning Pumilio2 protein levels. *EMBO J* 28 (6):697–710. doi: 10.1038/emboj.2009.10.

Fornito, A., A. Zalesky, C. Pantelis, and E. T. Bullmore. 2012. Schizophrenia, neuroim-aging and connectomics. *Neuroimage* 62 (4):2296–314. doi: 10.1016/j.neuroimage.2011.12.090.

Gardiner, E., N. J. Beveridge, J. Q. Wu et al. 2012. Imprinted DLK1-DIO3 region of 14q32 defines a schizophrenia-associated miRNA signature in peripheral blood mononuclear cells. *Mol Psychiatry* 17 (8):827–40. doi: 10.1038/mp.2011.78.

Gardiner, E., A. Carroll, P. A. Tooney, and M. J. Cairns. 2014. Antipsychotic drug-associated gene-miRNA interaction in T-lymphocytes. *Int J Neuropsychopharmacol* 17 (6):929–43. doi: 10.1017/S1461145713001752.

Garrido, M. I., J. M. Kilner, K. E. Stephan, and K. J. Friston. 2009. The mismatch negativity: A review of underlying mechanisms. *Clin Neurophysiol* 120 (3):453–63. doi: 10.1016/j.clinph.2008.11.029.

Gauthier, J., N. Champagne, R. G. Lafreniere et al. 2010. *De novo* mutations in the gene encoding the synaptic scaffolding protein SHANK3 in patients ascertained for schizo-phrenia. *Proc Natl Acad Sci USA* 107 (17):7863–8. doi: 10.1073/pnas.0906232107.

Gershon, E. S., N. Alliey-Rodriguez, and C. Liu. 2011. After GWAS: Searching for genetic risk for schizophrenia and bipolar disorder. *Am J Psychiatry* 168 (3):253–6. doi: 10.1176/appi.ajp.2010.10091340.

Green, E. K., D. Grozeva, I. Jones et al. 2010. The bipolar disorder risk allele at CACNA1C also confers risk of recurrent major depression and of schizophrenia. *Mol Psychiatry* 15 (10):1016–22. doi: 10.1038/mp.2009.49.

Green, M. J., M. J. Cairns, J. Wu et al. 2013. Genome-wide supported variant MIR137 and severe negative symptoms predict membership of an impaired cognitive subtype of schizophrenia. *Mol Psychiatry* 18 (7):774–80. doi: 10.1038/mp.2012.84.

Guella, I., A. Sequeira, B. Rollins et al. 2013. Analysis of miR-137 expression and rs1625579 in dorsolateral prefrontal cortex. *J Psychiatr Res* 47 (9):1215–21. doi: 10.1016/j.jpsychires.2013.05.021.

Guo, A. Y., J. Sun, P. Jia, and Z. Zhao. 2010. A novel microRNA and transcription factor mediated regulatory network in schizophrenia. *BMC Syst Biol* 4:10. doi: 10.1186/1752-0509-4-10.

Hansen, K. F., K. Karelina, K. Sakamoto et al. 2013. MiRNA-132: A dynamic regulator of cognitive capacity. *Brain Struct Funct* 218 (3):817–31. doi: 10.1007/s00429-012-0431-4.

Hansen, K. F., K. Sakamoto, G. A. Wayman, S. Impey, and K. Obrietan. 2010. Transgenic miR132 alters neuronal spine density and impairs novel object recognition memory. *PLoS One* 5 (11):e15497. doi: 10.1371/journal.pone.0015497.

Hansen, T., L. Olsen, M. Lindow et al. 2007. Brain expressed microRNAs implicated in schizophrenia etiology. *PLoS One* 2 (9):e873. doi: 10.1371/journal.pone.0000873.

Harrison, P. J., and D. R. Weinberger. 2005. Schizophrenia genes, gene expression, and neuropathology: On the matter of their convergence. *Mol Psychiatry* 10 (1):40–68; image 5. doi: 10.1038/sj.mp.4001558.

Haukvik, U. K., C. B. Hartberg, and I. Agartz. 2013. Schizophrenia—What does structural MRI show? *Tidsskr Nor Laegeforen* 133 (8):850–3. doi: 10.4045/tidsskr.12.1084.

Heck, A., M. Fastenrath, S. Ackermann et al. 2014. Converging genetic and functional brain imaging evidence links neuronal excitability to working memory, psychiatric disease, and brain activity. *Neuron* 81 (5):1203–13. doi: 10.1016/j.neuron.2014.01.010.

Hickie, I. B., R. Banati, C. H. Stewart, and A. R. Lloyd. 2009. Are common childhood or adolescent infections risk factors for schizophrenia and other psychotic disorders? *Med J Aust* 190 (4 Suppl):S17–21.

Hill A., J. Doecke, R.A. Sharples et al. 2014. Exosomal miRNA as biomarkers for diagnosing neurodegenerative diseases(presentation, Third International Meeting of International Society for Extracellular Vesicles, Rotterdam, the Netherlands, April 30–May 3, 2014).

Howes, O. D., and S. Kapur. 2009. The dopamine hypothesis of schizophrenia: Version III—The final common pathway. *Schizophr Bull* 35 (3):549–62. doi: 10.1093/schbul/sbp006.

Hu, R., J. Wallace, T. J. Dahlem, D. J. Grunwald, and R. M. O'Connell. 2013. Targeting human microRNA genes using engineered Tal-effector nucleases (TALENs). *PLoS One* 8 (5):e63074. doi: 10.1371/journal.pone.0063074.

Huntzinger, E., and E. Izaurralde. 2011. Gene silencing by microRNAs: Contributions of translational repression and mRNA decay. *Nat Rev Genet* 12 (2):99–110. doi: 10.1038/nrg2936.

Jackson, D., J. Kirkbride, T. Croudace et al. 2013. Meta-analytic approaches to determine gender differences in the age-incidence characteristics of schizophrenia and related psychoses. *Int J Methods Psychiatr Res* 22 (1):36–45. doi: 10.1002/mpr.1376.

Jin, X. F., N. Wu, L. Wang, and J. Li. 2013. Circulating microRNAs: A novel class of potential biomarkers for diagnosing and prognosing central nervous system diseases. *Cell Mol Neurobiol* 33 (5):601–13. doi: 10.1007/s10571-013-9940-9.

Kim, Y., G. Gabbine, J. Park et al. 2014. TALEN-based knockout library for human microRNAs. *Nat Struc & Mol Biol* 20 (12):1458–64. doi: 10.1038/nsmb.2701.

Kim, A. H., E. K. Parker, V. Williamson et al. 2012. Experimental validation of candidate schizophrenia gene ZNF804A as target for hsa-miR-137. *Schizophr Res* 141 (1):60–4. doi: 10.1016/j.schres.2012.06.038.

Kim, A. H., M. Reimers, B. Maher et al. 2010. MicroRNA expression profiling in the prefrontal cortex of individuals affected with schizophrenia and bipolar disorders. *Schizophr Res* 124 (1–3):183–91. doi: 10.1016/j.schres.2010.07.002.

Kocerha, J., M. A. Faghihi, M. A. Lopez-Toledano et al. 2009. MicroRNA-219 modulates NMDA receptor-mediated neurobehavioral dysfunction. *Proc Natl Acad Sci USA* 106 (9):3507–12. doi: 10.1073/pnas.0805854106.

Krichevsky, A. M., K. C. Sonntag, O. Isacson, and K. S. Kosik. 2006. Specific microRNAs modulate embryonic stem cell-derived neurogenesis. *Stem Cells* 24 (4):857–64. doi: 10.1634/stemcells.2005-0441.

Krol, J., V. Busskamp, I. Markiewicz et al. 2010a. Characterizing light-regulated retinal microRNAs reveals rapid turnover as a common property of neuronal microRNAs. *Cell* 141 (4):618–31. doi: 10.1016/j.cell.2010.03.039.

Krol, J., I. Loedige, and W. Filipowicz. 2010b. The widespread regulation of microRNA biogenesis, function and decay. *Nat Rev Genet* 11 (9):597–610. doi: 10.1038/nrg2843.

Kwon, E., W. Wang, and L. H. Tsai. 2013. Validation of schizophrenia-associated genes CSMD1, C10orf26, CACNA1C and TCF4 as miR-137 targets. *Mol Psychiatry* 18 (1):11–2. doi: 10.1038/mp.2011.170.

Kyriakopoulos, M., D. Dima, J. P. Roiser et al. 2012. Abnormal functional activation and connectivity in the working memory network in early-onset schizophrenia. *J Am Acad Child Adolesc Psychiatry* 51 (9):911–20 e2. doi: 10.1016/j.jaac.2012.06.020.

Lai, C. Y., S. L. Yu, M. H. Hsieh et al. 2011. MicroRNA expression aberration as potential peripheral blood biomarkers for schizophrenia. *PLoS One* 6 (6):e21635. doi: 10.1371/journal.pone.0021635.

Le, M. T., H. Xie, B. Zhou et al. 2009. MicroRNA-125b promotes neuronal differentiation in human cells by repressing multiple targets. *Mol Cell Biol* 29 (19):5290–305. doi: 10.1128/MCB.01694-08.

Lett, T. A., M. M. Chakravarty, D. Felsky et al. 2013. The genome-wide supported microRNA-137 variant predicts phenotypic heterogeneity within schizophrenia. *Mol Psychiatry* 18 (4):443–50. doi: 10.1038/mp.2013.17.

Lewis, B. P., I. H. Shih, M. W. Jones-Rhoades, D. P. Bartel, and C. B. Burge. 2003. Prediction of mammalian microRNA targets. *Cell* 115 (7):787–98.

Liu, C., F. Zhang, T. Li et al. 2012. MirSNP, a database of polymorphisms altering miRNA target sites, identifies miRNA-related SNPs in GWAS SNPs and eQTLs. *BMC Genomics* 13:661. doi: 10.1186/1471-2164-13-661.

Lu, B., G. Nagappan, and Y. Lu. 2014. BDNF and synaptic plasticity, cognitive function, and dysfunction. *Handb Exp Pharmacol* 220:223–50. doi: 10.1007/978-3-642-45106-5_9.

Maffioletti, E., D. Tardito, M. Gennarelli, and L. Bocchio-Chiavetto. 2014. Micro spies from the brain to the periphery: New clues from studies on microRNAs in neuropsychiatric disorders. *Front Cell Neurosci* 8:75. doi: 10.3389/fncel.2014.00075.

Marenco, S., and D. R. Weinberger. 2000. The neurodevelopmental hypothesis of schizophrenia: Following a trail of evidence from cradle to grave. *Dev Psychopathol* 12 (3):501–27.

McIntosh, A., I. Deary, and D. J. Porteous. 2014. Two-back makes step forward in brain imaging genomics. *Neuron* 81 (5):959–61. doi: 10.1016/j.neuron.2014.02.023.

Mellios, N., H. S. Huang, S. P. Baker et al. 2009. Molecular determinants of dysregulated GABAergic gene expression in the prefrontal cortex of subjects with schizophrenia. *Biol Psychiatry* 65 (12):1006–14. doi: 10.1016/j.biopsych.2008.11.019.

Meyer, U., J. Feldon, and O. Dammann. 2011. Schizophrenia and autism: Both shared and disorder-specific pathogenesis via perinatal inflammation? *Pediatr Res* 69 (5 Pt 2):26R–33R. doi: 10.1203/PDR.0b013e318212c196.

Miller, B. H., Z. Zeier, L. Xi et al. 2012. MicroRNA-132 dysregulation in schizophrenia has implications for both neurodevelopment and adult brain function. *Proc Natl Acad Sci USA* 109 (8):3125–30. doi: 10.1073/pnas.1113793109.

Moghaddam, B., and D. Javitt. 2012. From revolution to evolution: The glutamate hypothesis of schizophrenia and its implication for treatment. *Neuropsychopharmacology* 37 (1):4–15. doi: 10.1038/npp.2011.181.

Moreau, M. P., S. E. Bruse, R. David-Rus, S. Buyske, and L. M. Brzustowicz. 2011. Altered microRNA expression profiles in postmortem brain samples from individuals with schizophrenia and bipolar disorder. *Biol Psychiatry* 69 (2):188–93. doi: 10.1016/j.biopsych.2010.09.039.

Murphy, K. C., L. A. Jones, and M. J. Owen. 1999. High rates of schizophrenia in adults with velo-cardio-facial syndrome. *Arch Gen Psychiatry* 56 (10):940–5.

Navarrete, K., I. Pedroso, S. De Jong et al. 2013. TCF4 (e2-2; ITF2): A schizophrenia-associated gene with pleiotropic effects on human disease. *Am J Med Genet B Neuropsychiatr Genet* 162B (1):1–16. doi: 10.1002/ajmg.b.32109.

Ouchi, Y., Y. Banno, Y. Shimizu et al. 2013. Reduced adult hippocampal neurogenesis and working memory deficits in the Dgcr8-deficient mouse model of 22q11.2 deletion-associated schizophrenia can be rescued by IGF2. *J Neurosci* 33 (22):9408–19. doi: 10.1523/JNEUROSCI.2700-12.2013.

Perkins, D. O., C. D. Jeffries, L. F. Jarskog et al. 2007. MicroRNA expression in the prefrontal cortex of individuals with schizophrenia and schizoaffective disorder. *Genome Biol* 8 (2):R27. doi: 10.1186/gb-2007-8-2-r27.

Piper, M., M. Beneyto, T. H. Burne et al. 2012. The neurodevelopmental hypothesis of schizophrenia: Convergent clues from epidemiology and neuropathology. *Psychiatr Clin North Am* 35 (3):571–84. doi: 10.1016/j.psc.2012.06.002.

Potkin, S. G., F. Macciardi, G. Guffanti et al. 2010. Identifying gene regulatory networks in schizophrenia. *Neuroimage* 53 (3):839–47. doi: 10.1016/j.neuroimage.2010.06.036.

Robinson, M. B., and J. T. Coyle. 1987. Glutamate and related acidic excitatory neurotransmitters: From basic science to clinical application. *FASEB J* 1 (6):446–55.

Rong, H., T. B. Liu, K. J. Yang et al. 2011. MicroRNA-134 plasma levels before and after treatment for bipolar mania. *J Psychiatr Res* 45 (1):92–5. doi: 10.1016/j.jpsychires.2010.04.028.

Santarelli, D. M., N. J. Beveridge, P. A. Tooney, and M. J. Cairns. 2011. Upregulation of dicer and microRNA expression in the dorsolateral prefrontal cortex Brodmann area 46 in schizophrenia. *Biol Psychiatry* 69 (2):180–7. doi: 10.1016/j.biopsych.2010.09.030.

Santarelli, D. M., B. Liu, C. E. Duncan et al. 2013. Gene-microRNA interactions associated with antipsychotic mechanisms and the metabolic side effects of olanzapine. *Psychopharmacology (Berl)* 227 (1):67–78. doi: 10.1007/s00213-012-2939-y.

Scarr, E., T. F. Cowie, S. Kanellakis et al. 2009. Decreased cortical muscarinic receptors define a subgroup of subjects with schizophrenia. *Mol Psychiatry* 14 (11):1017–23. doi: 10.1038/mp.2008.28.

Scarr, E., J. M. Craig, M. J. Cairns et al. 2013. Decreased cortical muscarinic M1 receptors in schizophrenia are associated with changes in gene promoter methylation, mRNA and gene targeting microRNA. *Transl Psychiatry* 3:e230. doi: 10.1038/tp.2013.3.

Schipper, H. M., O. C. Maes, H. M. Chertkow, and E. Wang. 2007. MicroRNA expression in Alzheimer blood mononuclear cells. *Gene Regul Syst Biol* 1:263–74.

Schizophrenia Psychiatric Genome-Wide Association Study, Consortium. 2011. Genome-wide association study identifies five new schizophrenia loci. *Nat Genet* 43 (10):969–76. doi: 10.1038/ng.940.

Schofield, C. M., R. Hsu, A. J. Barker et al. 2011. Monoallelic deletion of the microRNA biogenesis gene Dgcr8 produces deficits in the development of excitatory synaptic transmission in the prefrontal cortex. *Neural Dev* 6:11. doi: 10.1186/1749-8104-6-11.

Shi, J., D. F. Levinson, J. Duan et al. 2009. Common variants on chromosome 6p22.1 are associated with schizophrenia. *Nature* 460 (7256):753–7. doi: 10.1038/nature08192.

Shi, W., J. Du, Y. Qi et al. 2012. Aberrant expression of serum miRNAs in schizophrenia. *J Psychiatr Res* 46 (2):198–204. doi: 10.1016/j.jpsychires.2011.09.010.

Shibata, H., A. Tani, T. Chikuhara et al. 2009. Association study of polymorphisms in the group III metabotropic glutamate receptor genes, GRM4 and GRM7, with schizophrenia. *Psychiatry Res* 167 (1–2):88–96. doi: 10.1016/j.psychres.2007.12.002.

Shohamy, D., and N. B. Turk-Browne. 2013. Mechanisms for widespread hippocampal involvement in cognition. *J Exp Psychol Gen* 142 (4):1159–70. doi: 10.1037/a0034461.

Silber, J., D. A. Lim, C. Petritsch et al. 2008. MiR-124 and miR-137 inhibit proliferation of glioblastoma multiforme cells and induce differentiation of brain tumor stem cells. *BMC Med* 6:14. doi: 10.1186/1741-7015-6-14.

Song, H. T., X. Y. Sun, L. Zhang et al. 2014. A preliminary analysis of association between the down-regulation of microRNA-181b expression and symptomatology improvement in schizophrenia patients before and after antipsychotic treatment. *J Psychiatr Res* 54:134–40. doi: 10.1016/j.jpsychires.2014.03.008.

Stark, K. L., B. Xu, A. Bagchi et al. 2008. Altered brain microRNA biogenesis contributes to phenotypic deficits in a 22q11-deletion mouse model. *Nat Genet* 40 (6):751–60. doi: 10.1038/ng.138.

Steen, R. G., C. Mull, R. McClure, R. M. Hamer, and J. A. Lieberman. 2006. Brain volume in first-episode schizophrenia: Systematic review and meta-analysis of magnetic resonance imaging studies. *Br J Psychiatry* 188:510–8. doi: 10.1192/bjp.188.6.510.

Stefansson, H., R. A. Ophoff, S. Steinberg et al. 2009. Common variants conferring risk of schizophrenia. *Nature* 460 (7256):744–7. doi: 10.1038/nature08186.

Sun, G., J. Yan, K. Noltner et al. 2009. SNPs in human miRNA genes affect biogenesis and function. *RNA* 15 (9):1640–51. doi: 10.1261/rna.1560209.

Sung, Y. H., J. M. Kim, H. T. Kim et al. 2014. Highly efficient gene knockout in mice and zebrafish with RNA-guided endonucleases. *Genome Res* 24 (1):125–31. doi: 10.1101/gr.163394.113.

Tabares-Seisdedos, R., and J. L. Rubenstein. 2009. Chromosome 8p as a potential hub for developmental neuropsychiatric disorders: Implications for schizophrenia, autism and cancer. *Mol Psychiatry* 14 (6):563–89. doi: 10.1038/mp.2009.2.

Thompson, P. M., J. L. Stein, S. E. Medland et al. 2014. The ENIGMA Consortium: Large-scale collaborative analyses of neuroimaging and genetic data. *Brain Imaging Behav* 8 (2):153–82. doi: 10.1007/s11682-013-9269-5.

Vita, A., L. De Peri, G. Deste, and E. Sacchetti. 2012. Progressive loss of cortical gray matter in schizophrenia: A meta-analysis and meta-regression of longitudinal MRI studies. *Transl Psychiatry* 2:e190. doi: 10.1038/tp.2012.116.

Vita, A., L. De Peri, C. Silenzi, and M. Dieci. 2006. Brain morphology in first-episode schizophrenia: A meta-analysis of quantitative magnetic resonance imaging studies. *Schizophr Res* 82 (1):75–88. doi: 10.1016/j.schres.2005.11.004.

Volpicelli, Laura A., and Allan I. Levey. 2004. Muscarinic acetylcholine receptor subtypes in cerebral cortex and hippocampus. 145:59–66. doi: 10.1016/s0079-6123(03)45003-6.

Weickert, C. S., S. J. Fung, V. S. Catts et al. 2013. Molecular evidence of *N*-methyl-D-aspartate receptor hypofunction in schizophrenia. *Mol Psychiatry* 18 (11):1185–92. doi: 10.1038/mp.2012.137.

Whitfield-Gabrieli, S., H. W. Thermenos, S. Milanovic et al. 2009. Hyperactivity and hyperconnectivity of the default network in schizophrenia and in first-degree relatives of persons with schizophrenia. *Proc Natl Acad Sci USA* 106 (4):1279–84. doi: 10.1073/pnas.0809141106.

Wray, N. R., S. M. Purcell, and P. M. Visscher. 2011. Synthetic associations created by rare variants do not explain most GWAS results. *PLoS Biol* 9 (1):e1000579. doi: 10.1371/journal.pbio.1000579.

Xu, Y., F. Li, B. Zhang et al. 2010. MicroRNAs and target site screening reveals a pre-microRNA-30e variant associated with schizophrenia. *Schizophr Res* 119 (1–3):219–27. doi: 10.1016/j.schres.2010.02.1070.

Zhan, Y., R. C. Paolicelli, F. Sforazzini et al. 2014. Deficient neuron-microglia signaling results in impaired functional brain connectivity and social behavior. *Nat Neurosci* 17 (3):400–6. doi: 10.1038/nn.3641.

Zhang, F., Y. Chen, C. Liu et al. 2012. Systematic association analysis of microRNA machinery genes with schizophrenia informs further study. *Neurosci Lett* 520 (1):47–50. doi: 10.1016/j.neulet.2012.05.028.

Zhou, R., P. Yuan, Y. Wang et al. 2009. Evidence for selective microRNAs and their effectors as common long-term targets for the actions of mood stabilizers. *Neuropsychopharmacology* 34 (6):1395–405. doi: 10.1038/npp.2008.131.

Zhou, Y., M. Liang, L. Tian et al. 2007. Functional disintegration in paranoid schizophrenia using resting-state fMRI. *Schizophr Res* 97 (1–3):194–205. doi: 10.1016/j.schres.2007.05.029.

Zhu, Y., T. Kalbfleisch, M. D. Brennan, and Y. Li. 2009. A microRNA gene is hosted in an intron of a schizophrenia-susceptibility gene. *Schizophr Res* 109 (1–3):86–9. doi: 10.1016/j.schres.2009.01.022.

Ziats, M. N., and O. M. Rennert. 2014. Identification of differentially expressed microRNAs across the developing human brain. *Mol Psychiatry* 19 (7):848–52. doi: 10.1038/mp.2013.93.

# 4 Impact of MicroRNAs in Synaptic Plasticity, Major Affective Disorders, and Suicidal Behavior

*Gianluca Serafini, Yogesh Dwivedi, and Mario Amore*

## CONTENTS

Abstract.................................................................................................................. 101
4.1 Introduction ................................................................................................ 102
4.2 Methods ...................................................................................................... 103
4.3 MicroRNAs Biogenesis and Expression..................................................... 104
4.4 Role of MiRNAs in Neurogenesis and Synaptic Plasticity ........................ 105
4.5 MicroRNAs and Neurotrophic Factors: What Implications for Major
    Affective Disorders?................................................................................... 108
4.6 Impact of MiRNAs in Stress-Related Disorders ........................................ 109
4.7 Impact of MiRNAs in Major Depressive Disorder and Suicidal
    Behavior...................................................................................................... 110
4.8 Critical Considerations and Main Limitations ............................................ 113
4.9 Conclusive Remarks ................................................................................... 127
Acknowledgment ................................................................................................. 128
References............................................................................................................. 128

## ABSTRACT

Major affective disorders are common conditions associated with relevant disability, psychosocial impairment, and suicidal behavior. Both affective disorders and suicidal behavior have been associated with impairments in synaptic plasticity and cellular resilience. It has been suggested that small noncoding RNAs, especially microRNAs (miRNAs) may be implicated in the pathogenesis of major affective disorders playing a critical role in the translational regulation at the synapse. In the present chapter, we aimed to carefully review the current literature about the impact of miRNAs on neurogenesis, synaptic plasticity, pathological changes induced by chronic stress, and major affective disorders including suicidal behavior. MiRNAs played a critical role in the evolution of the most important brain functions; they represent a relevant class of gene expression regulators involved in the development, physiology, and diseases of the central nervous system. Consistent evidence suggested that abnormalities of some intracellular mechanisms together with impaired assembly, localization, and translational regulation of specific RNA-binding proteins may significantly contribute to the pathogenesis of major affective disorders. At a molecular level, measurements of miRNAs may comprehensively help to understand how gene expression networks are reorganized in both major depression and/or suicide. The present chapter aimed to discuss the main implications of the studies analyzing the association between miRNAs, major affective disorders, and suicidal behavior.

## 4.1  INTRODUCTION

Major depressive disorder (MDD) is a disabling psychiatric condition associated with negative psychosocial consequences, unemployment, and impaired quality of life (Andersen et al., 2006). Also, suicide is among the top 20 leading causes of death worldwide with approximately 1 million people died by suicide every year (World Health Organization, 2012). Although several studies have been conducted in order to investigate the molecular and cellular mechanisms underlying both major depression and suicide, multiple questions still remain unclear. Recently, it has been reported that abnormalities in the expression of some genes involved in the regulation of neural and structural plasticity may play a critical role in the pathogenesis of major affective disorders and suicide (Dwivedi, 2009; Serafini, 2011, 2012).

MicroRNAs (miRNAs) represent an important class of gene expression regulators implicated in the development, physiology, but also disorders of the central nervous system (Bushati and Cohen, 2007; Fiore et al., 2008; Mouillet-Richard et al., 2012). MiRNAs may be considered as metacontrollers of gene expression related to brain development, cognitive functioning, and synaptic plasticity.

These small noncoding RNAs regulate gene expression through several mechanisms such as ribosomal RNA modifications, repression of mRNA expression by RNA interference, alternative splicing, and regulatory mechanisms mediated by RNA–RNA interactions (Dwivedi, 2014).

Being important regulators of gene expression, miRNAs are relevant and easily accessible biomarkers for the diagnosis, treatment, and progression of multiple disorders in humans. As suggested by some researchers (Saugstad, 2010; Dwivedi, 2011), understanding the molecular mechanisms underlying the regulation of gene miRNAs expression is fundamental in order to investigate the pathophysiology of neuropsychiatric disorders. Flavell and Greenberg (2008) reported that an activity-regulated transcriptional program of hundreds of genes in neurons plays a crucial role in the development of neural circuits, activity-dependent changes, and neural connectivity. Dendritic and synaptic remodeling seems to be closely regulated by genes encoding retrograde signals to the presynaptic cell (e.g., brain-derived neurotrophic factor (BDNF)) or truncated, dominant-interfering forms of full-length proteins at the synapse level (e.g., homer1a). Newman and Hammond (2010) suggested that miRNAs modulated many gene expression patterns during development and tissue homeostasis and identified the specificity of the microprocessor (the protein complex essential for maturation of canonical miRNAs).

There are many evidence reporting that miRNAs are involved in the onset and maintenance of common neuropsychiatric disorders like Huntington (Johnson et al., 2008) and Parkinson (Kim et al., 2007) diseases as well as Tourette's syndrome (Abelson et al., 2005). Furthermore, based on recent postmortem evidence, altered levels of specific miRNAs have been found in the brains of schizophrenic subjects (Perkins et al., 2007; Beveridge et al., 2008, 2010; Kim et al., 2010a) with these abnormalities nondependent by antipsychotic treatment (Perkins et al., 2007). Hansen et al. (2007) also reported the existence of two brain-expressed miRNAs related to schizophrenia in a case–control study using single-nucleotide polymorphic analysis.

In addition, it has been suggested that stress, glucocorticoids, and mood stabilizers modulated the expression of selective miRNAs having specific key roles (Hunsberger et al., 2009). More recently, the same authors found that pretreatment with 3 mM lithium plus 0.8 mM valproate (about 3 and 1.5 times human therapeutic lithium and valproate levels, respectively) for 6 days in glutamate-exposed rat cerebellar granule cells was associated with neuroprotection from excitotoxicity and miR-222 upregulation (Hunsberger et al., 2013).

MiRNAs alterations may also be indirectly associated with psychiatric conditions. For example, altered miRNAs regulation may be associated with learning and memory impairments that are common in MDD.

Measurements of miRNAs may considerably help our understanding of how gene expression networks are reorganized in both major depression and/or suicidal behavior; however, whether miRNAs may be considered as reliable biomarkers of major depression/suicidal behavior is still a matter of debate. To date, knowledge concerning the impact of miRNAs in synaptic plasticity, major affective disorders, and suicidal behavior is still inconsistent.

The present review chapter aimed to summarize the most relevant biochemical information (according to the current literature) about the role of miRNAs in neurogenesis/neuroplasticity mechanisms and their possible involvement in the development of pathological stress changes, affective disorders, and suicidal behavior.

## 4.2   METHODS

Medline and ScienceDirect databases have been carefully analyzed using the following terms: "microRNAs" and "synaptic plasticity" and "neuroplasticity" and "neurogenesis" and "major affective disorders" and "major depression" or "major depressive disorder (MDD)" and "Suicid*" (including suicidal behavior or suicide ideation or suicidal thoughts or deliberate self-harm or suicidal attempt). The reference lists of all identified papers have also been reviewed. Papers were included if they were published in peer reviewed journals.

Whether a title or abstract appeared to describe a study eligible for inclusion, the full-text article was obtained and examined to assess its relevance. A two-step literature search has been conducted by two independent researchers and any discrepancies between the two reviewers who, blind to each other, examined the studies for their possible inclusion were resolved by consultations with a senior author.

## 4.3   MicroRNAs BIOGENESIS AND EXPRESSION

The discovery of miRNAs undoubtedly added a new intriguing dimension to our understanding of complex gene regulatory networks (Labermaier et al., 2013).

Gene expression may be regulated at multiple levels within the central nervous system. MiRNAs may act as regulators of developmental timing and cell fate (Kosik, 2006; Bushati and Cohen, 2007). Gene expression can be activated through transcriptional factors and alternative splicing but also noncoding RNA transcripts such as miRNAs, antisense RNAs, and other critical forms of RNAs (Dwivedi, 2011, 2014; Serafini et al., 2012). More than 2000 miRNAs have been identified within the human genome (miRBase, 2012) in both tissue- and cell-specific manner (Landgraf et al., 2007; Kapsimali et al., 2007; Choi et al., 2008).

He et al. (2007) suggested that miRNAs have been evolutionarily conserved as involved in critical processes such as the stress response. Some miRNAs are expressed in neurons (Smirnova et al., 2005), whereas others may be found in astrocytes (Mor et al., 2011). Many researchers (Presutti et al., 2006; Fiore et al., 2008; Liu and Zhao, 2009; Vreugdenhil and Berezikov, 2009) reported that miRNAs may locally modulate mRNA translation by inducing changes in neurogenesis, synaptic/ axon development, and neuronal plasticity. As able to induce abnormalities in synaptic plasticity/neurogenesis or regulate the expression of genes critically involved in MDD, miRNAs may significantly contribute to either the onset, and maintenance of major affective disorders (Dwivedi, 2011).

MiRNAs are encoded in primary miRNAs (pri-miRNA) and processed by RNA polymerase II, III, and RNase III enzyme Drosha within the nucleus to create miRNA precursors (pre-miRNAs). Pre-miRNAs are also processed in the cytoplasm by the RNase III enzyme Dicer requiring cofactors such as the human transactivation response (TAR), RNA-binding protein (TRBP), or RNA-activated protein kinase (PKR)-activating protein (PACT). RNA-induced silencing complex binds to specific sequences within the mRNAs 3′ untranslated region (3′-UTR) and may interfere with the translation and subsequent mRNA levels.

The inhibition of translation may be mediated by protein Argonaute (AGO; having a splicer function) or, alternatively, by exonuclease, decapase, and deadenylase. A downregulation of transcription factors leading to reduced protein synthesis is another possible mechanism by which miRNAs may regulate gene transcription (Michalak, 2006).

MiRNAs may present several sites of dysfunctions (McNeill and Van Vactor, 2012) depending on the threshold of the miRNA level, miRNA cellular location or compartment, presence of a target mRNA, stoichiometric miRNA/mRNA target ratio, RNA cofactors binding to the target mRNA, and eventually cellular activation events such as raised calcium levels or proteolytic cleavage/phosphorylation of miRNA pathway protein components (McNeill and Van Vactor, 2012; Serafini et al., 2012).

Also, genetic changes in the promoter region of pri-miRNA gene transcripts, pre-miRNA secondary hairpin structure, or mature miRNA structure may be other mechanisms inducing altered miRNA expression (Michalak, 2006). RNA editing (Alon et al., 2012) of transcripts or epigenetic suppression of the chromosomal region encoding miRNAs could also play a role in the miRNA control as well as miRNA concentrations may be influenced by complex changes in the levels of miRNA machinery (e.g., Dicer, Drosha, or RISC; Michalak, 2006).

Figure 4.1 summarizes how miRNAs are functionally related to each other and how different factors may be implicated in influencing their expression. The figure schematically shows the possible sites of dysfunctions and/or miRNAs abnormal functioning.

## 4.4 ROLE OF MiRNAs IN NEUROGENESIS AND SYNAPTIC PLASTICITY

There are multiple evidence reporting an active role of miRNAs in the regulation of synaptic plasticity, neurogenesis, and neuropsychiatric disorders.

Hippocampal neurogenesis may be enhanced by environmental stimuli, exercise, and antidepressant (AD) medications (Lazarov et al., 2010), whereas pathological stress and major depression are able to negatively affect it (DeCarolis and Eisch, 2010; Lucassen et al., 2010).

The temporal progression of adult neurogenesis may be significantly regulated by miR-124 (Cheng et al., 2009), miR-124a, and let-7. MiR-200 family variants seem to be expressed by olfactory tissues (Choi et al., 2008), whereas miR-137 that is regulated by sex-determining region (Sox9) has been reported to enhance both the proliferation and differentiation of adult neural stem cells (Szulwach et al., 2010).

The existence of a link between Ezh2 (a histone methyltransferase and polycomb group protein) and miR-137 seems to further demonstrate the association between miRNAs and epigenetic mechanisms regulating adult neurogenesis.

Small RNAs have been reported as important mediators of long-term plasticity by Malan-Müller et al. (2013), who hypothesized their role as translational regulators. Schratt et al. (2006) suggested that miRNAs may regulate the development of dendritic spines associated with postsynaptic signaling/plasticity in rats.

Pre-miRNAs, Dicer, and eukaryotic translation initiation factor 6 (eIF2c) protein may regulate protein synthesis by binding to specific mRNA sites presumably through RISC in the somatodendritic compartment (Caudy et al., 2002; Ishizuka et al., 2002; Qurashi et al., 2007). It has also been suggested that Armitage plays an important role for long-term potentiation and synaptic protein synthesis. Similarly, the proteasomal degradation of MOV10, an RNA helicase associated with Argonaute protein seems to closely modulate synaptic plasticity (Banerjee et al., 2009).

Additional evidence supported the association between miRNAs and synaptic plasticity. Protein synthesis in rats together with the size of postsynaptic sites in dendritic hippocampal spines have been found to be regulated by miR-134 (Schratt et al., 2006). Siegel et al. (2009) suggested that miR-134 suppression (mediated by the translation of mRNA encoding a specific protein kinase, Limdomain-containing protein kinase 1 (Limk1)) may modulate dendritic spine growth. Conversely,

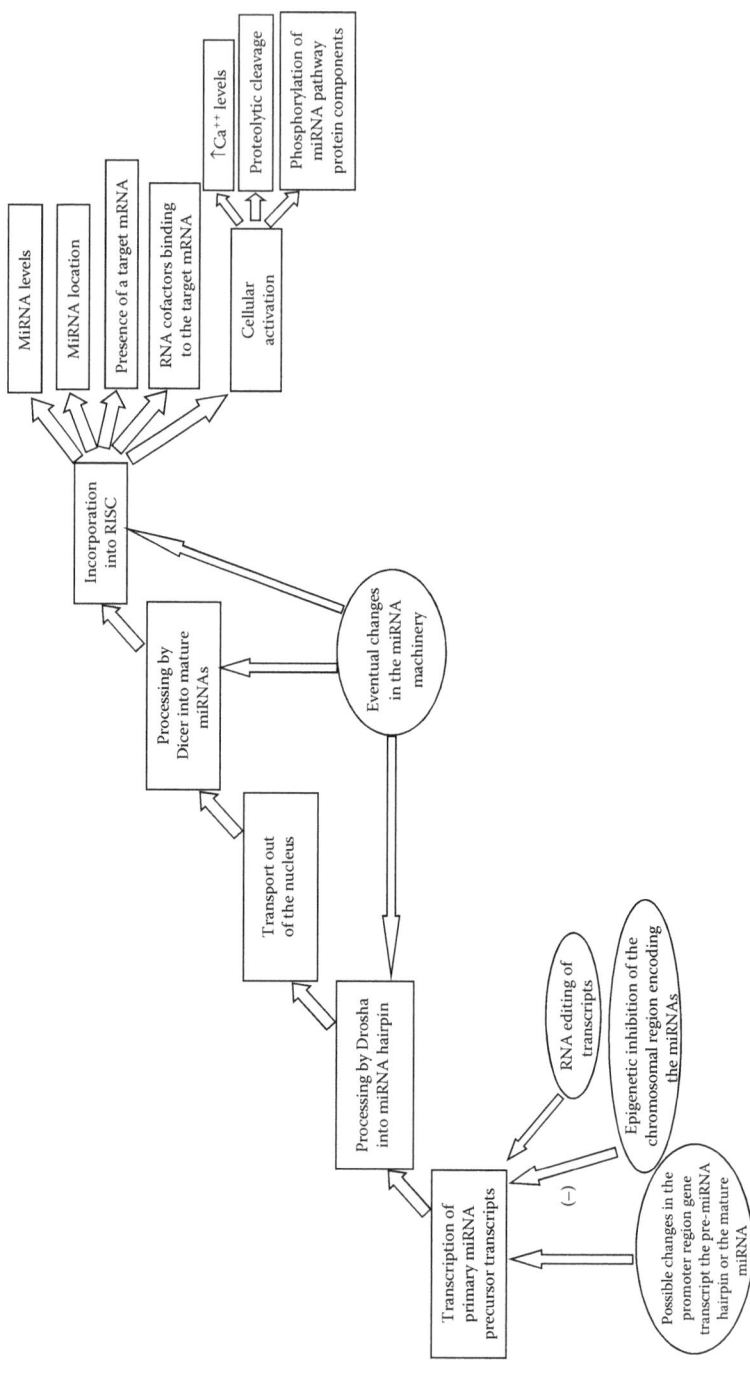

**FIGURE 4.1**　Altered miRNA expression: Sites of dysfunctions and/or miRNAs abnormal functioning.

miR-138 downregulated the size of dendritic spines in rat hippocampal neurons affecting the expression of acyl protein thioesterase 1 that is presumably implicated in the palmitoylation status of proteins having synaptic dendritic functions. In addition, the suppression of miR-138 induced spine growth that may be inhibited by interference-mediated knockdown of acyl protein thioesterase 1 and expression of membrane-localized Ga13.

Lugli et al. (2008) reported the critical role of miRNAs in local protein synthesis (as they are partially processed near synapses). Here, some miRNA genes such as miR-132 have binding sites for CREB1 (cAMP response element binding) and may be activated by BDNF through CREB (Vo et al., 2005; Wu and Xie, 2006). The increased miR-132 transcription has been reported to regulate spine formation (Impey et al., 2010), whereas neurite outgrowth/dendritic morphogenesis can be enhanced by miR-132 and miR-124 overexpression (Vo et al., 2005; Wayman et al., 2008). MiR-124 is fundamental for cell identity and is presumably regulated by 5-hydroxytriptamine (HT) derepressing CREB and enhancing 5-HT-dependent long-term facilitation, respectively.

Synaptic plasticity and memory formation have also been suggested to be regulated by the gene SIRT1 playing a fundamental role in oxidative stress and circadian rhythms (Nakahata et al., 2008; Finkel et al., 2009). SIRT1 reduction may induce an unchecked miR-134 expression with subsequent reduced CREB and BDNF expressions and impaired synaptic plasticity (Gao et al., 2010). Critical processes of memory formation are affected by miRNAs via synaptic tagging with the aim to establish a synaptic input specificity (Martin and Kosik, 2002; Schaeffer et al., 2003; Kim et al., 2004).

MiRNAs have been also proposed as important effectors in the pathophysiology of anxiety disorders considering their critical role in neurogenesis, neurite outgrowth, synaptogenesis, synaptic/neural plasticity, and stress-related conditions (Malan-Müller et al., 2013).

Lee et al. (2012) found increased levels of miR-206 in the brains of Tg2576 mice and temporal cortex of Alzheimer's disease mice. They also reported that a neutralizing inhibitor of miR-206 increased brain levels of BDNF into the mice brains with a subsequent improvement in memory function and prevention of the negative amyloid-$\beta$42 effects. Furthermore, AM206 hippocampal synaptic density as well as neurogenesis was enhanced by this neutralizing inhibitor of miR-206.

Thirty miRNAs were also found to be differentially expressed after 12 weeks of escitalopram treatment in the blood of 10 depressed subjects (Bocchio-Chiavetto et al., 2013). Interestingly, there was a significant enrichment in neuroactive ligand–receptor interaction, axon guidance, long-term potentiation, and depression confirming that miRNAs may be differentially involved in the regulation of AD mechanism.

Moreover, as miR-132 overexpression in transplanted neurons may enhance neuronal survival (Pathania et al., 2012), it has been suggested that it may represent the basis of a structural plasticity observed in postnatal neurogenesis. Specifically, a relevant increase in the survival of newborn neurons, specifically a timely directed overexpression of miR-132 at the onset of synaptic integration using *in vivo* electroporation has been reported.

According to the multiple evidence reported above, it is possible to summarize that miRNAs significantly affect both neurogenesis and neuroplasticity processes.

## 4.5   MicroRNAs AND NEUROTROPHIC FACTORS: WHAT IMPLICATIONS FOR MAJOR AFFECTIVE DISORDERS?

There are multiple evidence suggesting that neurotrophic factors are involved in neurogenesis, synaptic plasticity, and major affective disorders but there are also studies reporting the association between miRNAs levels and neurotrophic factors activity.

BDNF, a member of the neurotrophin family of structurally related proteins, induces neuronal differentiation and cell survival during development (Cowansage et al., 2010). BDNF is involved in the maintenance of various neuronal functions and neurogenesis (Sofroniew et al., 1990; Cooper et al., 1996) as well as regulation of morphological plasticity, neuronal proliferation, migration, and phenotypic differentiation (McAllister et al., 1999; Thoenen, 2000; McAllister, 2001). In detail, BDNF-TrkB signaling pathway is associated with cell survival, migration, outgrowth of axons and dendrites, synaptogenesis, and remodeling of synapses (Ohira and Hayashi, 2009). Epigenetic and post-translational mechanisms may induce changes in BDNF expression, release, and activity. MiRNAs have been also suggested to interact with BDNF.

For example, the X-linked transcriptional repressor methyl CpG-binding protein 2 (MeCP2) translation seems to be regulated by miR-132 (Klein et al., 2007). Several neurodevelopmental disorders such as Rett syndrome are associated with changes in MeCP2 levels. The blocking of miR-132-mediated repression increased MeCP2 and BDNF concentrations in cultured rat neurons, whereas the loss of MeCP2 reduced BDNF and miR-132 concentrations *in vivo*. Im et al. (2010) suggested that MeCP2 is a fundamental regulator of neuroplasticity in postmitotic neurons. The effects of cocaine on dorsal striatal BDNF levels seem to be controlled by MeCP2 through miR-212 (Im et al., 2010). The authors suggested that vulnerability to cocaine addiction is associated with the homeostatic interaction between MeCP2 and miR-212.

Glucocorticoids have been suggested to inhibit BDNF-induced synaptic maturation and excitatory neurotransmitter glutamate release (Kumamaru et al., 2008; Numakawa et al., 2009). Pariante and Miller (2001) reported that the expression of glucocorticoid receptors in the paraventricular nucleus of F344 rats is downregulated in depressed subjects (Uchida et al., 2008). The expression of miR-18a able to inhibit the translation of glucocorticoid receptors mRNA has been reported to be greater in these rats. Glucocorticoid receptors expression appearing downregulated in stress-related disorders seems to be controlled by miR-18a and miR-124a (Vreugdenhil et al., 2009). The expression of these two miRNAs may be reduced by glucocorticoid receptors concentrations via the induction of the gene leucine zipper. It is, however, important to note that the real effect of miR-124a expression on glucocorticoid receptor levels is uncertain as Vreugdenhil and colleagues (2009) reported an effect only in the case of miR-124a but not miR-18 overexpression after luciferase assays using the GR 3′-UTR. Glucocorticoids seem to be implicated in the neurogenic action of ADs (Huang and Herbert, 2006; David et al., 2009).

Consistent evidence (Duman, 2004; McEwen and Chattarji, 2004; Castrén et al., 2007; Sairanen et al., 2007; Bessa et al., 2009) suggested that neurotrophin signaling, promoting neuronal and synaptic remodeling and neurogenesis in key brain regions such as the hippocampus and prefrontal cortex, may be increased by the administration of AD medications. It has been reported that AD drugs may enhance both BDNF and glutamate release (Yagasaki et al., 2006).

Antidepressant-induced changes in neurogenesis seem to be directly dependent on the glucocorticoid receptors (Anacker et al., 2011). Kawashima et al. (2010) reported that the expression of miR-132 (associated with neurite outgrowth and dendritic morphogenesis) was regulated by BDNF through the action of CREB. Specifically, BDNF concentrations and ERK1/2 pathway activation (involved in pathological stress-induced changes and major depression) may induce the upregulation of miR-132. Mellios and colleagues (2011) hypothesized the existence of a correlation between miR-30a, miR-195, and BDNF levels associated with major depression.

As suggested by Huang et al. (2012), BDNF rapidly increased Dicer and subsequently mature miRNA levels together with RNA processing bodies in neurons and Lin28 associated with an upregulation in translation of Lin28-regulated target mRNAs. Lin28 deficiency or Lin28-resistant Let-7 precursor miRNA may inhibit BDNF translation specificity and BDNF-dependent dendrite arborization. The specificity in BDNF-regulated translation is linked to two-part post-transcriptional control of miRNA biogenesis enhancing mRNA repression together with GW182 (selectively derepressing and increasing translation of specific mRNAs).

In summary, multiple evidence reported a deep interaction between miRNAs activity and neurotrophic factors playing a fundamental role in the pathophysiology of major affective disorders.

## 4.6 IMPACT OF MiRNAs IN STRESS-RELATED DISORDERS

Several studies reported that miRNAs may play a major role in the development and maintenance of stress-related disorders.

MiRNAs expression and synaptic transmission in key brain regions such as the frontal cortex may be rapidly influenced by acute environmental stress able to regulate mRNA translation. Kagias et al. (2012) reported that stress-related processes may significantly affect neuronal differentiation and changes in the expression of several molecules such as transcription factors; miRNAs have been found to impact on stress resistance and adaptation as well.

After acute stress, the expression of several miRNAs was increased in the frontal cortex but not hippocampus of mice (Rinaldi et al., 2010). Specifically, the expression of Let-7a, miR-9, and miR-26a/b were significantly increased after repeated stress. Meerson et al. (2010) reported that chronic stress-induced relevant changes in the expression of miR-134, miR-183, miR-132, Let-7a-1, miR-9-1, and miR-124a-1 in the central amygdala and hippocampal CA1 region compared to acute stress, whereas both miR-376b and miR-208 levels were increased and miR-9 decreased in the hippocampal CA1 region after acute or chronic stress. The expression of miR-Let-7a-1 in the central amygdala seemed to be significantly modulated by acute and chronic stress.

Some miRNAs such as miR-134 and miR-183 may regulate alternative splicing. Knockdown of miR-183 expression and overexpression of miR-183 may increase and decrease, respectively, the splicing factor SC35 promoting the alternative splicing of acetylcholinesterase. The expression of profilin-2 mRNA associated with neurotransmitter homeostasis and dendritic spine morphology in neuronal cells may be targeted by miR-183 (Witke, 2004).

It has been suggested that the development of stress-related disorders in adulthood is significantly associated with repressor element 1 silencing transcription factor (REST4). Maternal separation increased the expression of REST4 and depression-like behaviors in the medial prefrontal cortex of neonatal mice (Uchida et al., 2010). The expression of several brain-enriched miRNAs involved in brain development and neural plasticity (Vo et al., 2005; Kosik, 2006; Rajasethupathy et al., 2009) may be regulated by REST (Conaco et al., 2006; Otto et al., 2007). REST has been reported to control the expression of pre-miRNAs-132 and pre-miRNAs-124 (Vo et al., 2005; Rajasethupathy et al., 2009) in the medial prefrontal cortex of stressed rats, reduce dendritic length (Pascual et al., 2007), and increase synaptic density (Ovtscharoff and Braun, 2001) together with interfering with basal neuronal activity (Stevenson et al., 2008).

The altered repression of miRNAs such as miR-137 seems to be directly mediated by REST (Soldati et al., 2013). The association between REST and cytoplasmatic Huntingtin is reported to be impaired in Huntington's disease and associated with increased nuclear REST and repression of neuronal-specific genes such as BDNF. Significant changes in the expression of more than 10 miRNAs, some predicted miRNAs targets, and functional clusters related to neural development have been reported by Nerini-Molteni et al. (2012). The authors suggested that pathway-oriented toxicity may be significantly mediated by miRNA expression analysis.

There are also evidence suggesting that homeostatic miRNA response after stress may affect the expression of transcriptional factors such as CREB (Uchida et al., 2008; Remenyi et al., 2010). Neurotrophins levels are able to regulate the production of specific miRNAs. For example, Dwivedi and colleagues (2003) reported that miR-124 regulated the expression of CREB (having a CREB site within its promoter region).

In overall, both preclinical and clinical evidence clearly suggested that miRNAs are extensively involved in stress-related disorders.

## 4.7    IMPACT OF MiRNAs IN MAJOR DEPRESSIVE DISORDER AND SUICIDAL BEHAVIOR

The potential of miRNAs as diagnostic markers for major psychiatric disorders and suicidal behavior is advancing rapidly (Dwivedi, 2011; Serafini et al., 2012; Chana et al., 2013; Dorval et al., 2013).

Dwivedi Y and Smalheiser NR (unpublished findings) reported that some miRNAs promoting brain-active transcription factors such as NOVA1, signaling proteins, and genes implicated in neurotransmitter release and synaptic plasticity, were significantly downregulated in the prefrontal cortex of depressed individuals. According to animal and human studies, some miRNAs downregulated the expression of VEGFA, signaling proteins, ion channels, and ubiquitin ligases.

MiRNA expression profiles in peripheral blood mononuclear cells have been investigated in a sample of 16 severe depressed patients and 13 matched controls at baseline, 2 and 8 weeks after treatment by Belzeaux et al. (2012) who found that

14 miRNAs were globally dysregulated (specifically, nine miRNAs were upregulated and five downregulated) but only two miRNAs (miR-941 and miR-589) demonstrated stable overexpression. The authors hypothesized the existence of a common RNA regulatory network in MDD suggesting the potential of miRNAs as biomarkers of major depressive episode evolution.

The possible association between miRNA polymorphisms and vulnerability to major depression has also been investigated (Xu et al., 2010). The authors reported that P300 latency resulted associated with miR-30e ss178077483 genotypes (C/T genotype) and frequency of miR-30e ss178077483 in depressed subjects after analyzing 1088 Chinese depressed patients and 1102 control subjects.

An association between miRNA-processing gene variants, vulnerability to depression and suicidal behavior has been recently identified by He et al. (2012) in a case–control study of 314 patients and 252 matched healthy controls. These authors found that subjects with variant allele of DGCR8 rs3757 were more likely to have suicidal tendencies and improvement in response to AD treatments compared to those with AGO1 rs636832 variant who showed reduced risk of suicidal tendency, suicidal behavior as well as recurrent depression. They concluded that DGCR8 rs3757 and AGO1 rs636832 variants were significantly associated with major depression, whereas GEMIN4 rs7813 variant did not affect vulnerability to depression.

Recently, Smalheiser et al. (2014b) suggested that treatment with the fluoroquinolone enoxacin for 1 week increased the expression of miRNAs in rat frontal cortex and decreased the proportion of animals exhibiting learned helplessness behavior following inescapable shock. However, the same authors reported that further additional studies are needed to test whether enoxacin may improve depressive behaviors in other animal paradigms as well as human clinical conditions.

MiRNAs have been also reported as effectors of the occurrence and development of stroke and having fundamental regulatory effects on poststroke depression. However, other factors such as anatomic location of lesions, gene polymorphism, inflammatory cytokines, abnormal circadian rhythms, and social/psychological factors may play a relevant role in the occurrence of poststroke depression (Yan et al., 2013).

MiRNA alterations have been also reported in patients with bipolar disorder (BD). Rong et al. (2011) found a significantly reduced expression of miR-134 after acute treatment together with its increased plasmatic levels after 2/4 weeks of treatment in the peripheral cells of 21 drug-free patients with BD type I compared to 21 matched healthy controls. Also, Kim et al. (2010b) suggested that the expression of 15 miRNAs are both up- and downregulated in the prefrontal cortex of BD individuals.

Psychoactive drugs may also significantly modify miRNAs and their effector targets levels (Zhou et al., 2009). The majority of studies examined the effect of lithium or valproate on miRNAs concentrations. MiR-15 and miR-16 seem to be inhibited by mood stabilizers (Marmol, 2008; Kim et al., 2010b); for example, chronic lithium and valproate treatment may alter hippocampal miRNA levels. Neurite outgrowth, neurogenesis, and important signaling pathways are reported to be regulated by the predicted effectors of these miRNAs. Lithium or valproate treatment is associated with a reduction of miR-34a and increased levels of the predicted effector metabotropic glutamate receptor 7. Furthermore, miR-34a precursor reduced and miR-34a increased the inhibitor metabotropic glutamate receptor 7 levels, respectively. Important signaling pathways like protein kinase C (PKC), phosphatase and tensin homolog (PTEN), extracellular signal-regulated kinase mitogen-activated protein (ERK-MAP kinase), Wnt/b-catenin (implicated in embryonic development and tumorigenesis), and adrenergic signaling pathways have been suggested as the identified miRNAs targets involved in both major depression and BD (Machado-Vieira et al., 2010). Chronic administration of lithium and valproate upregulated dipeptidylpeptidase 10 and glutamate receptor 7 (Saus et al., 2010) together with thyroid hormone receptor beta hippocampal protein levels and downregulated the expression of miR-128a, miR-24, and miR-34a, respectively. MiRNA-mediated effects at the synapse with a direct effect on transcripts through a stimulation-dependent translation have been reported (Kosik, 2006; Kuss and Chen, 2008). The expression of miR-221, miR-152, miR-15a, miR-494, miR-155, miR-181c, and miR-34a was persistently altered after chronic lithium treatment in 20 lymphoblastoid cell lines in culture (Chen et al., 2009).

Some miRNAs seem to be also involved in the effects of AD medications. MiR-16 may play a crucial role in regulating 5-HT transporter expression as well as mediating adaptive responses of serotonergic and noradrenergic neurons to fluoxetine treatment (Baudry et al., 2010). MiR-16 levels (expressed in noradrenergic neurons) and 5-HT transporter expression have been reported to be inversely correlated. Elevated miR-16 levels and decreased 5-HT transporter expression were reported in serotonergic raphe nuclei of mice after long-term treatment with fluoxetine (Baudry et al., 2010). Long-term fluoxetine induced a release of S100b associated with a reduction of miR-16 levels and an unlocked expression of serotonergic functions in the locus coeruleus of noradrenergic neurons.

O'Connor et al. (2012) suggested that ADs and mood stabilizers may use miRNAs as downstream effectors. Modified behaviors have been associated with modifying miRNA levels according to preclinical models. The fundamental role of miRNAs as critical effectors in regulating serotonergic transmission and major depression has been reported by Millan (2011). 5-HT is associated with a modulation of the miR-124 expression (Rajasethupathy et al., 2009).

Jensen et al. (2009) suggested the presence of the A-element able to confer repression by miR-96 within mRNA of 5HT1B receptors. A common human variant (G-element) associated with a reduced repressive activity of this element on miR-96 has been reported. Subjects who were homozygous for the ancestral A element suffered from more conduct disorder behaviors when compared with those with the G element (Jensen et al., 2009).

MiRNAs may be detected in many circulating biological fluids such as serum, plasma, urine, saliva, and cerebrospinal fluid (Cogswell et al., 2008; Hanke et al., 2010; Weber et al., 2010). Importantly, the profile of miRNAs is significantly modified in pathological conditions suggesting the possibility that peripheral miRNAs may be used as reliable biomarkers (Dwivedi, 2014).

Over the last years, several efforts have been carried out to identify developing circulating miRNAs as potential biomarkers for many illnesses. But what is the possible contribution of miRNAs studies in our understanding of suicidal behavior? The examination of postmortem brains to directly assess the status of miRNAs in psychiatric disorders was one of the most common approach to investigate suicidal behavior. MiRNAs have been suggested to play a prominent role in suicidal behavior based on studies on depressed suicide victims.

The first direct study primarily examining the impact of miRNAs alterations in patients with suicidal behavior is that of Smalheiser et al. (2012). MiRNA expression was globally found as downregulated, by 17% on average, in the prefrontal cortex of 18 depressed suicide victims compared to that of 17 well-matched nonpsychiatric controls died for other causes. The authors suggested that both the global decrease of miRNA expression and its lower variability were consistent with a hypoactivation of the frontal cortex observed in depressed suicidal subjects. Also, significant modifications of 21 miRNAs that are known to be involved in cellular growth and differentiation have been reported in this study.

In another study, Maussion et al. (2012) reported an increase of miR-185 expression levels regulating the TrkB-T1 decrease in the frontal cortex of 38 subjects who died by suicide. Interestingly, miR-185 and miR-491-3p were significantly and globally upregulated in individuals with low expression of TrkB-T1 who completed suicide. The relevant downregulation of TrkB-T1 levels in the subgroup of suicide completers was not explained by confounders such as age, pH, PMI, or suicide method. Importantly, the authors found that five possible binding sites for the DiGeorge syndrome linked miR-185 in the 3′-UTR of TrkB-T1 emerged at bioinformatic analyses. Based on these findings, deletions or duplications in miR-185 may potentially impact the expression levels of TrkB-T1 leading to a psychiatric phenotype.

Tables 4.1 and 4.2 summarize the most relevant studies reporting the association between miRNAs, major depression, and suicidal behavior.

## 4.8 CRITICAL CONSIDERATIONS AND MAIN LIMITATIONS

MiRNAs may be investigated in different circulating fluids such as serum, plasma, urine, saliva, and cerebrospinal fluid (Cogswell et al., 2008; Hanke et al., 2010; Weber et al., 2010).

The role of miRNAs in post-transcriptional regulation of an array of genes and, in particular, their impact in the pathophysiology of major neuropsychiatric disorders has been recognized worldwide. We have to learn a lot about the mechanisms by which synaptic miRNAs are dysregulated and we did not know to what extent they are involved into the pathogenesis as well as the way they may be modeled in humans. Another exciting perspective is that miRNAs have been described as potential molecular intermediaries of therapeutic response in neuropsychiatric disorders.

A recent study suggested that miR-1202 is associated with the pathophysiology of depression as able to mediate response to AD treatment (Lopez et al., 2014a). Specifically, an upregulation of miR-1202 after chronic treatment with imipramine and citalopram has been reported. Interestingly, when cells are treated with other psychoactive drugs such as lithium or valproate that do not act on the reuptake of serotonin or serotonin transporter, these effects have been not observed. Therefore, the increase in miR-1202 expression after chronic administration of imipramine/ citalopram seems to be exclusively based on the effect on SERT and the relative

**TABLE 4.1**

**Most Relevant Studies about the Association between MiRNAs and Major Depressive Disorders**

| Author(s), Year | Study Design | Sample | Results | Type of Treatment | Conclusions |
|---|---|---|---|---|---|
| Liu et al. (2015) | Animal model in which the TrkB antagonist K252a, the ERK phosphorylation inhibitor U0126, the Akt phosphorylation inhibitor LY294002 or vehicle was given intracerebroven-tricularly to mice 30 min before 7-CTKA or vehicle intraperitoneal injection | ICR mice (22–26 g; 4 weeks old) | The reduction in sucrose preference and multiple hippocampal miRNAs changes were reversed by 7-CTKA which also mediated 15 common miRNAs via TrkB-ERK/Akt pathways. Overall, only the expression levels of miR-34a-5p, miR-200a-3p, miR-144-3p, miR-1894-5p were validated by quantitative real-time PCR that generally supports findings from microarray analysis (with the exception of miR-1894-5p expression that resulted nonsignificant) | K252a; U0126; LY294002; 7-CTKA | The miRNA identified targets shared by TrkB-ERK/Akt pathways could be involved in the molecular mechanism underlying the rapid AD effects of 7-CTKA |
| Issler et al. (2014) | Animal model in which male mice were used | ePet-EYFP, ePet-Cre, C57BL/6 male mice | After administration of ADs, miR-135a levels resulted upregulated. Higher/lower levels of miR-135a expressions may be found in genetically modified mouse models who also demonstrated major anxiety- and depressive-like behaviors. MiR-135a levels in blood and brain of depressed human patients resulted significantly lower | ADs | A potential endogenous AD role for miR-135 has been suggested |
| Lopez et al. (2014a) | Postmortem human brain samples, cellular assays and samples from clinical trials of subjects with MDD | 32 patients with MDD and 18 HC | MiR-1202 may be differentially expressed in patients with depression. Also, it has been reported that miR-1202 regulates expression of the gene encoding GRM4 and is able to predict AD response at baseline | Citalopram | MiR-1202 is associated with the pathophysiology of depression and is a potential target for new AD treatments |

*(Continued)*

**TABLE 4.1 (*Continued*)**

**Most Relevant Studies about the Association between MiRNAs and Major Depressive Disorders**

| Author(s), Year | Study Design | Sample | Results | Type of Treatment | Conclusions |
|---|---|---|---|---|---|
| Yang et al. (2014) | Animal study in which the rat hippocampus was examined after ketamine injection | Male SD rats (50 days old) | Overall, 18 miRNAs were significantly decreased, and 22 significantly increased in the rat hippocampus after ketamine injection. Notably, miR-206 was downregulated in ketamine-treated rats. BDNF was a direct target gene of miR-206 in both cultured neuronal cells *in vitro* and hippocampus *in vivo*. MiR-206 modulated the expression of BDNF through this target gene. Also, overexpression of miR-206 significantly reduced ketamine-induced upregulation of BDNF | Ketamine | MiRNA-206 may be one of the novel therapeutic target for the antidepressive effect of ketamine |
| Bai et al. (2014) | Animal study in which rats were investigated with MD and CUPS | 10 MD and CUPS SD rats at the age of 3 months | MD rats demonstrated significantly lower Htr4 mRNA and protein expression and significantly higher Let-7a level in the hippocampus when compared to control rats. Let-7a expression negatively correlated with Htr4 mRNA and protein expression. Sucrose preference rate positively correlated with Htr4 mRNA expression and sucrose preference rate negatively correlated with Let-7a expression | No active treatment | Anhedonia can be associated with upregulation of Let-7a and downregulation of Htr4 expression in the rat hippocampus |

(*Continued*)

**TABLE 4.1 (Continued)**

**Most Relevant Studies about the Association between MiRNAs and Major Depressive Disorders**

| Author(s), Year | Study Design | Sample | Type of Treatment | Results | Conclusions |
|---|---|---|---|---|---|
| Ryan et al. (2013) | Animal study in which rat brain and blood were used after either acute (×1) or chronic (×10) ECS | Male SD rats | ECT/ECS | Using qRT-PCR, 14 BDNF-associated miRNA species were found. Levels of miR-212 appeared significantly increased in rat dentate gyrus after both acute and chronic ECS. Increased miR-212 levels were also observed in whole blood after chronic ECS. Elevated miR-212 levels were reported to be positively correlated with chronic ECS in the dentate gyrus | MiRNA altered expression may provide interesting information about the mechanism of action of ECT/ECS |
| Li et al., 2013 | Serum levels of miR-132 and miR-182 | 40 patients with MDD and 40 HC | No active treatment | MDD patients demonstrated lower serum BDNF levels (through the enzyme-linked immunosorbent assays) and higher serum miR-132 and miR-182 levels (through the real-time PCR) relative to HC. A significant negative association between SDS scores and serum BDNF levels, and a positive association between SDS scores and miR-132 levels were also reported. Finally, a reverse relationship between serum BDNF levels and the miR-132/miR-182 levels in MDD was observed | MiR-182 is a putative BDNF-regulatory miRNA. Serum BDNF together with related miRNAs may be used as possible biomarkers in the diagnosis or as therapeutic targets of MDD |

*(Continued)*

**TABLE 4.1 (*Continued*)**

**Most Relevant Studies about the Association between MiRNAs and Major Depressive Disorders**

| Author(s), Year | Study Design | Sample | Results | Type of Treatment | Conclusions |
|---|---|---|---|---|---|
| Oved et al. (2013) | Genome-wide expression profiles of LCLs lines derived by unrelated subjects treated with 1 µM paroxetine for 21 days and untreated control cells | Human LCLs of MDD subjects and HC | After chronic paroxetine exposure, ITGB3 showed the most important abnormal expression. Also, using genome-wide miRNA arrays, a corresponding reduction in the expression of miR-221 and miR-222, (both predicting to target ITGB3) was reported. Also, the expression of SERT or serotonin receptors was not modified | Paroxetine | SSRIs may act promoting neuronal synaptogenesis/ neurogenesis together with enhancing 5-HT neurotransmission |
| O'Connor et al. (2013) | Animal model in which SD rats were used | SD rats | Early-life stress influenced the expression of multiple hippocampal miRNAs based on microarray analysis. AD reversed stress-related changes concerning miR-451. Ketamine and ECT had the highest number of common targets. Thus, the existence of common pathways has been reported. MiR-598-5p was a common target of all three treatments | Fluoxetine; ketamine; ECT | Changes in hippocampal miRNA expression may exert an important role in stress-induced disorders that may be reversed using ADs |

*(Continued)*

**TABLE 4.1 (Continued)**
**Most Relevant Studies about the Association between MiRNAs and Major Depressive Disorders**

| Author(s), Year | Study Design | Sample | Results | Type of Treatment | Conclusions |
|---|---|---|---|---|---|
| Belzeaux et al. (2012) | Follow-up study in which assessments were performed at baseline, 2 and 8 weeks after treatment | 16 severe MDD patients and 13 HC | 14 miRNAs were globally dysregulated (specifically, nine miRNAs were upregulated and five downregulated) but only two miRNAs (miR-941 and miR-589) demonstrated stable overexpression | Current antidepressive treatments | MiRNAs have been identified as biomarkers of major depressive episode evolution and a common RNA regulatory network has been specifically recognized |
| Bai et al. (2012) | Animal study in which rats showed depression-like behaviors related to both MD and chronic CUPS | SD rats at the age of 3 months | Reduced BDNF mRNA and higher miR-16 expression were showed in MD but not CUPS rats compared with control rats. BDNF mRNA expression negatively correlated with the expression of miR-16 in MD rats. BDNF expression positively correlated with the total distance rats crawled and vertical activity in the OFT while miR-16 expression negatively correlated with these behaviors. BDNF positively correlated with sucrose preference rate whereas miR-16 negatively correlated with sucrose preference rate of the sucrose consumption test | No active treatment | Depression-like behavior induced by MD (but not CUPS) was significantly associated with upregulation of miR-16 and possible downregulation of BDNF in the hippocampus |

*(Continued)*

**TABLE 4.1 (*Continued*)**
**Most Relevant Studies about the Association between MiRNAs and Major Depressive Disorders**

| Author(s), Year | Study Design | Sample | Results | Type of Treatment | Conclusions |
|---|---|---|---|---|---|
| Bocchio-Chiavetto et al. (2013) | Blood of depressed subjects after 12 weeks of treatment | 10 MDD patients | Overall, 28 miRNAs were upregulated, and 2 were downregulated after AD treatment. After miRNA target gene prediction and functional annotation analysis, a relevant enrichment of neuroactive ligand–receptor interaction, axon guidance, long-term potentiation, and depression was found | Escitalopram | AD mechanism may be differentially regulated by miRNAs |
| Oved et al. (2012) | Human LCLs screened from healthy adult female subjects for growth inhibition by paroxetine | 80 human LCLs | MiR-151-3p had 6.7-fold higher basal expression in paroxetine-sensitive LCLs corresponding with reduced expression of CHL1 (a target of miR-151-3p). Also, miR-212, miR-132, miR-30b*, Let-7b, and Let-7c significantly differed by >1.5-fold between LCLs with high or low sensitivities to paroxetine | Paroxetine | MiRNAs may serve as biomarkers of SSRI response |

(*Continued*)

**TABLE 4.1 (*Continued*)**

**Most Relevant Studies about the Association between MiRNAs and Major Depressive Disorders**

| Author(s), Year | Study Design | Sample | Results | Type of Treatment | Conclusions |
|---|---|---|---|---|---|
| Launay et al. (2011) | Animal study in which Swiss-Kunming mice were used | Adult 6- to 8-week-old male Swiss-Kunming mice (25–30 g) | MiR-16 seems to mediate adult neurogenesis in the mouse hippocampus. Fluoxetine reduced miR-16 levels in the hippocampus, which in turn increased SERT levels (the target of SRI), bcl-2, and neuronal maturation. The hippocampal activity of fluoxetine is related to S100β, secreted by raphe and acting through LC. Also, serotonergic neurons released BDNF, Wnt2, and 15-deoxy-delta12,14-prostaglandin J2 acting synergistically to modulate hippocampal miR-16 levels | Fluoxetine | MiR-16 seems to regulate fluoxetine activity by acting as a micromanager of hippocampal neurogenesis |
| Baudry et al. (2010) | Animal study in which mice were investigated | Mice | SERT is a target of miR-16 which is expressed at higher levels in noradrenergic cells; its decrease in noradrenergic neurons induces *de novo* SERT expression. Chronic treatment with fluoxetine increases miR-16 levels in serotonergic raphe nuclei, which reduces SERT expression in mice. The effect of fluoxetine on raphe induces the release of neurotrophic factor S100β, which is able to act on noradrenergic cells of LC | Stereotaxic injection (2 mL/min) of fluoxetine (1 mM) into raphe in combination or not with activators of the canonical Wnt pathway | MiR-16 may contribute to the therapeutic action of SSRIs |

*(Continued)*

**TABLE 4.1 (*Continued*)**
**Most Relevant Studies about the Association between MiRNAs and Major Depressive Disorders**

| Author(s), Year | Study Design | Sample | Results | Type of Treatment | Conclusions |
|---|---|---|---|---|---|
| Xu et al. (2010) | An association analyses from the Han Chinese population | 1088 MDD patients and 1102 control subjects | A positive association between miR-30e ss178077483 and MDD was reported. Also, P300 latency was associated with miR-30e ss178077483 genotypes and subjects with the C/T genotype had a longer P300 latency compared to those carrying the C/C genotype | No active treatment | MiRNA polymorphisms may play a critical role in determining MDD susceptibility |
| Lambert et al. (2010) | Animal study using cultured mouse hippocampal neurons | Neurons isolated from the hippocampi of P0–P1 wild-type (C57BL/6;129SvJ) mice | Overexpression of miR132 is associated with increased paired-pulse ratio and reduced synaptic depression in cultured mouse hippocampal neurons without influencing the initial likelihood of neurotransmitter release, the calcium sensitivity of release, the amplitude of excitatory postsynaptic currents or the size of the readily releasable pool of synaptic vesicles | No active treatment | MicroRNAs can regulate short-term plasticity in neurons |

(*Continued*)

**TABLE 4.1 (*Continued*)**
**Most Relevant Studies about the Association between MiRNAs and Major Depressive Disorders**

| Author(s), Year | Study Design | Sample | Results | Type of Treatment | Conclusions |
|---|---|---|---|---|---|
| Uchida et al. (2010) | Animal study in which rat pups were separated from the dams for 180 min/day from postnatal day 2 through day 14 to create HMS180 rats | Timed-pregnant SD female rats | HMS180 rats demonstrated a greater HPA axis response to acute restraint stress compared to nonseparated control rats. Also, repeatedly restrained HMS180 rats demonstrated increased depression-like behavior as well as an anhedonic response relative to nonrestrained HMS180 rats. Moreover, increased expression of REST4 and a variety of miRNAs in the mPFC were observed in HMS180 rats. Furthermore, REST4 overexpression in the mPFC of neonatal mice via polyethyleneimine-mediated gene transfer induced the expression of its target genes as well as behavioral vulnerability to repeated restraint stress | No active treatment | Stress vulnerability may be influenced by the activation of REST4-mediated gene regulation in the mPFC during postnatal development |
| Rajasethupathy et al. (2009) | Animal study in which RNA amounts were derived by the whole animal, central nervous system, pleural and abdominal ganglia | Aplysia | It has been reported that miR-124, the most abundant and well-conserved brain-specific miRNA resulted present presynaptically in a sensorymotor synapse where it constrained serotonin-induced synaptic facilitation through regulation of the transcriptional factor CREB | No active treatment | A regulatory neurotransmitter critical for learning may modulate miR-124 levels involved in long-term plasticity of synapses |

*(Continued)*

**TABLE 4.1 (*Continued*)**
**Most Relevant Studies about the Association between MiRNAs and Major Depressive Disorders**

| Author(s), Year | Study Design | Sample | Results | Type of Treatment | Conclusions |
|---|---|---|---|---|---|
| Jensen et al. (2009) | Cross-sectional study in which HeLa cells were grown in Dulbecco's modified eagle medium | 574 Caucasian subjects of which 62.5% had a conduct-disorder behavior | The presence of A-element was able to confer repression by miR-96 throughout mRNA of 5-HT$_{1B}$ receptors. A common human variant (G element) associated with a reduced repressive activity of this element on miR-96 has been found. Subjects who were homozygous for the ancestral A element suffered from more conduct disorder behaviors when compared with those with the G element | No active treatment | Functional variants of A element may help to develop the search for genes involved in disabling behavioral disorders |

*Abbreviations:*    5-HT = 5-hydroxytryptamine; 7-CTKA = 7-chlorokynurenic acid; AD = antidepressant; BCL2 = B-cell lymphoma 2; BDNF = brain-derived neurotrophic factor; CREB = cAMP response element-binding protein; CHL1 = cell adhesion molecule L1; Cre = Cre recombinase; CUPS = chronic unpredictable stress; HMS18 = daily maternal separation for 18 min; ECS = electroconvulsive stimulation; ECT = electroconvulsive therapy; ERK = extracellular signal-regulated kinase; EYFP = enhanced yellow fluorescent protein; HC = healthy controls; HPA = hypothalamic-pituitary-adrenal; LCLs = human lymphoblastoid cell; ICR = imprinting control region; ITGB3 = integrin, beta 3; LC = locus coeruleus; MD = maternal deprivation; MDD = major depressive disorder; mPFC = medial prefrontal cortex; GRM4 = metabotropic glutamate receptor 4; MiRNAs = MicroRNAs; OFT = open field test; ePet = Pet-1 enhancer; qRT-PCR = real-time quantitative reverse transcription polymerase chain reaction; REST = repressor element 1 silencing transcription factor; SDS = self-rating depression scale; Akt = serine-threonine kinase; SERT = serotonin transporter; SD = Sprague–Dawley; SRI = serotonin reuptake inhibitor; SSRI = selective serotonin reuptake inhibitor; TrkB = tropomyosin-related kinase receptor B; Mnt2 = wingless-type MMTV integration site family, member 2 = Mnt2.

**TABLE 4.2**

**Most Relevant Studies about the Association between miRNAs and Suicidal Behavior**

| Author(s), Year | Study Design | Sample | Results | Conclusion |
|---|---|---|---|---|
| Smalheiser et al. (2014b) | Animal study in which rats were exposed to treatment for 1 week | Rat frontal cortex | 10 or 25 mg/kg fluoroquinolone enoxacin increased the expression of miRNAs in rat frontal cortex and reduced the proportion of animals exhibiting LH behavior following inescapable shock | Fluoroquinolone enoxacin may improve depressive behaviors |
| Lopez et al. (2014b) | Postmortem study | 15 suicide victims and 16 HC | The study profiled the expression of 10 miRNAs in the prefrontal cortex (BA44) of all participants using qRT-PCR. Several miRNAs demonstrated significant upregulation in the prefrontal cortex of suicide victims relative to psychiatric HC. Furthermore, a significant correlation between miRNAs and SAT1 and SMOX expression levels was found | A correlation between miRNAs and polyamine gene expression in the brain of suicide victims together with SAT1 and SMOX downregulation by post-transcriptional activity of miRNAs have been reported |
| Smalheiser et al. (2012) | Postmortem study in which the expression of miRNAs was measured in the prefrontal cortex (BA 9) using multiplex RT-PCR plates | 18 AD-free depressed suicide and 17 HC | MiRNA expression was found downregulated, by 17% on average, in the prefrontal cortex of depressed suicide completers relative to that of HC. This global reduction of miRNA expression and its lower variability may be positively associated with a hypoactivation of the frontal cortex in depressed suicidal subjects. Also, significant changes of 21 miRNAs that are supposed to be implicated in cellular growth and differentiation were found | Widespread changes in miRNA expression seem to be implicated in the pathogenesis of MDD and/or suicide |
| Maussion et al. (2012) | Postmortem study | 38 subjects who died by suicide | The study found increased expression of miR-185 regulating the TrkB-T1 reduction in the frontal cortex of suicide completers. MiR-185 and miR-491-3p resulted significantly and globally upregulated in suicide victims with low expression of TrkB-T1. Also, after bioinformatic analyses, five possible binding sites for the DiGeorge syndrome linked miR-185 in the 3'-UTR of TrkB-T1 were reported | Deletions or duplications in miR-185 may potentially affect the expression of TrkB-T1 leading to a psychiatric phenotype |

*(Continued)*

# TABLE 4.2 (Continued)

## Most Relevant Studies about the Association between miRNAs and Suicidal Behavior

| Author(s), Year | Study Design | Sample | Results | Conclusion |
|---|---|---|---|---|
| He et al. (2012) | Case–control study in which three polymorphisms from miRNA-processing genes were investigated | 314 patients and 252 matched HC | Individuals with variant allele DGCR8 rs3757 may be at higher suicidal risk and may have higher improvements related to AD treatments when compared to those with AGO1 rs636832 variant who demonstrated a reduced suicidal risk, and recurrent depression | A correlation between miRNA processing gene variants, vulnerability to depression, and suicidal behavior has been reported. DGCR8 rs3757 and AGO1 rs636832 but not GEMIN4 rs7813 may influence vulnerability to depression |
| Smalheiser et al. (2011) | Animal study in which the frontal cortex of male Holtzman rats subjected to repeated inescapable shocks at days 1 and 7, tested for LH at days 2 and 8, and sacrificed at day 15 was analyzed | Rats that did vs. did not exhibit LH | A relevant adaptive miRNA response to inescapable shock was demonstrated for NLH but not for LH rats who demonstrated a blunted response. Mir-96, 141, 182, 183, 183*, 298, 200a, 200a*, 200b, 200b*, 200c, 429 exhibited significant downregulation in NLH rats compared to HC. The majority of the mentioned miRNAs were enriched in synaptic fractions. Notably, half of these miRNAs was predicted to hit Creb1 as a target. Also, a specific coexpression module of 36 miRNAs which were highly correlated with each other among the LH group (but not among NLH or HC groups) was found | MiRNAs showed a specific role in the alterations of gene expression networks underlying NLH and altered LH response to repeated shocks |

*Abbreviations:* BA = Brodmann area; miRNAs = microRNAs; CREB = cAMP response element-binding protein; DGCR8 = DiGeorge syndrome critical region gene 8; GEMIN4 = Gem-associated protein 4; LH = learned helplessness; NLH = nonlearned helpless; AGO1 = protein Argonaute 1; qRT-PCR = real-time quantitative reverse transcription polymerase chain reaction; SAT1 = spermidine/spermine N1-acetyltransferase 1; SMOX = spermine oxidase; TrkB = tropomyosin-related kinase receptor B; 3′-UTR = the three prime untranslated region.

reuptake inhibition related to the AD activity. This has been further confirmed by the fact that changes in depression severity negatively correlated with changes in miR-1202 expression. Based on these findings, miR-1202 seems to represent a valid biomarker of treatment prediction or response in major depression. More generally, targeting miRNAs directly may result a useful therapeutic strategy as some miRNAs such as miR-1202 have been hypothesized as directly implicated in the pathophysiology of MDD and this may have stimulating implications for the future.

However, what is the possible interaction between the peripheric and brain miRNA levels? Some researchers (Bocchio-Chiavetto et al., 2013) have hypothesized that peripheral and brain miRNAs may interact during AD treatment, as miRNAs could be able to actively cross the blood–brain barrier. Also, some miRNA changes that have been observed in blood could reflect neuroendocrine or neuroimmune responses in the brain. It is, however, important to state that some miRNAs may modulate the interaction between immune and neuronal processes (Soreq and Wolf, 2011).

MiRNAs may be increasingly depicted as having a fundamental role in MDD outcome considering their utility as potential markers of major depressive episode evolution (Belzeaux et al., 2012).

We firmly believe that miRNAs may participate in the pathogenesis of MDD; also, alterations in targeted miRNAs play, in our opinion, a critical role in both major depression and suicidal behavior. Notably, miRNAs may actively participate in developing the MDD phenotype. One of the most difficult tasks is to identify the combination of dysregulated miRNAs in MDD that may help to understand the nature of altered pathways implicated in the pathogenesis of this illness. It would be helpful to identify the common transcription/epigenetic factors influencing the expression of a specific miRNAs network and the corresponding network of mRNAs that are both significantly affected in MDD across the different brain regions. Unfortunately, there are a variety of possible reasons related to altered miRNA expression.

Overall, genetic changes in the promoter region upstream of pri-miRNA gene transcripts, the pre-miRNA hairpin, or the mature miRNA, or RNA editing of transcripts or epigenetic suppression of the chromosomal region encoding the miRNAs have been reported, but many more possible dysfunctions are expected to exist (for more details, see Serafini et al., 2012 and Figure 4.1). Future studies have to elucidate their impact in terms of diagnostic tools, preventive strategies, and effective pharmacological treatment for major affective disorders. This has been clearly demonstrated in cancer patients where miRNAs appeared not only as useful indicators of different types of cancer, but according to miRNA profiling, it was possible to identify which patient may respond better to a specific type of treatment compared to others (Budhu and Wang, 2010). Detecting reliable biomarkers in psychiatric disorders instead represents a challenging task given the highly heterogeneous and complex nature of these conditions (Dwivedi, 2014).

Some limitations should be discussed when analyzing most of the studies investigating the association between miRNAs, synaptic plasticity, major affective disorders, and suicidal behavior. First, as reported by Smalheiser et al. (2014a), disease-related miRNA changes in the whole tissue often derive from a mixture of cell types (e.g., changes in the cell types related to cell shrinkage, gliosis, or invasion by microglia). Expression patterns of miRNAs differ according to various brain regions and may be cell-type specific. The authors (Smalheiser et al., 2014a) who assessed miRNA expression in prefrontal cortex (Brodmann area 10) in a cohort of 45 major depressed, BD, and schizophrenia subjects, reported a significant down-regulation of both mir-219 (highly enriched in brain compared to other tissues) and miR-219-5p (the most highly (fivefold) enriched miRNA in synaptic fractions). They reported that synaptosomes are more selectively informative regarding neurons and the synaptic compartment. Identifying biomarkers in neuropsychiatric diseases is a challenging task due to the heterogeneity and complex nature of psychiatric disorders. Also, expression studies of miRNAs are often difficult to interpret and some expression patterns of miRNAs may significantly differ in various brain regions, they may be cell-type specific, or alternatively can derive by a change of specific cell types (Hommers et al., 2015). As reported by the authors, expression studies could be available soon in the early stages of illness, although they would be presumably limited by some technical issues and lack of replication.

In addition, single mRNA changes detected by gene expression profiling usually require independent validation as miRNAs regulation occurred at multiple levels. The eventual search for miRNA polymorphisms as well as any efforts to replicate the predicted functional targets of miRNAs associated with depressive phenotypes may be crucial in future studies.

Moreover, our current technologies are not able to detect miRNAs or other small noncoding RNAs together with the complex spectrum of miRNA targets and the link to neuropsychiatric disorders. Biological processes may be affected by ncRNAs and differential changes in their expression levels. Interestingly, multiple evidence (Hüttenhofer et al., 2005; Verdel et al., 2009; Taft et al., 2010) suggested the existence of specific noncoding RNAs such as siRNAs and piRNAs that may induce epigenetic regulation through chromatin remodeling events.

It is also unclear how miRNAs regulatory networks control cellular signaling networks and how this regulation may finally impact on biological phenotypes. Currently, the variability of miRNA targets and combination of this regulatory network with existing signal transduction networks represent unclear critical issues that need to be further addressed.

Future studies are needed in order to examine the molecular mechanisms by which synaptic miRNAs are dysregulated, test whether they directly or indirectly contribute to the pathophysiology of major neuropsychiatric disorders, and identify whether they can be modified in animal studies.

## 4.9  CONCLUSIVE REMARKS

According to the existing literature and based on our point of view, miRNAs play a critical role not only in synaptic plasticity, neurogenesis, and stress responses, but also in disabling psychiatric conditions such as major depression and suicidal behavior. It has been widely recognized that multiple miRNA targets are involved in neural development, neurogenesis, and synaptic plasticity. Notably, the expression of specific miRNAs may be reversed by some psychoactive drugs such as ADs and mood stabilizers suggesting interesting therapeutic implications for the future.

There are multiple evidence suggesting the association between miRNA levels and neurotrophic factors activity. The homeostasis of important neural and synaptic pathways involved in stress-related disorders and major depression seems to be regulated by CREB-BDNF pathways via miRNAs in a complex manner.

The development of miRNAs as potential biomarkers for the pathogenesis of MDD represents an important step in detecting the pathophysiological mechanisms underlying this complex disease. As differential expression levels of peripheral miRNAs have been associated with multiple disease processes, using miRNAs as a new generation of biomarkers in neuropsychiatric conditions is a promising and innovative option for the treatment of major affective disorders (Maffioletti et al., 2014).

But we are only at the beginning of a fascinating journey and much remains to be clarified about the role of miRNAs in major neuropsychiatric conditions.

## ACKNOWLEDGMENT

Part of the study was funded by grants from National Institute of Mental Health (R01MH107183 and R01MH100616) to Dr. Yogesh Dwivedi.

## REFERENCES

Abelson JF, Kwan KY, O'Roak BJ et al., Sequence variants in SLITRK1 are associated with Tourette's syndrome, *Science* 310;2005:317–20.

Alon S, Mor E, Vigneault F et al., Systematic identification of edited microRNAs in the human brain, *Genome Research* 22;2012:1533–40.

Anacker C, Zunszain PA, Cattaneo A et al., Antidepressants increase human hippocampal neurogenesis by activating the glucocorticoid receptor, *Molecular Psychiatry* 16;2011:738–50.

Andersen I, Thielen K, Bech P, Nygaard E, Diderichsen F, Increasing prevalence of depression from 2000 to 2006, *Scandinavian Journal of Public Health* 39;2011:857–63.

Bai M, Zhu X, Zhang Y et al., Abnormal hippocampal BDNF and miR-16 expression is associated with depression-like behaviors induced by stress during early life, *PLoS One* 7;2012:e46921.

Bai M, Zhu XZ, Zhang Y et al., Anhedonia was associated with the dysregulation of hippocampal HTR4 and microRNA. Let-7a in rats, *Physiology & Behavior* 129;2014:135–41.

Banerjee S, Neveu P, Kosik KS, A coordinated local translational control point at the synapse involving relief from silencing and MOV10 degradation, *Neuron* 64;2009:871–84.

Baudry A, Mouillet-Richard S, Schneider B, Launay JM, Kellermann O, miR-16 targets the serotonin transporter: A new facet for adaptive responses to antidepressants, *Science* 329;2010:1537–41.

Belzeaux R, Bergon A, Jeanjean V et al., Responder and nonresponder patients exhibit different peripheral transcriptional signatures during major depressive episode, *Translational Psychiatry* 2;2012:e185.

Bessa JM, Ferreira D, Melo I et al., The mood-improving actions of antidepressants do not depend on neurogenesis but are associated with neuronal remodeling, *Molecular Psychiatry* 14;2009:764–73.

Beveridge NJ, Gardiner E, Carroll AP, Tooney PA, Cairns MJ, Schizophrenia is associated with an increase in cortical microRNA biogenesis, *Molecular Psychiatry* 15;2010:1176–89.

Beveridge NJ, Tooney PA, Carroll AP et al., Dysregulation of miRNA 181b in the temporal cortex in schizophrenia, *Human Molecular Genetics* 17;2008:1156–68.

Bocchio-Chiavetto L, Maffioletti E, Bettinsoli P et al., Blood microRNA changes in depressed patients during antidepressant treatment, *European Neuropsychopharmacology* 23;2013:602–11.

Budhu A, Ji J, Wang XW, The clinical potential of microRNAs, *Journal of Hematology & Oncology* 3;2010:37.

Bushati N, Cohen SM, microRNA functions, *Annual Review of Cell and Developmental Biology* 23;2007:175–205.

Castrén E, Voikar V, Rantamaki T, Role of neurotrophic factors in depression, *Current Opinion in Pharmacology* 7;2007:18–21.

Caudy AA, Myers M, Hannon GJ, Hammond SM, Fragile X-related protein and VIG associate with the RNA interference machinery, *Genes & Development* 16;2002:2491–96.

Chana G, Bousman CA, Money TT et al., Biomarker investigations related to pathophysiological pathways in schizophrenia and psychosis, *Frontiers in Cellular Neuroscience* 7;2013:95.

Chen H, Wang N, Burmeister M, McInnis MG, MicroRNA expression changes in lymphoblastoid cell lines in response to lithium treatment, *International Journal of Neuropsychopharmacology* 2;2009:1–7.

Cheng LC, Pastrana E, Tavazoie M, Doetsch F, miR-124 regulates adult neurogenesis in the subventricular zone stem cell niche, *Nature Neuroscience* 12;2009:399–408.

Choi PS, Zakhary L, Choi WY et al., Members of the miRNA-200 family regulate olfactory neurogenesis, *Neuron* 57;2008:41–55.

Cogswell JP, Ward J, Taylor IA et al., Identification of miRNA changes in Alzheimer's disease brain and CSF yields putative biomarkers and insights into disease pathways, *Journal of Alzheimer's Disease* 14;2008:27–41.

Conaco C, Otto S, Han JJ, Mandel G, Reciprocal actions of REST and a microRNA promote neuronal identity, *Proceedings of the National Academy of Sciences of the United States of America* 103;2006:2422–7.

Cooper JD, Skepper JN, Berzaghi MD, Lindholm D, Sofroniew MV, Delayed death of septal cholinergic neurons after excitotoxic ablation of hippocampal neurons during early postnatal development in the rat, *Experimental Neurology* 139;1996:143–55.

Cowansage KK, LeDoux JE, Monfils MH, Brain-derived neurotrophic factor: A dynamic gatekeeper of neural plasticità, *Current Molecular Pharmacology* 3;2010:12–29.

David DJ, Samuels BA, Rainer Q et al., Neurogenesis-dependent and -independent effects of fluoxetine in an animal model of anxiety/depression, *Neuron* 62;2009:479–93.

DeCarolis NA, Eisch AJ, Hippocampal neurogenesis as a target for the treatment of mental illness: A critical evaluation, *Neuropharmacology* 58;2010:884–93.

Dorval V, Nelson PT, Hébert SS, Circulating microRNAs in Alzheimer's disease: The search for novel biomarkers, *Frontiers Molecular Neuroscience* 6;2013:24.

Duman RS, Introduction: Theories of depression—from monoamines to neuroplasticity In Olié, JP, Costa e Silva JA, Macher JP, *Neuroplasticity: A New Approach to the Pathophysiology of Depression*, 1–11, London: Science Press Ltd, 2004.

Dwivedi Y, Brain-derived neurotrophic factor: Role in depression and suicide, *Neuropsychiatric Disease and Treatment* 5;2009:433–49.

Dwivedi Y, Evidence demonstrating role of microRNAs in the etiopathology of major depression, *Journal of Chemical Neuroanatomy* 42;2011:142–56.

Dwivedi Y, Emerging role of microRNAs in major depressive disorder: Diagnosis and therapeutic implications, *Dialogues in Clinical Neuroscience* 16;2014:43–61.

Dwivedi Y, Rao JS, Rizavi HS et al., Abnormal expression and functional characteristics of cyclic adenosine monophosphate response element binding protein in postmortem brain of suicide subjects, *Archives of General Psychiatry* 60;2003:273–82.

Finkel T, Deng CX, Mostoslavsky R, Recent progress in the biology and physiology of sirtuins, *Nature* 460;2009:587–91.

Fiore R, Siegel G, Schratt G, MicroRNA function in neuronal development, plasticity and disease, *Biochimica et Biophysica Acta* 1779;2008:471–8.

Flavell SW, Greenberg ME, Signaling mechanisms linking neuronal activity to gene expression and plasticity of the nervous system, *Annual Review of Neuroscience* 31;2008:563–90.

Gao J, Wang WY, Mao YW et al., A novel pathway regulates memory and plasticity via SIRT1 and miR-134, *Nature* 466;2010:1105–9.

Hanke M, Hoefig K, Merz H et al., A robust methodology to study urine microRNA as tumor marker: MicroRNA-126 and microRNA-182 are related to urinary bladder cancer, *Urologic Oncology* 28;2010:655–61.

Hansen T, Olsen L, Lindow M et al., Brain expressed microRNAs implicated in schizophrenia etiology, *PLoS One* 2;2007:e873.

He L, He X, Lim LP et al., A microRNA component of the p53 tumour suppressor network, *Nature* 447;2007:1130–4.

He Y, Zhou Y, Xi Q et al., Genetic variations in microRNA processing genes are associated with susceptibility in depression, *DNA and Cell Biology* 31;2012:1499–506.

Hommers LG, Domschke K, Deckert J, Heterogeneity and individuality: MicroRNAs in mental disorders, *Journal of Neural Transmission* 122;2015:79–97.

Huang GJ, Herbert J, Stimulation of neurogenesis in the hippocampus of the adult rat by fluoxetine requires rhythmic change in corticosterone, *Biological Psychiatry* 59;2006:619–24.

Huang YW, Ruiz CR, Eyler EC, Lin K, Meffert MK, Dual regulation of miRNA biogenesis generates target specificity in neurotrophin-induced protein synthesis, *Cell* 148;2012:933–46.

Hunsberger JG, Austin DR, Chen G, Manji HK, MicroRNAs in mental health: From biological underpinnings to potential therapies, *Neuromolecular Medicine* 11;2009:173–82.

Hunsberger JG, Fessler EB, Chibane FL et al., Mood stabilizer-regulated miRNAs in neuropsychiatric and neurodegenerative diseases: Identifying associations and functions, *American Journal of Translational Research* 5;2013:450–64.

Hüttenhofer A, Schattner P, Polacek N, Non-coding RNAs: Hope or hype? *Trends in Genetics* 21;2005:289–97.

Im HI, Hollander JA, Bali P, Kenny PJ, MeCP2 controls BDNF expression and cocaine intake through homeostatic interactions with microRNA-212, *Nature Neuroscience* 13;2010:1120–7.

Impey S, Davare M, Lasiek A et al., An activity-induced micro-RNA controls dendritic spine formation by regulating Rac1-PAK signaling, *Molecular and Cellular Neuroscience* 43;2010:146–56.

Ishizuka A, Siomi MC, Siomi H, A *Drosophila* fragile X protein interacts with components of RNAi and ribosomal proteins, *Genes Development* 16;2002:2497–508.

Issler O, Haramati S, Paul ED et al., MicroRNA 135 is essential for chronic stress resiliency, antidepressant efficacy, and intact serotonergic activity, *Neuron* 83;2014:344–60.

Jensen KP, Covault J, Conner TS, Tennen H, Kranzler HR, Furneaux HM, A common polymorphism in serotonin receptor 1B mRNA moderates regulation by miR-96 and associates with aggressive human behaviors, *Molecular Psychiatry* 14;2009:381–9.

Johnson R, Zuccato C, Belyaev ND, Guest DJ, Cattaneo E, Buckley NJ, A microRNA-based gene dysregulation pathway in Huntington's disease, *Neurobiology of Disease* 29;2008:438–45.

Kagias K, Nehammer C, Pocock R, Neuronal responses to physiological stress, *Frontiers in Genetics* 3;2012:222.

Kapsimali M, Kloosterman WP, de Bruijn E, Rosa F, Plasterk RH, Wilson SW, MicroRNAs show a wide diversity of expression profiles in the developing and mature central nervous system, *Genome Biology* 8;2007:R173.

Kawashima H, Numakawa T, Kumamaru E et al., Glucocorticoid attenuates brain-derived neurotrophic factor-dependent upregulation of glutamate receptors via the suppression of microRNA-132 expression, *Neuroscience* 165;2010:1301–11.

Kim AH, Reimers M, Maher B, Williamson V, McMichael O, McClay JL, MicroRNA expression profiling in the prefrontal cortex of individuals affected with schizophrenia and bipolar disorders, *Schizophrenia Research* 124;2010a:183–91.

Kim HW, Rapoport SI, Rao JS, Altered expression of apoptotic factors and synaptic markers in postmortem brain from bipolar disorder patients, *Neurobiology of Disease* 37;2010b:596–603.

Kim J, Inoue K, Ishii J et al., A MicroRNA feedback circuit in midbrain dopamine neurons, *Science* 317;2007:1220–4.

Kim J, Krichevsky A, Grad Y et al., Identification of many microRNAs that copurify with polyribosomes in mammalian neurons, *Proceedings of the National Academy of Sciences of the United States of America* 101;2004:360–5.

Klein ME, Lioy DT, Ma L, Impey S, Mandel G, Goodman RH, Homeostatic regulation of MeCP2 expression by a CREB-induced microRNA, *Nature Neuroscience* 10;2007:1513–4.

Kosik KS, The neuronal microRNA system, *Nature Reviews Neuroscience* 7;2006:911–20.

Kumamaru E, Numakawa T, Adachi N et al., Glucocorticoid prevents brain-derived neurotrophic factor mediated maturation of synaptic function in developing hippocampal neurons through reduction in the activity of mitogen-activated protein kinase, *Molecular Endocrinology* 22;2008:546–58.

Kuss AW, Chen W, MicroRNAs in brain function and disease, *Current Neurology and Neuroscience Reports* 8;2008:190–7.

Labermaier C, Masana M, Müller MB, Biomarkers predicting antidepressant treatment response: How can we advance the field? *Disease Markers* 35;2013:23–31.

Lambert TJ, Storm DR, Sullivan JM, MicroRNA132 modulates short-term synaptic plasticity but not basal release probability in hippocampal neurons, *PLoS One* 5;2010:e15182.

Landgraf P, Rusu M, Sheridan R et al., A mammalian microRNA expression atlas based on small RNA library sequencing, *Cell* 129;2007:1401–14.

Launay JM, Mouillet-Richard S, Baudry A, Pietri M, Kellermann O, Raphe-mediated signals control the hippocampal response to SRI antidepressants via miR-16, *Transl Psychiatry* 1;2011:e56.

Lazarov O, Mattson MP, Peterson DA, Pimplikar SW, van Praag H, When neurogenesis encounters aging and disease, *Trends in Neurosciences* 33;2010:569–79.

Lee ST, Chu K, Jung KH et al., MiR-206 regulates brain-derived neurotrophic factor in Alzheimer disease model, *Annals of Neurology* 72;2012:269–77.

Li YJ, Xu M, Gao ZH et al., Alterations of serum levels of BDNF-related miRNAs in patients with depression, *PLoS One* 8;2013:e63648.

Liu BB, Luo L, Liu XL, Geng D, Liu Q, Yi LT, 7-Chlorokynurenic acid (7-CTKA) produces rapid antidepressant-like effects: Through regulating hippocampal microRNA expressions involved in TrkB-ERK/Akt signaling pathways in mice exposed to chronic unpredictable mild stress, *Psychopharmacology (Berl)* 232;2015:541–550.

Liu C, Zhao X, MicroRNAs in adult and embryonic neurogenesis, *Neuromolecular Medicine* 11;2009:141–52.

Lopez JP, Fiori LM, Gross JA et al., Regulatory role of miRNAs in polyamine gene expression in the prefrontal cortex of depressed suicide completers, *International Journal of Neuropsychopharmacology* 17;2014a:23–32.

Lopez JP, Lim R, Cruceanu C et al., miR-1202 is a primate-specific and brain-enriched microRNA involved in major depression and antidepressant treatment. *Nature Medicine* 20;2014b:764–8.

Lucassen PJ, Meerlo P, Naylor AS et al., Regulation of adult neurogenesis by stress, sleep disruption, exercise and inflammation: Implications for depression and antidepressant action, *European Neuropsychopharmacology* 20;2010:1–17.

Lugli G, Torvik VI, Larson J, Smalheiser NR, Expression of microRNAs and their precursors in synaptic fractions of adult mouse forebrain, *Journal of Neurochemistry* 106;2008:650–61.

Machado-Vieira R, Salvadore G, DiazGranados N et al., New therapeutic targets for mood disorders, *The Scientific World Journal* 10;2010:713–26.

Maffioletti E, Tardito D, Gennarelli M, Bocchio-Chiavetto L, Micro spies from the brain to the periphery: New clues from studies on microRNAs in neuropsychiatric disorders, *Frontiers in Cellular Neuroscience* 8;2014:75.

Malan-Müller S, Hemmings SM, Seedat S, Big effects of small RNAs: A review of microRNAs in anxiety, *Molecular Neurobiology* 47;2013:726–39.

Marmol F, Lithium: Bipolar disorder and neurodegenerative diseases. Possible cellular mechanisms of the therapeutic effects of lithium, *Progress in Neuro-Psychopharmacology & Biological Psychiatry* 232;2008:1761–71.

Martin KC, Kosik KS, Synaptic tagging—Who's it? *Nature Reviews Neuroscience* 3;2002:813–20.

Maussion G, Yang J, Yerko V et al., Regulation of a truncated form of tropomyosin-related kinase B (TrkB) by Hsa-miR-185* in frontal cortex of suicide completers, *PLoS One* 7;2012:e39301.

McAllister AK, Neurotrophins and neuronal differentiation in the central nervous system, *Cellular and Molecular Life Sciences* 58;2001:1054–60.

McAllister AK, Katz LC, Lo DC, Neurotrophins and synaptic plasticità, *Annual Review of Neuroscience* 22;1999:295–318.

McEwen BS, Chattarji S, Molecular mechanisms of neuroplasticity and pharmacological implications: The example of tianeptine, *European Neuropsychopharmacology* 14;2004:S497–502.

McNeill E, Van Vactor D, MicroRNAs shape the neuronal landscape, *Neuron* 75;2012:363–79.

Meerson A, Cacheaux L, Goosens KA, Sapolsky RM, Soreq H, Kaufer D, Changes in brain MicroRNAs contribute to cholinergic stress reactions, *Journal of Molecular Neuroscience* 40;2010:47–55.

Mellios N, Huang HS, Grigorenko A, Rogaev E, Akbarian S, A set of differentially expressed miRNAs, including miR-30a-5p, act as post-transcriptional inhibitors of BDNF in prefrontal cortex, *Human Molecular Genetics* 17;2011:3030–42.

Michalak P, RNA world—The dark matter of evolutionary genomics, *Journal of Evolutionary Biology* 19;2006:1768–74.

"miRBase," Accessed on August 25, 2014. Available at http://www.mirbase.org/, 2012.

Millan MJ, MicroRNA in the regulation and expression of serotonergic transmission in the brain and other tissues, *Current Opinion in Pharmacology* 11;2011:11–22.

Mor E, Cabilly Y, Goldshmit Y et al., Species-specific microRNA roles elucidated following astrocyte activation, *Nucleic Acids Research* 39;2011:3710–23.

Mouillet-Richard S, Baudry A, Launay JM, Kellermann O, MicroRNAs and depression, *Neurobiology of Disease* 46;2012:272–8.

Nakahata Y, Kaluzova M, Grimaldi B et al., The NAD + -dependent deacetylase SIRT1 modulates CLOCK-mediated chromatin remodeling and circadian control, *Cell* 134;2008:329–40.

Nerini-Molteni S, Mennecozzi M, Fabbri M et al., MicroRNA profiling as a tool for pathway analysis in a human *in Vitro* model for neural development, *Current Medical Chemistry* 19;2012:6214–23.

Newman MA, Hammond SM, Emerging paradigms of regulated microRNA processing, *Genes & Development* 24;2010:1086–92.

Numakawa T, Kumamaru E, Adachi N, Yagasaki Y, Izumi A, Kunugi H, Glucocorticoid receptor interaction with TrkB promotes BDNF-triggered PLCgamma signaling for glutamate release via a glutamate transporter, *Proceedings of the National Academy of Sciences of the United States of America* 106;2009:647–52.

O'Connor RM, Dinan TG, Cryan JF, Little things on which happiness depends: MicroRNAs as novel therapeutic targets for the treatment of anxiety and depression, *Molecular Psychiatry* 17;2012:359–76.

O'Connor RM, Grenham S, Dinan TG, Cryan JF, microRNAs as novel antidepressant targets: Converging effects of ketamine and electroconvulsive shock therapy in the rat hippocampus, *International Journal of Neuropsychopharmacology* 16;2013:1885–92.

Ohira K, Hayashi M, A new aspect of the TrkB signaling pathway in neural plasticity, *Current Neuropharmacology* 7;2009:276–85.

Otto SJ, McCorkle SR, Hover J et al., A new binding motif for the transcriptional repressor REST uncovers large gene networks devoted to neuronal functions, *Journal of Neurosciences* 27;2007:6729–39.

Oved K, Morag A, Pasmanik-Chor M et al., genome-wide miRNA expression profiling of human lymphoblastoid cell lines identifies tentative SSRI antidepressant response biomarkers, *Pharmacogenomics* 13;2012:1129–39.

Oved K, Morag A, Pasmanik-Chor M, Rehavi M, Shomron N, Gurwitz D, genome-wide expression profiling of human lymphoblastoid cell lines implicates integrin beta-3 in the mode of action of antidepressants, *Translational Psychiatry* 3;2013:e313.

Ovtscharoff Jr, W, Braun K, Maternal separation and social isolation modulate the postnatal development of synaptic composition in the infralimbic cortex of Octodon degus, *Neuroscience* 104;2001:33–40.

Pariante CM, Miller AH, Glucocorticoid receptors in major depression: Relevance to pathophysiology and treatment, *Biological Psychiatry* 49;2001:391–404.

Pascual R, Zamora-León P, Catalán-Ahumada M, Valero-Cabré A, Early social isolation decreases the expression of calbindin D-28k and dendritic branching in the medial prefrontal cortex of the rat, *International Journal of Neuroscience* 117;2007:465–76.

Pathania M, Torres-Reveron J, Yan L et al., MiR-132 enhances dendritic morphogenesis, spine density, synaptic integration, and survival of newborn olfactory bulb neurons, *PLoS One* 7;2012:e38174.

Perkins DO, Jeffries CD, Jarskog LF et al., MicroRNA expression in the prefrontal cortex of individuals with schizophrenia and schizoaffective disorder, *Genome Biology* 8;2007:R27.

Presutti C, Rosati J, Vincenti S, Nasi S, Non coding RNA and brain, *BMC Neuroscience* Suppl 1;2006:S5.

Qurashi A, Chang S, Peng J, Role of microRNA pathway in mental retardation, *The Scientific World Journal* 7;2007;146–54.

Rajasethupathy P, Fiumara F, Sheridan R et al., Characterization of small RNAs in Aplysia reveals a role for miR-124 in constraining synaptic plasticity through CREB, *Neuron* 63;2009:803–17.

Remenyi J, Hunter CJ, Cole C et al., Regulation of the miR-212/132 locus by MSK1 and CREB in response to neurotrophins, *Biochemistry Journal* 428;2010:281–91.

Rinaldi A, Vincenti S, De Vito F et al., Stress induces region specific alterations in microRNAs expression in mice, *Behavioural Brain Research* 208;2010:265–9.

Rong H, Liu TB, Yang KJ et al., MicroRNA-134 plasma levels before and after treatment for bipolar mania, *Journal of Psychiatric Research* 45;2011:92–5.

Ryan KM, O'Donovan SM, McLoughlin DM, Electroconvulsive stimulation alters levels of BDNF-associated microRNAs, *Neurosci Lett* 549;2013:125–9.

Sairanen M, O'Leary OF, Knuuttila JE, Castren E, Chronic antidepressant treatment selectively increases expression of plasticity-related proteins in the hippocampus and medial prefrontal cortex of the rat, *Neuroscience* 144;2007:368–74.

Saugstad JA, MicroRNAs as effectors of brain function with roles in ischemia and injury, neuroprotection, and neurodegeneration, *Journal of Cerebral Blood Flow & Metabolism* 30;2010:1564–76.

Saus E, Brunet A, Armengol L et al., Comprehensive copy number variant (CNV) analysis of neuronal pathways genes in psychiatric disorders identifies rare variants within patients, *Journal of Psychiatric Research* 44;2010:971–8.

Schaeffer C, Beaulande M, Ehresmann C, Ehresmann B, Moine H, The RNA binding protein FMRP: New connections and missing links, *Biology of the Cell* 95;2003:221–8.

Schratt GM, Tuebing F, Nigh EA et al., A brain-specific microRNA regulates dendritic spine development, *Nature* 439;2006:283–9.

Serafini G, Pompili M, Innamorati M et al., Glycosides, depression and suicidal behaviour: The role of glycoside-linked proteins, *Molecules* 16;2011:2688–713.

Serafini G, Pompili M, Innamorati M et al., The role of microRNAs in synaptic plasticity, major affective disorders and suicidal behavior, *Neuroscience Research* 73;2012:179–90.

Siegel G, Obernosterer G, Fiore R et al., A functional screen implicates microRNA-138-dependent regulation of the depalmitoylation enzyme APT1 in dendritic spine morphogenesis, *Nature Cell Biology* 11;2009:705–16.

Smalheiser NR, Lugli G, Rizavi HS et al., microRNA expression in rat brain exposed to repeated inescapable shock: Differential alterations in learned helplessness vs. non-learned helplessness, *International Journal of Neuropsychopharmacology* 14;2011:1315–25.

Smalheiser NR, Lugli G, Rizavi HS, Torvik VI, Turecki G, Dwivedi Y, MicroRNA expression is down-regulated and reorganized in prefrontal cortex of depressed suicide subjects, *PLoS One* 7;2012:e33201.

Smalheiser NR, Lugli G, Zhang H et al., Expression of microRNAs and other small RNAs in prefrontal cortex in schizophrenia, bipolar disorder and depressed subjects, *PLoS One* 9 ;(2014a):e86469.

Smalheiser NR, Zhang H, Dwivedi Y, Enoxacin elevates microRNA levels in rat frontal cortex and prevents learned helplessness, *Frontiers in Psychiatry* 5;2014b:6.

Smirnova L, Gräfe A, Seiler A, Schumacher S, Nitsch R, Wulczyn FG, Regulation of miRNA expression during neural cell specification, *European Journal of Neuroscience* 21;2005:1469–77.

Sofroniew MV, Galletly NP, Isacson O, Svendsen CN, Survival of adult basal cholinergic neurons after loss of target neurons, *Science* 247;1990:338–42.

Soldati C, Bithell A, Johnston C, Wong KY, Stanton LW, Buckley NJ, Dysregulation of REST-regulated coding and non-coding RNAs in a cellular model of Huntington's disease, *Journal of Neurochemistry* 124;2013:418–30.

Soreq H, Wolf Y. NeurimmiRs: MicroRNAs in the neuroimmune interface, *Trends Mol Med* 17;2011:548–55.

Stevenson CW, Marsden CA, Mason R, Early life stress causes FG-7142- induced corticolimbic dysfunction in adulthood, *Brain Research* 1193;2008:43–50.

Szulwach KE, Li X, Smrt RD et al., Cross talk between microRNA and epigenetic regulation in adult neurogenesis, *Journal of Cell Biology* 189;2010:127–41.

Taft RJ, Pang KC, Mercer TR, Dinger M, Mattick JS, Non-coding RNAs: Regulators of disease, *Journal of Pathology* 220;2010:126–39.

Thoenen H, Neurotrophins and activity-dependent plasticità, *Progress in Brain Research* 128;2000:183–91.

Uchida S, Hara K, Kobayashi A et al., Early life stress enhances behavioral vulnerability to stress through the activation of REST4-mediated gene transcription in the medial prefrontal cortex of rodents, *Journal of Neuroscience* 30;2010;15007–18.

Uchida S, Nishida A, Hara K et al., Characterization of the vulnerability to repeated stress in Fischer 344 rats: Possible involvement of microRNA-mediated down-regulation of the glucocorticoid receptor, *European Journal of Neuroscience* 27;2008:2250–61.

Verdel A, Vavasseur A, Le Gorrec M, Touat-Todeschini L, Common themes in siRNA-mediated epigenetic silencing pathways, *International Journal of Developmental Biology* 53;2009; 245–57.

Vo N, Klein ME, Varlamova O et al., A cAMP-response element binding protein-induced microRNA regulates neuronal morphogenesis, *Proceedings of the National Academy of Sciences of the United States of America* 102;2005:16426–31.

Vreugdenhil E, Berezikov E, Fine-tuning the brain: MicroRNAs, *Frontiers in Neuroendocrinology* 31;2009:128–33.

Vreugdenhil E, Verissimo CS, Mariman R et al., MicroRNA 18 and 124a down-regulate the glucocorticoid receptor: Implications for glucocorticoid responsiveness in the brain, *Endocrinology* 150;2009:2220–8.

Wayman GA, Davare M, Ando H et al., An activity regulated microRNA controls dendritic plasticity by down-regulating p250GAP, *Proceeding of the National Academy of Sciences USA* 105;2008:9093–8.

Weber JA, Baxter DH, Zhang S et al., The microRNA spectrum in 12 body fluids, *Clinical Chemistry* 56;2010:1733–41.

Witke W, The role of profilin complexes in cell motility and other cellular processes, *Trends in Cellular Biology* 14;2004:461–9.

World Health Organization, *Mental Health: Suicide Prevention*, Geneva: World Health Organization, accessed on August 25, 2014. Available at http://www.who.int/mental_health/prevention/en/, 2012.

Wu J, Xie X, Comparative sequence analysis reveals an intricate network among REST, CREB and miRNA in mediating neuronal gene expression, *Genome Biology* 7;2006:R85.

Xu Y, Liu H, Li F et al., A polymorphism in the microRNA-30e precursor associated with major depressive disorder risk and P300 waveform, *Journal of Affective Disorders* 127;2010:332–6.

Yagasaki Y, Numakawa T, Kumamaru E, Hayashi T, Su TP, Kunugi H, Chronic antidepressants potentiate via sigma-1 receptors the brain-derived neurotrophic factor-induced signaling for glutamate release, *Journal of Biological Chemistry* 281;2006:12941–9.

Yan H, Fang M, Liu XY, Role of microRNAs in stroke and poststroke depression, *The ScientificWorld Journal* 2013;2013:459692.

Yang X, Yang Q, Wang X et al., MicroRNA expression profile and functional analysis reveal that miR-206 is a critical novel gene for the expression of BDNF induced by ketamine, *Neuromolecular Medicine* 16;2014:594–605.

Zhou R, Yuan P, Wang Y et al., Evidence for selective microRNAs and their effectors as common long-term targets for the actions of mood stabilizers, *Neuropsychopharmacology* 34;2009:1395–405.

# Section III

---

*MicroRNAs in the Nervous System and Neurological Diseases (Basic)*

*MicroRNAs in Neurological Diseases*

# 5 MicroRNAs in Prion Diseases

*Daniela Zimbardi and Tiago Campos Pereira*

## CONTENTS

Abstract ................................................................................................................ 139
5.1 Prion: An Unconventional Infectious Agent ................................................ 140
    5.1.1 Introduction ......................................................................................... 140
    5.1.2 Basic Historic Events ......................................................................... 140
    5.1.3 Mode of Replication ........................................................................... 141
5.2 MicroRNA Signature in the Preclinical and Clinical Stages of Prion Diseases ........................................................................................................ 143
5.3 Occurrence of Pathogenic Prions and MicroRNAs in Exosomes and Its Relevance to the Spread of Disease and Diagnosis ...................................... 144
5.4 Prion Protein Interactions with MicroRNA Machinery .............................. 145
5.5 Use of Artificial MicroRNAs to Knockdown Prion Protein ....................... 146
    5.5.1 Combating Pathogenic Prions with RNA Interference ................... 146
    5.5.2 Mouse Lines Expressing an Artificial MicroRNA Against Prion.... 147
    5.5.3 Dual Targeting of Prion Transcripts with Artificial miRNAs ......... 147
5.6 Mutations within Prion-Like Domains of MicroRNA-Binding Proteins ..... 148
5.7 Conclusions ................................................................................................. 148
References ............................................................................................................ 149

## ABSTRACT

MicroRNAs (miRNAs) encompass a class of small RNA molecules (18–22 nucleotides) able to regulate gene expression posttranscriptionally, controlling virtually all processes in the cell in an unforeseen new manner. On the other hand, prions are unique pathogens composed solely of proteins and cause a class of intriguing fatal neurodegenerative diseases, such as bovine spongiform encephalopathy (also known as mad cow disease) and Creutzfeldt–Jakob disease in humans, among many others.

Recently, a series of reports has revealed unexpected and curious associations between prions and miRNAs. This chapter discusses on them, as follows: (i) the concept of prion as an infectious agent, (ii) miRNA signature in the preclinical and clinical stages of prion diseases, (iii) occurrence of pathogenic prions and miRNAs in exosomes, (iv) prion interactions with miRNA machinery, (v) use of artificial miRNAs to knockdown prions, and (vi) prion-like domains of miRNA-binding proteins.

This chapter provides an unparalleled connection between two of the most fascinating concepts in modern molecular biology: prions and miRNAs, both of which have considerably changed our understanding of the cell. Investigations on the cross-talk between these two fields started only very recently, thus representing a new, exciting, and very promising research area.

## 5.1   PRION: AN UNCONVENTIONAL INFECTIOUS AGENT

### 5.1.1   INTRODUCTION

Prions encompass a singular class of intriguing pathogens, better known as the causative agent of the mad cow disease (reviewed in Prusiner, 2013). They are unlike any other conventional infectious agents such as (i) parasitic worms (multicellular organisms), (ii) yeasts (unicellular eukaryotes), (iii) bacteria (unicellular prokaryotes), (iv) viruses (noncellular structures composed basically of proteins and nucleic acids), or (v) viroids (composed solely of RNAs). A prion ("proteinaceous infectious particle") is thought to be composed of a single protein (Prusiner, 1982).

This "protein-only theory" is a dramatic paradigm shift, since the central dogma of molecular biology states that information in living organisms flows in an unidirectional way (from DNA to protein). Therefore, the dogma also dictates that any known life form depends on nucleic acids to replicate. This is not the case for prions: they are proteins capable of replicating their *structural conformations*.

### 5.1.2   BASIC HISTORIC EVENTS

The discovery of prions is as unique as its nature. Scrapie, a fatal disease in sheep which causes them to "scrape off" their wool against fences, as if they suffered of an intense itch, was known to occur in England since 1732. For a period, scrapie was thought to be caused by the protozoan *Sarcocystis tenella*, while others attributed it to inbreeding depression (M'Gowan, 1918).

Clinical and neuropathological links between scrapie and a mysterious human disease named "kuru" was observed by William J. Hadlow, Carleton Gajdusek, and others in the late 1950s (reviewed in Alpers, 2008). Kuru is an encephalopathy with a very long period of incubation which affected isolated tribes in Papua (New Guinea) and was propagated within the community due to cannibalism. Cerebellar ataxia, tremors, dementia, choreiform, and athetoid movements were distinctive and prominent clinical signs (Imran and Mahmood, 2011).

Both scrapie and kuru are slow, progressive fatal neurodegenerative diseases of the central nervous system marked by the spongiform aspect of the brain. Many other conditions revealed to be similar: bovine spongiform encephalopathy (BSE, commonly known as mad cow disease, in cattle), chronic wasting disease (in elk and deer), Creutzfeldt–Jakob disease (CJD), Gerstmann–Sträussler–Scheinker syndrome, and fatal familial insomnia (all in humans).

A large effort performed by several research groups during the twentieth century, including of Daniel Carleton Gajdusek and Stanley Benjamin Prusiner (both Nobel Laureates in 1976 and 1997, respectively) helped to unveil the nature of these diseases. The initial attempts to characterize and isolate the infectious agent proved that it was exquisitely small in size (somewhere between a virus and a protein; Alper et al., 1966), extremely resistant to nucleic acid damaging factors (UV light, radiation), and sensitive to some proteases. These findings lead to the (then) heretical hypothesis that the etiological agent was a novel class of pathogen deprived of nucleic acids (Alper et al., 1967). After several decades of biochemical (isolation of the protein; Prusiner et al., 1981, 1984), genetic (production and characterization of mice strains with mutation in the corresponding gene; Hsiao et al., 1990; Büeler et al., 1993), and cellular studies (among many others), this "protein-only model" was accept by the vast majority of scientific community.

### 5.1.3 Mode of Replication

Mammals present a gene encoding a protein named "cellular prion," whose structure is rich in alpha-helices, is protease-sensitive and exerts a natural function in the cell (Pan et al., 1993). However, mutations in this gene may lead to an altered "pathogenic prion," whose structure is rich in β-sheets, is protease-resistant, and tends to form self-aggregates. The resulting amyloid plaques may increase in size and act as seeds to form other aggregates, which is associated with neuronal death and the vacuoles observed in spongiform brains.

The most intriguing aspect of the pathogenic prion is its ability to convert a cellular prion into a pathogenic one (Kocisko et al., 1994; Figure 5.1). In other words, it is able to change the folding of cellular prion, from *rich in alpha-helices* into *rich in beta-sheets*. By doing so, pathogenic prion is "propagating" its tridimensional conformation. Another interpretation of the phenomenon is that pathogenic prion is self-replicating its *folding information*. Therefore, this *biological information* is flowing from protein to protein, violating the central dogma of molecular biology. Since the cell is naturally producing normal prions, if it gets in contact with the pathogenic version (e.g., via ingestion of contaminated food), a chain reaction initiates, however, of slow progression.

One can develop a prionic disease by three different ways (Figure 5.2). The first one is the hereditary form, where the gene encoding prion is mutated and transmitted to the offspring (Montagna et al., 2003). Thus, the organism will produce pathogenic prions endogenously.

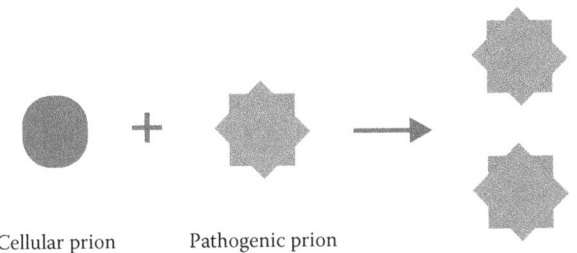

Cellular prion        Pathogenic prion

**FIGURE 5.1** Conversion of cellular into pathogenic prion. Apparently, the simple contact between isoforms is sufficient to promote the transition to scrapie (pathogenic) version.

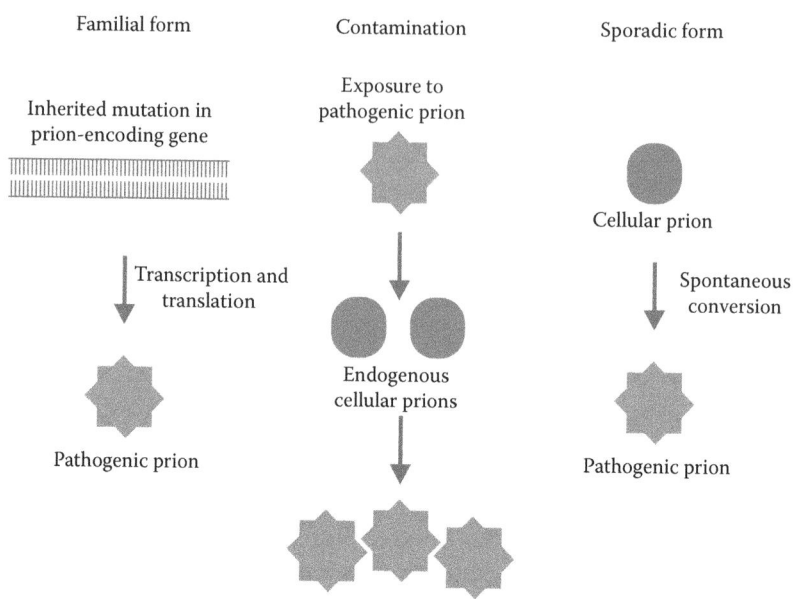

**FIGURE 5.2**  Three forms of developing prionic diseases. Pathogenic prions may be passed on to offspring due to inherited mutations in prion-encoding gene (familial form) or due to exposure to contaminated surgical instruments (or ingestion of contaminated food, among others). Prionic diseases may also emerge due to rare and spontaneous transition of cellular prion into pathogenic one.

The second form is contamination. Ingestion of food derived from cattle with BSE caused a sharp increase in the incidence of CJD in the United Kingdom in the turn of the century (Prusiner, 1997; Belay and Schonberger, 2005). Contamination may also occur by surgical instruments (Gibbs et al., 1994; Fichet et al., 2004) and in rare cases by blood transfusion (Llewelyn et al., 2004; Castilla et al., 2005; Andréoletti et al., 2012). In this form, the organism is producing cellular prions endogenously, which serve as templates (or substrates) for the acquired pathogenic prion to propagate (i.e., amplify the copies of pathogenic conformation).

Finally, one can develop the sporadic form of the disease. This case is rare and is not due to mutations in the prion gene or contamination. Sporadic form is thought to occur because the cellular prion may spontaneously convert itself into pathogenic one, by stochastic transition between conformational states (Montagna et al., 2003; Prusiner, 2013). Since the pathogenic conformation is more stable, it acts as an initial seed for the propagation.

Prions are also known to occur in yeasts, promoting heritable phenotypic variation (Halfmann et al., 2012). More recently, the prion paradigm has been proposed to explain the pathogenesis of several neurological diseases marked by proteic aggregates such as Alzheimer's, Parkinson's, and Huntington's diseases, frontotemporal dementia, amyotrophic lateral sclerosis (ALS), polyglutamine diseases, and others (Prusiner, 2012; Soto, 2012).

## 5.2  MicroRNA SIGNATURE IN THE PRECLINICAL AND CLINICAL STAGES OF PRION DISEASES

MiRNAs have extensively been reported as potent regulators of most cellular and molecular processes such as proliferation, differentiation, cell death, stress response among others, supporting the complex control involved in the maintenance of homeostasis, and function of the adult central nervous system, as well as during the neurodevelopment. In addition, some authors have further indicated that miRNAs present an expression profile specific to different neuroanatomical regions of the brain, modulating cell phenotypes in those areas (for review, Saugstad, 2010).

Therefore, once these small RNAs have the ability to modulate hundreds of genes, aberrant expression of these molecules could be dramatically detrimental to a broad range of cellular processes, underlying the etiology of an increasing number of diseases, including neurodegenerative conditions. In fact, conditional ablation of Dicer enzyme in Purkinje cells culminated in cell death, cerebellar degeneration, and ataxia in a mouse model (Schaefer et al., 2007).

In this context, altered expression of endogenous miRNAs has been reported in several neurodegenerative diseases, such as Parkinson's, Alzheimer's, and Huntington's diseases (Saugstad, 2010). However, only recently, unique miRNA signatures in prion-infected neuronal cells have been found compared to noninfected ones, suggesting these molecules as potential new diagnostic markers of prionic diseases, other than the pathogenic prion itself (PrP$^{Sc}$).

Thus, by using a genome-wide approach and an integrative bioinformatics strategy, Saba et al. (2008) identified 15 miRNAs aberrantly expressed in the neurodegenerating brains of prion-infected mice. These included upregulated miRNAs miR-342-3p, miR-320, let-7b, miR-328, miR-128, miR-139-5p, and miR-146a and downregulated miR-338-3p and miR-337-3p. Subsequently, this subset of small RNAs were computationally predicted as targeting 119 genes linked to biological pathways involved in neurogenesis, synaptic function, and cell death. Importantly, many of these target genes had previously been reported to be altered in prion-infected mice. Additionally, this subset of miRNAs was found to be distinct from those sets previously found in other nonprion neurodegenerative disorders (except miR-128), thus suggesting these molecules as specific biomarkers in preclinical stages (Saba et al., 2008). It should be noted that the preclinical stage of prion diseases represents a long incubation period with minimal (or no) symptoms to the host and can reach many months or years; however, a growing dissemination of prion particles to the brain takes place. Therefore, identification of predictive markers of the disease in the preclinical period should contribute to a faster management of patients (Huzarewich et al., 2010).

In accordance to this, Majer et al. (2012) described specific and differential expression profiles in CA1 hippocampal neurons of mice, able to characterize both the preclinical and clinical stages of the prion disease. A dynamic and punctuated alteration in miRNA expression (including miR-132-3p, miR-124a-3p, miR-16-5p, miR-26a-5p, miR-29a-3p, and miR-140-5p) is observed between both phases: the upregulation noted in preclinical stage returned to the basal levels (or inverted) toward the clinical phase. Authors evidenced that prion replication in preclinical stage was characterized by a transcriptional response related to neuronal survival and dendrite and synapse formation, followed by a switch to a loss of the neuroprotective response, culminating in a progressive neuronal death in the clinical stage of the disease.

In addition, Montag et al. (2009) reported that the upregulation of hsa-miR-342-3p could be indicative of late clinical stages of prion disease. In their study, BSE-infected cynomolgus macaques (a nonhuman primate species) displayed clinical courses similar to CJD in humans. Two miRNAs, hsa-miR-342-3p, and hsa-miR-494, were found to be differentially expressed in this animal model compared to noninfected animals. Curiously, evaluation of two brain samples of human type 1 and type 2 sporadic CJD also revealed upregulation of hsa-miR-342-3p.

Therefore, the assumption that miRNAs could be considered as effective diagnostic and prognostic biomarkers has gained much more attention after the identification of intact molecules in body fluids such as serum, saliva, milk, and urine. In these fluids, small RNAs are usually found enveloped in microvesicles called exosomes, but in a minor extent, can also be found attached to circulating RNA-binding proteins or lipoproteins, wherein it is postulated to act in cell signaling (Chen et al., 2012).

As evidenced, studies investigating the crosstalk between the miRNA molecules and prion diseases are still scarce, but represent a promising research area.

## 5.3 OCCURRENCE OF PATHOGENIC PRIONS AND MicroRNAs IN EXOSOMES AND ITS RELEVANCE TO THE SPREAD OF DISEASE AND DIAGNOSIS

Exosomes are membranous vesicles reaching 50–130 nm in diameter found in body fluids such as blood, urine, saliva, breast milk, and cerebrospinal fluid, which are originated after a process of cell invagination and activation of the endocytic pathway that culminates with the formation of intracellular multivesicular bodies followed by the release to the extracellular environment as exosomes (Kalani et al., 2014).

Recent evidences have shown that these vesicles can transport a wide range of different molecules including mRNAs, proteins, lipids, and noncoding RNAs from an origin cell to the surrounding cells of an organism, suggesting an important involvement in cell-to-cell communication. Those molecules have been reported as retaining their functionality and being able to modulate biologic pathways of the recipient cells, wherein the mRNAs can be translated into proteins and the noncoding RNAs, especially miRNAs, are able to regulate target gene expression (Chen et al., 2012).

Exosomes could be identified and isolated from body fluids using a variety of internal and external markers, once these vesicles reflect the molecular and physiological status of the parental cells or tissues. The reported markers could be both the internal molecular content carried by, or also the proper membrane proteins and lipids presented at the surface of these vesicles, such as heat shock proteins, GTPases, annexins, tetraspanins, phospholipases, among others. Thus, in addition to the main function related to cell communication, the exosomes can also exhibit additional roles in the organism involving the induction and modulation of inflammatory properties (Kalani et al., 2014).

Besides the transference of normal and physiologically functional biomolecules to surrounding cells—which represents a process known as influencing the neuroprotection, synaptic activity, and regeneration—exosomes have also been reported as exhibiting biological relevance regarding the pathophysiology of neurodegenerative diseases, since they can also be a vehicle of altered and pathogenic molecules, including miRNAs and infectious particles, as prions proteins.

In this context, both forms of prion protein—the normal cellular (PrP$^C$) and the pathogenic misfolded PrP$^{Sc}$—have been isolated from exosomes, which facilitates the transference of infectious PrP$^{Sc}$ to normal cells, disseminating the infection, according to previous cellular (Fevrier et al., 2004) and animal-based studies (Alais et al., 2008). In addition, another hypothetical mechanism for the "spreading" of prion infection could be speculated here: the dissemination of some specific miRNAs molecules in circulating exosomes. These small RNA molecules could stimulate the production of cellular prion through an indirect induction of *PRNP* gene expression secondarily to *SP1* and *p53* gene expression (Bellingham et al., 2009; Vincent et al., 2009). This overexpression might facilitate the spontaneous transition from cellular to pathogenic form in a saturated intracellular environment (Figure 5.2).

In fact, some studies were able to characterize the presence of specific miRNAs in exosomes recovered from the peripheral blood (Hunter et al., 2008; Skog et al., 2008; Taylor and Gercel-Taylor, 2008) and saliva (Michael et al., 2010) suggesting a potential diagnostic role of these molecules for ovarian cancer (Taylor and Gercel-Taylor, 2008) and glioblastoma (Skog et al., 2008). Corroborating these findings, Bellingham et al. (2012) have recently reported the first study identifying a specific miRNA profile in exosomes derived from prion-infected neuronal cells after a cell culture–based assay followed by small RNA deep sequencing methodology. The miRNAs identified as aberrantly expressed in this pathologic condition encompassed let-7b, let-7i, miR-128a, miR-21, miR-222, miR-29b, miR-342-3p, and miR-424 found as upregulated and miR-146 as downregulated. Interestingly, some of these molecules were also reported as deregulated in mouse and primate models of prion disease and a small number of those were also altered in cancer and other neurological diseases as Alzheimer's and Huntington's diseases.

Taken together, these data strongly indicates that miRNAs molecules carried by the exosomes and easily isolated from body fluids of patients by a noninvasive technique could be a useful strategy for the diagnosis and prognosis of prion diseases.

## 5.4 PRION PROTEIN INTERACTIONS WITH MicroRNA MACHINERY

Mammalian endogenous mature miRNAs molecules are usually generated after a multi-step process that initiates with the transcription in the nucleus of a stem looped pri-miRNA molecule by RNA polymerase II, followed by RNase III Drosha cleavage into a smaller double-stranded precursor molecule (or pre-miRNA). Then, this molecule is actively transported by exportin-5 to the cytoplasm of the cell, where Dicer enzyme proceeds with the processing and maturation of the single-stranded miRNA molecule (Bartel, 2004).

After miRNA biogenesis, the molecule is committed to posttranscriptional silencing of targeted genes, according to well-established and characterized mechanisms of action. This biological pathway involves incorporation of the miRNA into miRISC complex (or miRNA-induced silencing complex), a ribonucleoprotein cluster which is guided to the target mRNA by a mechanism based on sequence complementarity with the 3'-UTR, culminating in the cleavage or repression of mRNA translation.

A number of studies have elucidated that the essential element of the miRISC is a member of the Argonaute protein family, besides some other well-characterized proteins (Meister, 2013). However, only recently a remarkable association of the cellular prion with the components of miRISC has been reported, suggesting that this protein is required for the effective repression of transcripts by the ribonucleoprotein complex associated with the miRNA. In this study, the authors identified motifs that specifically bind to Argonaute proteins and suggested that cellular prion could influence the assembly of the ribonucleoprotein complex in an effective manner or, at least, in an appropriate rate (Gibbings et al., 2012).

In a similar direction, Beaudoin et al. (2009) have also demonstrated that the cytosolic form of the cellular prion is usually found in association with RNA organelles, cell areas involved in the posttranscriptional regulation and characterized by a huge concentration of mRNA, RNase III Dicer, other RNA-binding proteins, and noncoding RNAs such as miRNAs. Thus, authors suggested that cellular prion has a strong involvement of in the formation of RNA organelles and noncoding RNAs biological pathways.

In addition, $Prnp^{-/-}$ mice lines are viable, most of them present no neurological phenotypic abnormalities and continue to exhibit miRNA activity (as judged by the fact that depletion of miRNA pathway is lethal; Aguzzi et al., 2008). Thus, while some studies presented robust evidences for links between cellular prion and miRNA machinery, the complete elucidation of the role(s) of this protein in miRNA pathways and consequently, to the posttranscriptional regulation of gene expression, remains to be elucidated.

## 5.5   USE OF ARTIFICIAL MicroRNAs TO KNOCKDOWN PRION PROTEIN

### 5.5.1   Combating Pathogenic Prions with RNA Interference

Due to its unique nature, there are no current commercially available vaccines or treatments for prionic diseases. Nevertheless, experimental procedures aiming the control of prions have been tested. One of the most promising ones is the use of RNA interference-based approaches.

RNA interference in the process by which double-stranded RNAs promote specific and potent posttranscriptional downregulation of genes (Fire et al., 1998). Several classes of double stranded RNAs may be used, such as long double stranded RNAs (ranging from 300 to 800 bp), small interfering RNAs (21-nt duplexes, known as siRNAs), as well as artificial miRNAs.

This therapeutic concept is based on the fact that pathogenic prion demands cellular prion in order to replicate. Therefore, if cellular prion is temporarily (or permanently) silenced by an artificial miRNA, prion propagation is controlled. In fact, RNAi has been used to generate genetically modified cattle, without cellular prion (Golding et al., 2006; Wongsrikeao et al., 2011). Such animals are expected to be fully resistant to do pathogenic prions, thus interrupting the transmission to humans.

### 5.5.2 Mouse Lines Expressing an Artificial MicroRNA Against Prion

The use of artificial miRNAs (amiRNAs) to knockdown specific genes is interesting, since it mimics a natural process. However, if this amiRNA is super-expressed, it will compete with endogenous miRNAs for the processing machinery, thus disturbing several miRNAs, a situation which may be lethal (Grimm et al., 2006).

Gallozzi and colleagues (2008) developed an artificial miRNA based on human pre-miR-30 structure and under control of human prion promoter and SV40 poly-A terminator (Figure 5.3). They first evaluated the ability of this amiRNA to inhibit ovine and murine prions in cell cultures. The strong downregulation of both proteins with the same amiRNA prompted them to assess its efficiency *in vivo* (mouse). They obtained nine transgenic founders, three of which revealed significant silencing of prion transcripts and proteins (up to 80%) throughout the brain. Apparently, the knockdown effect was directly proportional to the expression levels of the amiRNA, but even low amounts were able to efficiently downregulate prion protein. Therefore, it might be possible to control pathogenic prion amplification via low-expressing artificial miRNAs, without interfering with endogenous miRNA processing pathways.

### 5.5.3 Dual Targeting of Prion Transcripts with Artificial miRNAs

Some reports suggest that amiRNAs are more suitable for achieving RNAi in the mouse brain than short hairpin RNAs (shRNAs), in terms of toxicity (McBride et al., 2008). However, shRNAs promote higher silencing effects since they lead to targeted mRNA cleavage, instead of translational inhibition by amiRNAs.

In order to circumvent this situation, Kang et al. (2011) developed a dual-targeting amiRNA system against prion protein. They generated an expression cassette with two amiRNAs under the control of the cytomegalovirus promoter and thymidine kinase polyadenylation signal (Figure 5.4). These miRNAs target different regions of prion transcript (near the corresponding N and C termini). According to authors, this strategy enhanced knockdown efficacy more than threefold compared to single site targeting method and might provide a promising tool to investigate cellular prion function as well as a therapeutic intervention for prionic diseases.

```
                                            UGAAG
            A                 UC           G        C
5′ CGGAAUUCCG GGCUUUGGUGG    CUACAUGUUCU         C
          GGC CCGGAACCACC——GAUGUACGAGG       A
3′ CCUAUAG     C                       GUAGAC
```

**FIGURE 5.3** Artificial miRNA used to knockdown prion. The depicted structure refers to human pre-miR-30, used as a backbone to insert an artificial miRNA (nucleotides in bold) against prion protein.

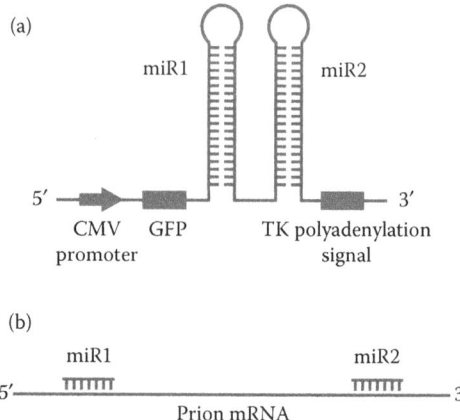

**FIGURE 5.4** Dual targeting miRNA cassette. Two miRNAs were inserted into an expression cassette under the control of cytomegalovirus promoter (a), targeting distinct regions of the same transcript (the prion mRNA, in its 5′ (N-terminus) and 3′ (C-terminus)) (b). According to authors, this dual-miRNA targeting strategy promotes a higher knockdown effect without the occasional toxicity effects that siRNAs may cause.

## 5.6 MUTATIONS WITHIN PRION-LIKE DOMAINS OF MicroRNA-BINDING PROTEINS

Curiously, many yeast and human proteins possess prion-like domains within their sequences (Alberti et al., 2009; Goldschmidt et al., 2010), some of which are RNA-binding proteins associated with neurodegenerative diseases. One of these polypeptides, named "heterogeneous nuclear ribonucleoprotein A2B1" (hnRNPA2B1) binds to determined miRNAs through the recognition of specific motifs, controlling their loading into exosomes (Villarroya-Beltri et al., 2013).

Mutations in the prion-like domain of hnRNPA2B1 (and hnRNPA1) enhance the propensity of these proteins to form self-seeding fibrils (a hallmark of prions), leading to inherited degeneration affecting muscle, brain, motor neuron and bone, and familial ALS. Intriguingly, it has been recently suggested that ALS is in fact a prionic disease (Polymenidou and Cleveland, 2011) and prions may be dispersed via exosomes (Klöhn et al., 2013; Pegtel et al., 2014). Therefore, hnRNPA2B1 represents an interesting case of very close relationship between prions and miRNAs: the same protein which is involved in the sorting of miRNAs into exosomes may also cause a prion disease, which in turn might be spread through these extracellular vesicles.

## 5.7 CONCLUSIONS

Once considered two remarkable but completely unrelated phenomena, recent reports have revealed unpredicted close links between miRNA-mediated gene regulation and prion biology. These initial findings evidence intriguing data and call attention of the scientific community for more studies. Since several important diseases are now thought to be prionic (Alzheimer's, Huntington's, and ALS) future works may provide valuable data, which could be used for specific diagnostics, prognostics, and therapeutics of prion diseases.

## REFERENCES

Aguzzi A., Baumann F., and Bremer J. 2008. The prion's elusive reason for being. *Annual Review of Neuroscience* 31: 439–477.

Alais S., Simoes S., Baas D. et al. 2008. Mouse neuroblastoma cells release prion infectivity associated with exosomal vesicles. *Biology of the Cell* 100: 603–615.

Alberti S., Halfmann R., King O. et al. 2009. A systematic survey identifies prions and illuminates sequence features of prionogenic proteins. *Cell* 137: 146–158.

Alpers M.P. 2008. The epidemiology of kuru: Monitoring the epidemic from its peak to its end. *Philosophical Transactions of the Royal Society B: Biological Sciences* 363: 3707–3713.

Alper T., Cramp W.A., Haig D.A. et al. 1967. Does the agent of scrapie replicate without nucleic acid? *Nature* 214: 764–766.

Alper T., Haig D.A., and Clarke M.C. 1966. The exceptionally small size of the scrapie agent. *Biochemical and Biophysical Research Communications* 22: 278–284.

Andréoletti O., Litaise C., Simmons H. et al. 2012. Highly efficient prion transmission by blood transfusion. *PLoS Pathogens* 8: e1002782.

Bartel D.P. 2004. MicroRNAs: Genomics, biogenesis, mechanism, and function. *Cell* 116: 281–297.

Beaudoin S., Vanderperre B., Grenier C. et al. 2009. A large ribonucleoprotein particle induced by cytoplasmic PrP shares striking similarities with the chromatoid body, an RNA granule predicted to function in posttranscriptional gene regulation. *Biochimica et Biophysica Acta* 1793: 335–345.

Belay E.D. and Schonberger L.B. 2005. The public health impact of prion diseases. *Annual Review of Public Health* 26: 191–212.

Bellingham S.A., Coleman B.M., and Hill A.F. 2012. Small RNA deep sequencing reveals a distinct miRNA signature released in exosomes from prion-infected neuronal cells. *Nucleic Acids Research* 40: 10937–10949.

Bellingham S.A., Coleman L.A., Masters C.L. et al. 2009. Regulation of prion gene expression by transcription factors SP1 and metal transcription factor-1. *The Journal of Biological Chemistry* 284: 1291–301.

Büeler H., Aguzzi A., Sailer A. et al. 1993. Mice devoid of PrP are resistant to scrapie. *Cell* 73: 1339–1347.

Castilla J., Saá P., and Soto C. 2005. Detection of prions in blood. *Nature Medicine* 11: 982–985.

Chen X., Liang H., Zhang J. et al. 2012. Horizontal transfer of microRNAs: Molecular mechanisms and clinical applications. *Protein Cell* 3: 28–37.

Fevrier B., Vilette D., Archer F. et al. 2004. Cells release prions in association with exosomes. *Proceedings of the National Academy of Sciences of the United States of America* 101: 9683–9688.

Fichet G., Comoy E., Duval C. et al. 2004. Novel methods for disinfection of prion-contaminated medical devices. *The Lancet* 364: 521–526.

Fire A., Xu S., Montgomery M.K. et al. 1998. Potent and specific genetic interference by double-stranded RNA in *Caenorhabditis elegans*. *Nature* 391: 806–811.

Gallozzi M., Chapuis J., Le Provost F. et al. 2008. Prnp knockdown in transgenic mice using RNA interference. *Transgenic Research* 17: 783–791.

Gibbings D., Leblanc P., Jay F. et al. 2012. Human prion protein binds Argonaute and promotes accumulation of microRNA effector complexes. *Nature Structural & Molecular Biology* 19: 517–524 (S1).

Gibbs C.J. Jr., Asher D.M., Kobrine A. et al. 1994. Transmission of Creutzfeldt–Jakob disease to a chimpanzee by electrodes contaminated during neurosurgery. *Journal of Neurology, Neurosurgery & Psychiatry* 57: 757–758.

Golding M.C., Long C.R., Carmell M.A. et al. 2006. Suppression of prion protein in livestock by RNA interference. *Proceedings of the National Academy of Sciences of the United States of America* 103: 5285–5290.

Goldschmidt L., Teng P.K., Riek R. et al. 2010. Identifying the amylome, proteins capable of forming amyloid-like fibrils. *Proceedings of the National Academy of Sciences of the United States of America* 107: 3487–3492.

Grimm D., Streetz K.L., Jopling C.L. et al. 2006. Fatality in mice due to oversaturation of cellular microRNA/short hairpin RNA pathways. *Nature* 441: 537–541.

Halfmann R., Jarosz D.F., Jones S.K. et al. 2012. Prions are a common mechanism for phenotypic inheritance in wild yeasts. *Nature* 482: 363–368.

Hsiao K.K., Scott M., Foster D. et al. 1990. Spontaneous neurodegeneration in transgenic mice with mutant prion protein. *Science* 250: 1587–1590.

Hunter M.P., Ismail N., Zhang X. et al. 2008. Detection of microRNA expression in human peripheral blood microvesicles. *PLoS One* 3: e3694.

Huzarewich R.L., Siemens C.G., and Booth S.A. 2010. Application of "omics" to prion biomarker discovery. *Journal of Biomedicine and Biotechnology* 2010: 613504.

Imran M. and Mahmood S. 2011. An overview of human prion diseases. *Virology Journal* 8: 559.

Kalani A., Tyagi A., and Tyagi N. 2014. Exosomes: Mediators of neurodegeneration, neuroprotection and therapeutics. *Molecular Neurobiology* 49: 590–600.

Kang S.G., Roh Y.M., Lau A., et al. 2011. Establishment and characterization of Prnp knockdown neuroblastoma cells using dual microRNA-mediated RNA interference. *Prion* 5: 93–102.

Kocisko D.A., Come J.H., Priola S.A. et al. 1994. Cell-free formation of protease-resistant prion protein. *Nature* 370: 471–474.

Klöhn P.C., Castro-Seoane R., and Collinge J. 2013. Exosome release from infected dendritic cells: A clue for a fast spread of prions in the periphery? *Journal of Infection* 67: 359–368.

Llewelyn C.A., Hewitt P.E., Knight R.S., et al. 2004. Possible transmission of variant Creutzfeldt-Jakob disease by blood transfusion. *The Lancet* 363: 417–421.

Majer A., Medina S.J., Niu Y. et al. 2012. Early mechanisms of pathobiology are revealed by transcriptional temporal dynamics in hippocampal CA1 neurons of prion infected mice. *PLoS Pathogens* 8: e1003002.

McBride J.L., Boudreau R.L., Harper S.Q. et al. 2008. Artificial miRNAs mitigate shRNA-mediated toxicity in the brain: Implications for the therapeutic development of RNAi. *Proceedings of the National Academy of Sciences of the United States of America* 105: 5868–5873.

Meister G. 2013. Argonaute proteins: Functional insights and emerging roles. *Nature Reviews Genetics* 14: 447–459.

M'Gowan J.P. 1918. Scrapie. *Journal of Comparative Pathology and Therapeutics* 31: 278–290.

Michael, A., Bajracharya, S.D., Yuen, P.S. et al. 2010. Exosomes from human saliva as a source of microRNA biomarkers. *Oral Diseases* 16: 34–38.

Montag J., Hitt R., Opitz L., et al. 2009. Upregulation of miRNA hsa-miR-342-3p in experimental and idiopathic prion disease. *Molecular Neurodegeneration* 4: 36.

Montagna P., Gambetti P., Cortelli P. et al. 2003. Familial and sporadic fatal insomnia. *The Lancet Neurology* 2: 167–176.

Pan K.M., Baldwin M., Nguyen J. et al. 1993. Conversion of alpha-helices into beta-sheets features in the formation of the scrapie prion proteins. *Proceedings of the National Academy of Sciences of the United States of America* 90: 10962–10966.

Pegtel D.M., Peferoen L., and Amor S. 2014. Extracellular vesicles as modulators of cell-to-cell communication in the healthy and diseased brain. *Philosophical Transactions of the Royal Society B: Biological Sciences* 369: pii 20130516.

Polymenidou M. and Cleveland D.W. 2011. The seeds of neurodegeneration: Prion-like spreading in ALS. *Cell* 147: 498–508.

Prusiner S.B. 1982. Novel proteinaceous infectious particles cause scrapie. *Science* 216: 136–144.

Prusiner S.B. 1997. Prion diseases and the BSE crisis. *Science* 278: 245–251.

Prusiner S.B. 2012. Cell biology: A unifying role for prions in neurodegenerative diseases. *Science* 336: 1511–1513.

Prusiner S.B. 2013. Biology and genetics of prions causing neurodegeneration. *Annual Review of Genetics* 47: 601–623.

Prusiner S.B., Groth D.F., Bolton D.C. et al. 1984. Purification and structural studies of a major scrapie prion protein. *Cell* 38: 127–134.

Prusiner S.B., McKinley M.P., Groth D.F. et al. 1981. Scrapie agent contains a hydrophobic protein. *Proceedings of the National Academy of Sciences of the United States of America* 78: 6675–6679.

Saba R., Goodman C.D., Huzarewich R.L. et al. 2008. A miRNA signature of prion-induced neurodegeneration. *PLoS One* 3: e3652.

Saugstad J.A. 2010. MicroRNAs as effectors of brain function with roles in ischemia and injury, neuroprotection, and neurodegeneration. *Journal of Cerebral Blood Flow & Metabolism* 30: 1564–1576.

Schaefer A., O'Carroll D., Tan CL. et al. 2007. Cerebellar neurodegeneration in the absence of microRNAs. *The Journal of Experimental Medicine* 204: 1553–1558.

Skog J., Wurdinger T., van Rijn S. et al. 2008. Glioblastoma microvesicles transport RNA and proteins that promote tumour growth and provide diagnostic biomarkers. *Nature Cell Biology* 10: 1470–1476.

Soto C. 2012. Transmissible proteins: Expanding the prion heresy. *Cell* 149: 968–977.

Taylor D.D. and Gercel-Taylor C. 2008. MicroRNA signatures of tumor-derived exosomes as diagnostic biomarkers of ovarian cancer. *Gynecologic Oncology* 110: 13–21.

Vincent B., Sunyach C., Orzechowski H.D. et al. 2009. p53-dependent transcriptional control of cellular prion by presenilins. *The Journal of Neuroscience* 29: 6752–6760.

Villarroya-Beltri C., Gutiérrez-Vázquez C., Sánchez-Cabo F. et al. 2013. Sumoylated hnRNPA2B1 controls the sorting of miRNAs into exosomes through binding to specific motifs. *Nature Communications* 4: 2980.

Wongsrikeao P., Sutou S., Kunishi M. et al. 2011. Combination of the somatic cell nuclear transfer method and RNAi technology for the production of a prion gene-knockdown calf using plasmid vectors harboring the U6 or tRNA promoter. *Prion* 5: 39–46.

# 6 MicroRNAs in Epileptogenesis and Epilepsy

*Eva M. Jimenez-Mateos, Tobias Engel,*
*Catherine Mooney, and David C. Henshall*

## CONTENTS

Abstract ........................................................................................................................ 154
6.1   Introduction ..................................................................................................... 154
6.2   Epilepsy ........................................................................................................... 154
    6.2.1   Diagnosis ............................................................................................. 155
    6.2.2   Treatment ............................................................................................. 155
        6.2.2.1   Nonpharmacological Treatments ....................................... 156
6.3   Mechanisms of Epilepsy .................................................................................. 156
    6.3.1   Genetic Mutations ................................................................................ 156
    6.3.2   Injury-Induced Epileptogenic Mechanisms ........................................ 157
        6.3.2.1   Neuronal Death ................................................................... 157
        6.3.2.2   Gliosis ................................................................................ 158
        6.3.2.3   Inflammation ....................................................................... 159
        6.3.2.4   Axonal/Dendritic Reorganization ....................................... 159
        6.3.2.5   Extracellular Matrix and Wound Repair ............................ 160
    6.3.3   Epigenetic Control of Gene Expression in Epileptogenesis ............. 160
6.4   MicroRNA Biogenesis Pathways in the Brain ................................................. 162
    6.4.1   MicroRNA Biogenesis Pathways in Epilepsy .................................... 163
    6.4.2   Altered MicroRNA Expression in Epilepsy ....................................... 164
    6.4.3   MicroRNA Profile after Status Epileptics: Early Epileptogenic
        Changes ................................................................................................ 165
    6.4.4   MicroRNA Profile in Experimental Temporal Lobe Epilepsy ......... 167
    6.4.5   MicroRNA Profile in Human Temporal Lobe Epilepsy .................... 167
    6.4.6   MicroRNAs Regulating Inflammation in Epilepsy: miR-21
        and miR-146a ....................................................................................... 170
    6.4.7   MicroRNAs Implicated in the Regulation of Dendritic Spines
        and Neuronal Activity: miR-132, miR-134, and miR-128 ............... 171
    6.4.8   MicroRNAs in Seizure-Induced Neuronal Death: miR-34a
        and miR-184 ......................................................................................... 172
6.5   Conclusions and Future Directions .................................................................. 173
Acknowledgments ...................................................................................................... 173
References .................................................................................................................. 173

## ABSTRACT

Epilepsy is the most common serious neurological disease affecting people of all ages. It is characterized by an enduring predisposition to seizures; a result of abnormal, excessive synchronization of neurons. The epileptogenic process features change to multiple signaling pathways including those regulating neuronal morphology and function, gliosis, neuroinflammation, and cell death. Recent work has identified select changes to microRNA levels within the hippocampus that may promote or oppose aberrant gene expression during epileptogenesis and in established epilepsy. Functional interrogation has been undertaken using intracerebral delivery of chemically modified antisense oligonucleotides (antagomirs) and genetic techniques. This has demonstrated roles for microRNAs in seizure-induced cell death (miR-34a, miR-184), inflammation (miR-146a), and neuronal microstructure (miR-128, miR-132, miR-134), in epilepsy. This chapter summarizes work that has characterized microRNA expression in experimental and human epilepsy and the evidence that functional manipulation of microRNAs may be a novel approach to treat or prevent epilepsy.

## 6.1   INTRODUCTION

Epilepsy is a chronic brain disease, characterized by recurrent unprovoked seizures. Current treatments are only successful in suppressing seizures in two-thirds of patients and they do not treat the underlying pathophysiology. An improved understanding of the molecular mechanisms coordinating gene expression during epilepsy development, and maintenance of the chronic state is required to identify novel therapeutic targets and potential antiepileptogenic treatments. MicroRNAs (miRNA) are small noncoding RNAs that function to reduce protein levels in cells through sequence-specific binding to target mRNAs. Genetic deletion of key components of the miRNA biogenesis pathway results in aberrant development and function of the brain that produces, in some cases, seizures. Specific research on miRNAs in epilepsy first appeared in 2010, with reports on individual miRNAs linked to brain inflammatory responses and neuronal microstructure. A number of large-scale miRNA profiling studies followed which defined more completely the alterations in miRNA expression in experimental and human epilepsy. Functional studies in rodents have since demonstrated that single miRNAs can exert powerful effects on seizures and epilepsy. In this chapter, we review the main discoveries on miRNA biogenesis and expression in experimental and human epilepsy and describe the functional studies of epilepsy-associated miRNAs that regulate inflammation, neuronal activity, microstructure, and cell death. Research on miRNA promises to provide both critical insight into the molecular mechanisms of gene expression and therapeutic targets to disrupt development or maintenance of the epileptic state.

## 6.2   EPILEPSY

Epilepsy is a neurological disease characterized by a permanent predisposition to generate seizures. Epilepsy affects around 65 million people worldwide. Seizures are defined as an abnormal excessive or synchronous neuronal activity in the brain. They are a result of transient deregulation of the normal excitatory and inhibitory balance. Epileptic seizures can be focal (limited to an area or one cerebral hemisphere), generalized (bilaterally distributed, affecting cortical, and subcortical areas of the brain, but not necessarily the entire cortex), and both focal generalized, where the affected area cannot be determined (Chang and Lowenstein, 2003; Fisher et al., 2014). Epilepsy is accompanied by profound neurobiologic, cognitive, psychological, and social consequences (Fisher et al., 2014). The clinical signs of focal seizures include one or more of the following features: aura, focal motor movements, autonomic movements, awareness, and responsiveness. Generalized seizures are associated with absence seizures, tonic-clonic, atonic and myoclonic seizures, and loss of consciousness (Fisher et al., 2014).

### 6.2.1 DIAGNOSIS

A diagnosis of epilepsy is based on having: (1) two epileptic seizures occurring more than 24 h apart; (2) one epileptic seizure associated with existing brain damage, such as injury from trauma or stroke, or (3) diagnosis of an epilepsy syndrome. These parameters alone often cannot differentiate between epileptic seizures and other disorders such as syncope or psychogenic nonepileptic attack disorders (Fisher et al., 2014). A correct diagnosis of epilepsy therefore requires multidimensional criteria, including family and personal history, age of seizure onset, neurological and cognitive status, and ictal and inter-ictal EEG monitoring (Moshe et al., 2014).

There is no cure for epilepsy. However, epilepsy is considered to be resolved (but not cured) in persons who either had an age-dependent epilepsy syndrome and is now past the symptomatic age, or has been seizure free for 10 years or without any anticonvulsant medication for the last 5 years (Fisher et al., 2014).

### 6.2.2 TREATMENT

In general, 70% of patients with epilepsy respond to antiepileptic drugs (AEDs), achieving seizure freedom with the first prescribed drug. The main mechanisms of action of AEDs are enhancement of inhibitory γ-amino butyric acid (GABA) transmission, inhibition of glutamatergic transmission, and blockage of voltage-dependent sodium channels (Bialer and White, 2010). However, AEDs with other mechanism of action are known, including levetiracetam that targets part of the synaptic vesicle release machinery (Moshe et al., 2014). Choice of first-line AED takes into account factors such as syndrome (focal vs. generalized), age, sex (contraception and childbearing potential), and drug interactions (other medications prescribed). Benzodiazepines and pentobarbital are most often prescribed for control of generalized seizure types, whereas levetiracetam and gabapentin are more effective against focal (partial) seizures (Bialer and White, 2010; Moshe et al., 2014). Phenytoin and carbamazepine are suitable for both focal and generalized seizures (Bialer and White, 2010; Moshe et al., 2014).

### 6.2.2.1 Nonpharmacological Treatments

For the approximately 30% of patients whose epilepsy is not controlled with single or combined AEDs, the main alternative therapies are surgical resection and neurostimulation. Surgical resection is generally effective for patients with drug-resistant focal epilepsy with at least half of these patients achieving seizure freedom (Thom et al., 2010). Multifocal epilepsies pose a greater challenge to postoperative seizure freedom. These include where damage affects the temporal lobe and neighboring regions, multifocal cortical dysplasias, and patients with tuberous sclerosis. Advances in noninvasive neuroimaging and neurophysiological techniques provide better resolution and delineation of the epileptogenic focus and may further improve diagnosis and outcome (de Tisi et al., 2011). A better understanding and identification of the epileptic network and removing the epileptogenic zones in one or several surgical steps can result in greater seizure-freedom rates (McGonigal et al., 2007; Cardinale et al., 2013).

Neurostimulation was mainly developed as a palliative treatment for patients with drug-resistant epilepsy who were not candidates for surgery. Vagus nerve stimulation is the most widely used nerve-stimulation technique, achieving up to 50% reduction in seizures, but only 5% of patients achieve seizure freedom. Novel, less invasive techniques include transcutaneous stimulation of the vagus and trigeminal nerve with evidence accumulating for the efficacy of these devices (Fisher, 2012).

Deep brain stimulation is used only in severe epilepsies; the main stimulated areas are the anterior nucleus of the thalamus and the cortex. This treatment results in small effects size and seizure-free rates. Further studies are necessary to establish risk-to-benefit ratio of invasive neurostimulation (Fisher et al., 2010).

## 6.3 MECHANISMS OF EPILEPSY

The causes of epilepsy can be classified as: (1) genetic, where genetic factors have a major causal role in the disease (with Mendelian or complex patterns of inheritance or the result of *de novo* mutations); (2) structural or metabolic, where a nongenetic factor is the main cause, such as stroke, brain tumor, or developmental malformations; and (3) unknown, where no cause has been identified (Moshe et al., 2014; Thomas and Berkovic, 2014).

### 6.3.1 GENETIC MUTATIONS

A number of generalized epilepsies have a Mendelian inheritance pattern. Such pure epilepsy syndromes are rare but studies of the mutations that underlie them have been pivotal in our understanding of the molecular mechanisms of epilepsy (Chang and Lowenstein, 2003; Hanna, 2006). Most familial cases involve mutations in genes encoding ion channels and transmitter receptors. This fits well with our concept of epilepsy being a disorder of imbalanced excitation and inhibition. An example is benign familial neonatal epilepsy which results from an autosomal dominant mutation in the *KCNQ2/3* genes. These encode potassium channels with the mutation resulting in reduced channel function. The affected channels display reduced outward potassium flow during activation preventing repolarization of the plasma membrane of neurons and thereby increasing intrinsic excitability (Chang and Lowenstein, 2003; Maljevic and Lerche, 2014).

*De novo* mutations are those present in an affected child only and include epileptic encephalopathies such as Dravet syndrome. As with inherited forms of epilepsy, mutations are found in genes associated with neuronal excitability, such as *SCN1A*, which encodes a voltage-gated sodium channel (Allen et al., 2013). Increasingly, however, mutations have been discovered in genes without obvious links to excitability including topoisomerase II (*TOP2*), which is involved in DNA stress and repair (Gomez-Herreros et al., 2014), and *ALG13*, which encodes an enzyme involved in *N*-linked glycosylation (Allen et al., 2013).

A number of disorders associated with seizures have been traced to mutations in genes that relate closely to miRNA biogenesis. The *FMR1* gene is mutated in patients with Fragile X syndrome, a disease which features seizures in up to 15% of sufferers. The *FMR1* gene product interacts with proteins involved in miRNA biogenesis and function (Jin et al., 2004). Likewise, deletion of 22q11.2, which includes DiGeorge syndrome critical region 8 (*DCGR8*), another component of the miRNA biogenesis machinery, is associated with seizures and childhood epilepsy (Piccione et al., 2011; Cheung et al., 2014). Last, research has shown a role for malin in miRNA processing (Singh et al., 2012), a gene (NHLRC1) frequently mutated in a rare progressive myoclonic epilepsy (Lafora disease; Merwick et al., 2012).

### 6.3.2 Injury-Induced Epileptogenic Mechanisms

Acquired epilepsy is the result of a series of changes in neuronal network excitability that follow an initial precipitating insult to the brain (e.g., traumatic brain injury, stroke, prolonged febrile seizures). Transcriptome profiling as well as studies focusing on single genes has provided comprehensive knowledge of the pathways altered following epileptogenic insults. The most consistently affected processes are neuronal death, gliosis, neuroinflammation, reorganization of the microarchitecture of neurons including axons and dendrites, and reorganization of the extracellular matrix. For comprehensive reviews of these different aspects, the reader is referred elsewhere (McNamara et al., 2006; Pitkanen and Lukasiuk, 2011; Vezzani et al., 2011; Goldberg and Coulter, 2013; Henshall and Engel, 2013). Questions remain, however, as to the relative importance of the individual processes. Moreover, it is likely that multiple aspects of epileptogenic signaling need to be targeted for successful antiepileptogenesis. This view is upheld by the rather modest effects reported for functional manipulation of the individual components, such as reducing neuronal death, gliosis, and reorganization of neuronal networks, on the frequency or severity of spontaneous seizures (Pitkanen and Lukasiuk, 2011). A fully effective antiepileptogenic treatment has not yet been reported in a preclinical animal model. Here, we summarize several key processes implicated in epileptogenesis.

#### 6.3.2.1 Neuronal Death

Neuron loss is a common finding within the hippocampus in patients with mesial temporal lobe epilepsy (TLE). The process of epileptogenesis is frequently associated with acute and delayed neuronal cell death, particularly where the precipitant is a brain injury such as trauma or prolonged seizures (status epilepticus). Indeed, cell death–signaling pathways often feature among gene lists in epileptogenesis profiling studies (Pitkanen and Lukasiuk, 2011). Neuronal death may contribute to the epileptic phenotype via various mechanisms. First, neuron loss creates an imbalance between excitation and inhibition, either by directly removing inhibitory neurons or by removing excitatory neurons which activate inhibitory neurons and thereby decreasing excitatory drive onto inhibitory neurons (Sloviter and Bumanglag, 2013). Neuronal death may be secondarily epileptogenic by provoking reactive gliosis and inflammation (Vezzani et al., 2011).

Significant support for the importance of neuronal cell death in epileptogenesis has emerged from studies of apoptosis-associated signaling pathways. Mice lacking the proapoptotic gene *Puma* display reduced hippocampal damage after status epilepticus and go on to develop fewer spontaneous seizures (Engel et al., 2010b). Conversely, mice lacking the transcription factor *Chop* display enhanced hippocampal injury after status epilepticus and go on to develop more frequent spontaneous seizures (Engel et al., 2013). Similar proportionate effects of damage on the epileptic phenotype were reported in models of trauma (Kharatishvili and Pitkanen, 2010). However, some studies have failed to observe an antiepileptogenic effect of neuroprotection (Ebert et al., 2002) and damage to some brain regions may even help disrupt epileptic neuronal networks (Andre et al., 2000). Additionally, work in the developing brain generally suggests neuronal loss is not a key precipitant of later-life epilepsy (Baram et al., 2002). Thus, although neuroprotection for antiepileptogenesis is intuitively the right thing to do, the precise networks and developmental stage may be critical to observe an antiepileptogenic effect. Last, it has been suggested that the biochemical pathways of apoptosis and other forms of programmed cell death, rather than the cell death they control, are causally important for epileptogenesis (Dingledine et al., 2014).

### 6.3.2.2 Gliosis

Astrocytes perform critical roles in maintaining the correct functioning of neuronal networks but they undergo structural and metabolic changes in epilepsy that are implicated in generating seizures (Wetherington et al., 2008; Clasadonte and Haydon, 2012). Gliosis is a common pathological finding in resected brain tissue from patients with epilepsy, even where overt neuron loss is absent. Indeed, triggering astrocyte hypertrophy alone, in the absence of cell death, is sufficient to decrease inhibitory drive within brain networks (Ortinski et al., 2010). The expansion of astrocytes and accompanying functional adaptations can promote seizures through a variety of mechanisms. Astrocytes are the main cell type expressing adenosine kinase which breaks down the brain's endogenous anticonvulsant, adenosine (Boison, 2008). Thus, astrogliosis lowers the threshold for seizures. Astrocytes also release neuromodulators including glutamate and ATP which can have pro- as well as anticonvulsive effects (Wetherington et al., 2008). Altered astrocyte handling of neurotransmitters and ions has also been reported in epileptic brain (Wetherington et al., 2008). However, functioning astrocytes after brain injury are critical to recover as demonstrated by the exacerbation of damage and inflammation after trauma in mice in which astrocyte proliferation was prevented (Myer et al., 2006).

The role of microglia in epilepsy is less well understood (Devinsky et al., 2013). Microglia perform important functions in maintenance of synapses in the normal brain and their phagocytic properties are critical for the restitution of injury (Tremblay et al., 2011). However, microglia release pro-inflammatory cytokines including interleukin-1β (IL-1β) which is a pro-ictogenic molecule that may promote seizures (Vezzani et al., 2011).

### 6.3.2.3 Inflammation

Growing experimental and clinical evidence links the immune system with seizure onset and generation or maintenance of epilepsy. Anti-inflammatory drugs (e.g., steroids and intravenous immunoglobulins) are often useful in selected drug-resistant epileptic syndromes, and fever and infection can precipitate seizures. Moreover, altered inflammatory signaling in genetically modified mice or pharmacological studies in animal seizure models demonstrates the involvement of inflammation in seizure generation and epileptogenesis (Dube et al., 2007; Vezzani et al., 2011). The inflammatory state is driven by both innate and adaptive arms of the immune response. Astrocyte and microglial activation, which are both common characteristics of epileptic foci release pro- and anti-inflammatory cytokines to maintain homeostasis and limit injury after an insult to the brain. Pro-inflammatory cytokines include IL-1β and transforming growth factor β, products of the COX-2 pathway, and danger signals such as the high mobility group box 1 (HMGB1; Vezzani et al., 2011). These have potent ictogenic effects and can further contribute to epileptogenesis by impairing blood–brain barrier (BBB) integrity (Ivens et al., 2007). Ineffective secretion of anti-inflammatory mediators after seizures or during epileptogenesis has been proposed to prolong brain injury–induced inflammation and its adverse effects on brain homeostasis (Devinsky et al., 2013). Accordingly, inhibition of cytokine release either by specific inhibitors or in genetic models has provided compelling evidence for their involvement in seizure generation; IL-1β inhibition or the inactivation of HMGB1 resulted in a significant delay in the onset time of seizures (Ravizza et al., 2006; Maroso et al., 2010). Inflammation is also implicated in producing long-lasting changes in brain excitability following early-life brain injuries such as febrile seizures (Vezzani, 2013). In contrast, inhibition of COX-2 has shown some inconsistent results against epileptogenesis (Rojas et al., 2014). Together, these data reveal inflammation is an integral aspect of epileptogenesis and chronic epilepsy.

### 6.3.2.4 Axonal/Dendritic Reorganization

Dendritic and axonal reorganization is thought to be an important contributor to the development of aberrant neuronal networks in epilepsy. The most intensely researched process is mossy fiber sprouting (MFS), which refers to the sprouting of axons from the dentate granule neurons (mossy fibers) and is a common finding in experimental and human epilepsy (Cronin and Dudek, 1988; Sutula et al., 1989). These axon collaterals innervate the supragranular region and inner molecular layer of the dentate gyrus (Pitkanen and Sutula, 2002). This is a potentially elegant explanation for the development of hyperinnervated and recurrent excitatory circuitry within the epileptic brain (Dudek et al., 1994). However, several findings have challenged whether this is pro-epileptogenic. First, sprouted mossy fibers may innervate inhibitory interneurons, increase GABA signaling, and reduce hippocampal excitability (Frotscher et al., 2006). Histological studies have found no relationship between MFS and seizure rates in animal models and epilepsy can develop in a time frame too brief for MFS to play a role (Nissinen et al., 2001; Bumanglag and Sloviter, 2008; Mouri et al., 2008). Recent data have identified the mammalian target of rapamycin (mTOR) pathway as responsible for MFS and treatment with rapamycin attenuates poststatus epilepticus MFS (Zeng et al., 2009). However, rapamycin has failed to provide antiepileptogenic effects in several TLE models (Buckmaster and Lew, 2011; Heng et al., 2013). Altogether, these data suggest MFS may serve both pro- and antiseizure roles but is not a requirement for epileptogenesis. Epileptogenic injuries also cause dramatic and rapid dendritic changes, including spine retraction and the appearance of new sites of innervation such as on hilar basal dendrites (Zeng et al., 2007). These may also alter the normal inhibitory tone in the hippocampus and increase excitability (Pitkanen and Sutula, 2002).

### 6.3.2.5   Extracellular Matrix and Wound Repair

Extracellular matrix reorganization and wound healing are consistently among the most altered processes in microarray analyses of the hippocampus in animal models of epileptogenesis (Lukasiuk and Pitkanen, 2007; Pitkanen and Lukasiuk, 2011; He et al., 2014). This is not unexpected since epileptogenic injuries cause cell death and gliosis and therefore reorganization is necessary to accommodate the remodeled extracellular environment. Altering such extracellular matrix reorganization may influence neuronal network hyperexcitability. Restitution of the extracellular matrix is obviously critical for recovery following an epileptogenic injury. However, the reorganization process may permit aberrant rewiring or create a permissive environment in which glia can expand and proliferate. There is emerging functional evidence that this process is important in epileptogenesis. Loss of matrix metalloproteinase 9 (*MMP-9*) protects against kindling-induced epileptogenesis (Mizoguchi et al., 2011). Similarly, kindling progression was impaired in mice lacking extracellular matrix glycoprotein tenascin-R (Hoffmann et al., 2009). However, mice lacking the urokinase-type plasminogen activator receptor, which is critical for extracellular matrix reorganization, develop worse epilepsy after kainate-induced status epilepticus, indicating a functioning wound repair system is beneficial (Ndode-Ekane and Pitkanen, 2013). Together, these results suggest an important influence of genes associated with remodeling of the extracellular space during epileptogenesis and indicate focused targeting of specific genes may be required for therapeutic benefit.

### 6.3.3   Epigenetic Control of Gene Expression in Epileptogenesis

Gene expression during epileptogenesis occurs in temporally coordinated waves (Gorter et al., 2006). There is increasing interest in how such programs are coordinated since they may offer novel targets for antiepileptogenesis. A number of transcription factors have emerged as potential master regulators. These include the neuron-restrictive silencing factor (Hu et al., 2011b; McClelland et al., 2014), cAMP response element-binding protein (Zhu et al., 2012), p53 (Engel et al., 2010a), and the CCAAT/enhancer-binding protein homologous protein CHOP (Engel et al., 2013).

Upstream of these transcription factors lies various layers of epigenetic regulation which influence the compactness of the chromatin state and thus the degree to which a permissive transcriptional environment exists. The main epigenetic processes are DNA methylation and histone modification although noncoding RNA is often included (Graff et al., 2011). Early work identified an increase in methylation of *RELN* which was associated with a reduction in reelin protein levels and granule cell dispersion (GCD) in patients with TLE (Kobow et al., 2009). The majority of DNA methylation in brain is highly static and studies from the author's laboratory demonstrated changes to DNA methylation after status epilepticus in mice affected less than 300 genes (Miller-Delaney et al., 2012). The main response was promoter hypomethylation, predicted to enable gene transcription, and this occurred within the promoters of genes involved in processes including protein binding, catalytic activity, and transcription (Miller-Delaney et al., 2012). Analyses of DNA methylation in experimental epilepsy reveal hypermethylation of genes becomes dominant (Kobow et al., 2013; Williams-Karnesky et al., 2013). Epileptogenic injuries also alter expression of epigenetic components including DNA methyltransferases (Zhang et al., 2007; Lundberg et al., 2009). Experimental treatments that counter hypermethylation of DNA, including the ketogenic diet and adenosine augmentation, are both associated with antiseizure effects (Kobow et al., 2013; Williams-Karnesky et al., 2013). However, it is likely that much methylation change represents homeostatic events that function to oppose expression of genes that drive hyperexcitability. Accordingly, broad treatments to reduce all methylation changes may not be sufficiently specific to be safe or effective.

The second major epigenetic process investigated in epilepsy is histone modification (Graff et al., 2011). These are generally shorter lived epigenetic modifications relative to DNA methylation. Numerous histone posttranslational modifications have been identified, including phosphorylation, acetylation, and methylation. Each modification functions as a potential molecular switch, either increasing or decreasing the degree of chromatin compaction (Jakovcevski and Akbarian, 2012). Acetylation of histones is typically associated with gene turn-on. In contrast, methylation can either promote or repress transcription depending on the site and number of methylations. For example, H3K4me and H4K12ac promote a transcriptionally active state, whereas H3K27/9me3 is an example of a repressive mark. Ultimately, it is the specific combinations of histone modifications that decide the effect on transcription in combination with other epigenetic marks (Graff et al., 2011; Jakovcevski and Akbarian, 2012). Posttranslational modification of histone variants have been reported by a number of groups. Histone modifications occur in regionally and temporally specific patterns following seizures (Sng et al., 2006). Seizures also alter expression of the molecular machinery involved in posttranslational modification of histones. Here, most work has focused on acetylation of histones that are predicted to affect genes involved in synaptic function (Huang et al., 2002, 2012; Park et al., 2014). Another finding that has driven interest in histone modifications was the discovery that the common AED valproate is a histone deacetylase (HDAC) inhibitor. There has been significant excitement that valproate may work by promoting histone acetylation. However, while some beneficial effects of valproate may be due to HDAC activity, other HDAC inhibitors are not seizure suppressive (Hoffmann et al., 2008).

## 6.4 MicroRNA BIOGENESIS PATHWAYS IN THE BRAIN

MiRNAs are small noncoding RNA (~20–22 nucleotides, nt) that regulate post-transcriptional gene expression in a sequence-specific manner. A single miRNA is potentially able to regulate the expression of hundreds of genes, making them key regulators of cellular functions and attractive targets for antiepileptogenesis. About half of all identified miRNAs are expressed in the mammalian brain and there is significant cell- and region-specific distribution reflecting roles in gene expression that direct the functional specialization of neurons and the morphological responses that are required to adapt to their continuously changing activity state (Siegel et al., 2011; O'Carroll and Schaefer, 2013). MiRNAs regulate gene expression via translational inhibition, mRNA degradation, or a combination of both mechanisms (Bartel, 2009; O'Carroll and Schaefer, 2013). Recently published work suggests that destabilization of mRNA accounts for most of the effect of miRNAs with translational inhibition being less important (Eichhorn et al., 2014). In the brain, however, miRNA targeting is frequently not associated with reduced mRNA levels of targets (Klein et al., 2007; Lee et al., 2012). MiRNAs and their biogenesis components display select localization within neurons, with significant enrichment in dendrites, enabling local, activity-dependent miRNA regulation of protein levels (Lugli et al., 2005, 2008; O'Carroll and Schaefer, 2013). Recent work demonstrated that certain pre-miRs have localization signals which ensure delivery to the synaptic sites and processing to the mature miRNA (Bicker et al., 2013).

Most miRNA genes are transcribed by RNA polymerase II (Cai et al., 2004; Lee et al., 2004). MiRNAs are generally expressed as a cluster; within a single polycistronic transcript (Baskerville and Bartel, 2005). Approximately 40% of miRNAs are localized within intronic regions and are coexpressed with their host gene. Another 10% of miRNAs are expressed from exons, following the transcription pattern of the host gene (Rodriguez et al., 2004; Baskerville and Bartel, 2005). The remaining 50% of miRNAs are expressed from noncoding regions of the genome. As mentioned, miRNAs often share functions of the host gene, thereby contributing to the operation of a genetic network. For example, transcriptional activation of *AATK*, which is essential for neuronal differentiation, leads to the expression of its intronic miR-338, which inhibits the expression of negative regulators of neuronal differentiation (Barik, 2008). Also, intronic miRNAs can negatively regulate the expression of their host gene, such as miR-128 which regulates the expression of its host gene *Arpp21* in the brain (Lin et al., 2011).

Transcription of the miRNA generates a primary sequence, which is recognized by the nuclear microprocessor complex containing Drosha and DGCR8 (Lee et al., 2003). This complex cleaves the primary miRNA to generate the precursor-miRNA (pre-miR) of ~60–70 nt in length. Next, pre-miRs are exported into the cytoplasm by exportin-5 in a GTP-dependent manner (Bohnsack et al., 2004). Once in the cytoplasm, the pre-miR is incorporated into the loading complex for the RISC (RNA-induced silencing complex), where it is cleaved by the type III ribonuclease Dicer into a 20- to 22-nt long miRNA (Bernstein et al., 2001). Finally, the mature miRNA is loaded into the RISC complex and bound to proteins of the Argonaute (Ago) family. This miRNA-bound-RISC complex is the functional and active form of the miRNA which then targets the mRNA. Binding of a 7- to 8-nt seed region, usually within the 3'UTR of the mRNA, results in either inhibition of translation or mRNA degradation, which requires the activation of de-cap enzymes and removal of the poly-A tail (Kwak and Tomari, 2012).

Evidence of the essential role of miRNAs in the brain has been obtained using a range of genetic tools, including mice with constitutive and conditional deletion of key biogenesis enzymes. Deletion of DGCR8 results in a reduction in brain size and loss of inhibitory synaptic neurotransmission (Babiarz et al., 2011; Hsu et al., 2012). Conditional deletion of Drosha in neuronal progenitors did not affect neurogenesis in the developing brain, but did affect differentiation and migration of neurons (Knuckles et al., 2012). Deletion of Dicer from neurons produces severe brain abnormalities, including microencephaly and defects in dendritic arborization in cortex and hippocampus (Davis et al., 2008; De Pietri Tonelli et al., 2008; Babiarz et al., 2011; Dorval et al., 2012). Mice lacking Dicer in astrocytes develop spontaneous seizures and many die prematurely (Tao et al., 2011). This indicates that the miRNA biogenesis system is essential for brain development and function. Surprisingly, one study reported that specific deletion of Dicer in the adult mouse forebrain transiently enhanced learning and memory, although these animals later displayed degeneration of neurons in the cortex and hippocampus (Konopka et al., 2010).

There are four Ago proteins (Ago1–4) encoded in the mouse and human genome (Burroughs et al., 2011). Of these, studies suggest Ago2 is critical for miRNA-mediated repression of mRNAs (Czech and Hannon, 2011) and is the most abundant form in various tissues (Liu et al., 2004). Deficiency in Ago2 results in death of mice during early embryogenesis or mid-gestation (Morita et al., 2007). This reflects not only the essential role of Ago2 in embryonic development but perhaps an effect of impaired miRNA generation since Ago2 also processes some miRNAs (Morita et al., 2007). Studies in conditional mutants show individual deficiencies in Ago1, Ago3, and Ago4 genes do not produce obvious effects in mice, suggesting a potential redundancy among Ago family members (O'Carroll et al., 2007).

### 6.4.1  MicroRNA Biogenesis Pathways in Epilepsy

Available data show there are select changes to the expression of miRNA biogenesis components in epilepsy (Table 6.1).

Drosha levels are normal in the resected hippocampus of TLE patients with hippocampal sclerosis and in adjacent neocortex (McKiernan et al., 2012a). Drosha levels are also normal in mice shortly after status epilepticus (Jimenez-Mateos et al., 2011; McKiernan et al., 2012a). In contrast, a reduction in Dicer protein levels was found in a subgroup of resected sclerotic hippocampi from patients with drug-resistant TLE (McKiernan et al., 2012a). Dicer levels were also lower in the lesioned hippocampus in a mouse model of TLE (McKiernan et al., 2012a). In TLE patient samples of adjacent neocortex that lacked pathology, Dicer levels were normal and Dicer levels were normal in mice subject to nonharmful seizures (McKiernan et al., 2012a). The mechanism of Dicer loss is unknown but Dicer is a substrate for caspase-3 (Nakagawa et al., 2010), and caspase-3 is activated in sclerotic hippocampi from TLE patients (Schindler et al., 2006). Thus, Dicer may be vulnerable to degradation in epilepsy.

**TABLE 6.1**
**MiRNA Biogenesis Components in Experimental and Human Epilepsy**

|  | Drosha | Dicer | Ago2 | Reference |
|---|---|---|---|---|
| Human TLE |  |  |  |  |
| Hpp | = | ↓ | ↑ | McKiernan et al. (2012a) |
| Cortex | = | = | ↑ |  |
| Murine TLE |  |  |  |  |
| Hpp | = | ↓ | = | Li et al. (2014) |
| Cortex | n.d. | n.d. | n.d. | McKiernan et al. (2012a) |
| Murine SE |  |  |  |  |
| Hpp | = | = | = | Jimenez-Mateos et al. (2011) |
| Cortex | n.d. | n.d. | n.d. |  |
| Murine nondamaging SZ |  |  |  |  |
| Hpp | = | = | ↓ | McKiernan et al. (2012b) |
| Cortex | n.d. | n.d. | n.d. |  |

*Note:* TLE, temporal lobe epilepsy; SE, status epilepticus; Hpp, hippocampus; SZ, seizure; =, no change; ↑, increase; ↓, decrease; n.d., not detected/reported.

Ago2 levels were found to be higher in the resected hippocampi that displayed loss of Dicer, which may be a result of loss of Ago2-targeting miRNAs (McKiernan et al., 2012a). Since Ago2 can perform some miRNA-processing functions, this could mean there is an adaptive upregulation to maintain miRNA biogenesis and function. Increased Ago2 protein levels were also found in resected neocortex from TLE patients suggesting this is a result of repeated seizures rather than a response to pathological changes such as neuron loss (McKiernan et al., 2012a). However, animal data have showed less congruence. Ago2 was not altered by status epilepticus or in experimental TLE in mice (Jimenez-Mateos et al., 2011; McKiernan et al., 2012a; Bicker et al., 2014). However, at least one study did detect an effect of seizures on Ago2 levels (McKiernan et al., 2012b). In summary, experimental and human data show alterations in several components of the miRNA biogenesis pathway in brain which may reflect pathology-related changes or adaptive changes in epilepsy.

### 6.4.2 Altered MicroRNA Expression in Epilepsy

Figure 6.1 provides a time line plotting the key seminal papers in the area of miRNA and epilepsy.

The first studies to analyze miRNA responses to seizures were published in 2010. Among these was the first profiling effort in a model of status epilepticus (Liu et al., 2010) and two reports on individual miRNAs in experimental models of status epilepticus (Aronica et al., 2010; Nudelman et al., 2010). The next studies on miRNA and epilepsy appeared in 2011 which included additional miRNA profiling studies as well as the first functional manipulation of a miRNA in a seizure model (Jimenez-Mateos et al., 2011). There have now been about a dozen miRNA

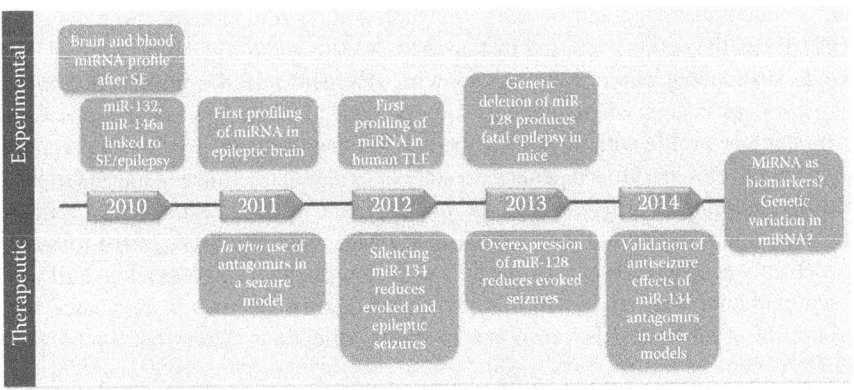

**FIGURE 6.1**    Time line of epilepsy research on microRNA.

profiling studies in experimental and human epilepsy. The studies that have analyzed miRNA profiles in the brain in the acute wake of status epilepticus featured
both chemically induced status epilepticus using kainic acid and pilocarpine and
electrically induced seizures, for example, using stimulation of the amygdala. Both
experimental models result in the development of spontaneous seizures and are
associated with pathologic changes in the hippocampus including neuron loss and
gliosis. A further series of studies has profiled miRNA changes in human epilepsy.
MiRNA profiles have been reported for other brain injuries known to precipitate
epilepsy but these are not reviewed presently (Jeyaseelan et al., 2008; Lei et al.,
2009; Liu et al., 2010).

### 6.4.3   MicroRNA PROFILE AFTER STATUS EPILEPTICS: EARLY EPILEPTOGENIC CHANGES

For a recent review on miRNA profiling after status epilepticus, the reader is
referred elsewhere (Henshall, 2013). The first miRNA profiling study in a seizure
model was reported by Liu and coworkers (2010). They detected a 1.5-fold increase
in 104 miRNAs in the hippocampus of rats 24 h after status epilepticus, while 179
miRNAs were downregulated. However, only four significantly regulated miRNAs
remained after statistical correction in this study (Liu et al., 2010). A second rat
study published the miRNA profile of the hippocampus 24 h after pilocarpine-
induced status epilepticus (Hu et al., 2011a). The authors reported upregulation of
19 miRNAs in the hippocampus. Among these were miR-132, which is involved in
dendritic spine morphology (Bicker et al., 2014), and miR-34a, which was linked
to control of apoptosis (Hermeking, 2010). This study also identified seven downregulated miRNAs, including miR-21 which was recently shown to have potent
protective effects against brain injury (Ge et al., 2014). Bioinformatics analyses
predicted effects on the mitogen-activated protein kinase pathway and long-term
potentiation, a model of synaptic plasticity (Hu et al., 2011a). In the same year, the
first mouse model profile was published, defining miRNA changes 24 h following

status epilepticus triggered by intra-amygdala kainic acid (Jimenez-Mateos et al., 2011). Here, the authors focused just on the CA3 subfield of the hippocampus. This avoids introducing other brain regions with divergent miRNA profiles and injury responses, as occurs when subfields are "pooled." The study also juxtaposed the normal injury profile with miRNA responses in mice that were previously preconditioned with brief seizures to generate a state of epileptic tolerance. Analysis showed that upregulation of miRNA was the predominant response after status epilepticus with, again, increased miR-132 (Jimenez-Mateos et al., 2011). In the tolerance model, only around 20% of the miRNAs were upregulated, whereas about half were downregulated. This suggested that protection against seizures is associated with a dampening of the miRNA response to status epilepticus. Moreover, a number of miRNAs showed bidirectional responses, suggesting there may be reprogramming of the miRNA response to injury in epileptic tolerance (Jimenez-Mateos et al., 2011). In contrast to Hu et al., there was strong upregulation of miR-21 after intra-amygdala kainic acid–induced status epilepticus in mice, suggestive of a protective response (Jimenez-Mateos et al., 2011).

A number of additional studies have since appeared that characterized the miRNA response after status epilepticus in rodents. Risbud and Porter (2013) profiled miRNA changes after pilocarpine-induced status epilepticus. At an early time point, they found 67 upregulated and none downregulated, whereas by 48 h, there were 10 upregulated and 188 downregulated. Among these, there was overlap with previous studies, including upregulation of miR-21 and miR-132. Most recently, miRNA profiles were analyzed 24 h after tetanic stimulation of the hippocampus in rats (Gorter et al., 2014). In this study, the authors analyzed three different regions of the brains; the CA1 and dentate gyrus subfields of the hippocampus subfield, and the parahippocampal region. In the CA1 and dentate gyrus, the main response was miRNA upregulation, consistent with many of the earlier studies. Interestingly, the parahippocampal gyrus showed significant downregulation of miRNAs which may reflect the different severity of the original insult and brain-region-specific miRNA responses. Again, miR-21 and miR-132 were among the miRNAs upregulated (Gorter et al., 2014).

Although profiling work typically focuses on just one or two time points, validation has often included more extensive time courses for individual miRNAs (Sano et al., 2012; McKiernan et al., 2012b; Peng et al., 2013). These studies reveal there are often sharp, temporal changes to miRNA expression in the wake of status epilepticus and these are brain-region specific. This implies there is a tight transcriptional control of miRNA biogenesis as well, presumably, as mechanisms for their rapid degradation (Zhang et al., 2012). This of course may lead to underestimation of the degree of overlap between miRNA profiles between models because some will be missed due to subtle differences in the timing of sampling relative to seizure initiation. Such tight control has also been observed for miRNA responses during brain development (Krichevsky et al., 2003). The abrupt stepwise miRNA expression patterns may exert strong influences on the timing of temporal waves of protein-coding genes that ultimately direct the cell and molecular reorganization that culminates in later epilepsy (Henshall, 2013).

### 6.4.4   MicroRNA Profile in Experimental Temporal Lobe Epilepsy

The first study of a miRNA in epilepsy was reported by Aronica and coworkers (2010) who found miR-146a levels were increased in the hippocampus of epileptic rats, particularly in astrocytes. Table 6.2 lists studies to date in which miRNAs have been profiled in experimental epilepsy.

The first profiling study of miRNA expression in established epilepsy was published in the following year (Song et al., 2011). The authors studied the miRNA profile in the hippocampus of rats that had undergone pilocarpine-induced status epilepticus 2 months earlier. Of 23 miRNAs regulated in the epileptic hippocampus, 18 were upregulated and 5 miRNAs were downregulated. Expression of miR-146a was also increased in their study. Notably, some of the same miRNAs altered acutely by status epilepticus remained elevated in established epilepsy, including miR-132 and miR-134 (Song et al., 2011). Risbud and Porter's study also found predominantly upregulated miRNAs in experimental epilepsy (Risbud and Porter, 2013). Again, miR-21 featured among those upregulated as well as miR-146a (Risbud and Porter, 2013). Increased expression of miR-21 has also been reported in models of prolonged seizures in immature rats (Peng et al., 2013). In the study by Gorter and colleagues (2014), the major response in both CA1 and dentate gyrus was also upregulation of miRNAs in the hippocampus, with 37 of 42 miRNAs showing an increased level. Li and colleagues were the first to use RNAseq to analyze miRNA expression in experimental epilepsy. Surprisingly, they only reported six significantly different miRNAs in the hippocampus of epileptic rats. Again, upregulation was the main miRNA response (up: miR-455-3p, miR-345-3p, miR-423-3p, miR-54, miR-365-5p; down: miR-296-5p; Li et al., 2014).

In a few reports, the direction of miRNA response has been more toward miRNA downregulation. Hu and colleagues found over 60% of deregulated miRNAs were downregulated in rats 2 months after pilocarpine. Again, miR-146a featured among those upregulated (Hu et al., 2012). Bot and colleagues (2013) also reported predominantly miRNA downregulation in epileptic rats. Recent work also compared miRNA profiles between seizure models in a single study (Kretschmann et al., 2014), explored the effects of sleep deprivation (Matos et al., 2014), and identified miRNAs specific to experimental drug-resistant epilepsy (Moon et al., 2014).

Overall, the experimental data show there are consistent miRNA changes in epilepsy (Table 6.2). Where there are discrepancies, these likely arise from differences between the animal models used, the choice of sampling time and sacrifice time, brain region studied, and profiling analysis platform.

### 6.4.5   MicroRNA Profile in Human Temporal Lobe Epilepsy

There have been three studies that have profiled miRNA expression in resected human brain tissue from patients with epilepsy. In 2012, two independent studies appeared that defined miRNA in sclerotic and nonsclerotic samples (Kan et al., 2012; McKiernan et al., 2012a). Kan and colleagues reported on 20 resected hippocampi from epilepsy patients and compared findings to 10 autopsy control samples.

**TABLE 6.2**

**List of MiRNAs Up- and Downregulated in Experimental Epilepsy Triggered by Status Epilepticus**

| | Song et al. (2011) | Hu et al. (2012) | Risbud and Porter (2013) | Bot et al. (2013) | Gorter et al. (2014) | Moon et al. (2014) | Matos et al. (2014) | Li et al. (2014) | Kretschmann et al. (2014) | MiRNA Count |
|---|---|---|---|---|---|---|---|---|---|---|
| **Upregulated** | | | | | | | | | | |
| MiR-146a | * | * | * | * | * | | | | | 5 |
| MiR-132-3p | * | | * | | * | | | | | 3 |
| MiR-23a | * | * | | | * | | | | | 3 |
| MiR-34a | | * | | | * | | * | | | 3 |
| MiR-126 | * | | | | * | | | | | 2 |
| MiR-132-5p | | | | * | * | | | | | 2 |
| MiR-135a | | | | | * | | * | | | 2 |
| MiR-140-3p | * | | | | * | | | | | 2 |
| MiR-152 | | * | | | * | | | | | 2 |
| MiR-193 | | | | | * | | * | | | 2 |
| MiR-21 | | * | | | * | | | | | 2 |
| MiR-210 | | | | * | * | | | | | 2 |
| MiR-211 | | | | * | | * | | | | 2 |
| MiR-212-3p | | | | * | * | | | | | 2 |
| MiR-212-5p | | | | * | * | | | | | 2 |
| MiR-23b | * | | | | * | | | | | 2 |
| MiR-24 | * | | | | * | | | | | 2 |
| MiR-27a | | * | | | * | | | | | 2 |
| MiR-27b | * | | | | * | | | | | 2 |
| MiR-423-3p | | | * | | | | | * | | 2 |

(Continued)

**TABLE 6.2 (Continued)**
List of MiRNAs Up- and Downregulated in Experimental Epilepsy Triggered by Status Epilepticus

| | Song et al. (2011) | Hu et al. (2012) | Risbud and Porter (2013) | Bot et al. (2013) | Gorter et al. (2014) | Moon et al. (2014) | Matos et al. (2014) | Li et al. (2014) | Kretschmann et al. (2014) | MiRNA Count |
|---|---|---|---|---|---|---|---|---|---|---|
| MiR-455-3p | | | | | | | | * | * | 2 |
| MiR-466f-3p | | | | | | * | | | * | 2 |
| MiR-467e-3p | | | | | | * | | | * | 2 |
| | | | | Downregulated | | | | | | |
| MiR-130a-3p | | | | * | | | | | * | 2 |
| MiR-138-1-3p | | * | | * | | | | | | 2 |
| MiR-139-5p | | | | * | * | | | | | 2 |
| MiR-144 | | * | | | | * | | | | 2 |
| MiR-296-5p | | * | | | | | | * | | 2 |
| MiR-551b | | | | | * | * | | | | 2 |
| MiR-935 | | | | * | * | | | | | 2 |

*Note:* A list of the miRNAs found in at least two studies.

They found 51 miRNAs were differentially expressed by over twofold between epileptic tissue and autopsy control, with similar numbers up- and downregulated. In contrast, a profiling analysis of miRNA in resected sclerotic hippocampus found major downregulation of miRNAs, which was attributed to loss of Dicer in the tissue (McKiernan et al., 2012a). Notably, pri-miRNA levels were normal in this tissue suggesting microprocessor function was not different in human TLE (McKiernan et al., 2012a). The miRNA profile of the dentate gyrus was recently reported for 14 samples with and without GCD (Zucchini et al., 2014), a frequent pathological finding in resected TLE hippocampus (Houser, 1990). The study identified altered expression of 12 miRNAs in the samples without GCD, with half showing lower levels in GCD samples including miR-487b. Further analysis identified higher levels of a potential target, *ANTXR1* which promotes cell spreading (Zucchini et al., 2014). Last, recent work has documented increased (including miR-21) as well as decreased miRNA in cortical dysplasia in humans, another common cause of epilepsy (Lee et al., 2014).

In summary, profiling analysis of status epilepticus and experimental and human epilepsy reveals a select group of miRNAs undergo expression changes. The predominant response after status epilepticus is upregulation while there is more of a balance with downregulation in epilepsy. Sufficient studies have now been published to see the emergence of a set of conserved miRNAs in epilepsy. This naturally paves the way to functional interrogation of these miRNAs to establish whether they have causal roles in epileptogenesis or maintenance of the chronic epileptic state.

### 6.4.6   MicroRNAs Regulating Inflammation in Epilepsy: miR-21 and miR-146a

The importance of inflammation in the genesis and maintenance of the epileptic state is increasingly understood (Vezzani et al., 2011). A large number of miRNAs have been linked to control of inflammation (Sheedy and O'Neill, 2008), of which miR-21 and miR-146a have repeatedly appeared in epilepsy studies (see above and Table 6.2). Inflammation-induced upregulation of miR-21 was first reported in a study by Moschos et al. (2007). Since then, many studies have reported miR-21 responses to inflammatory insults and a number of targets have been identified, including ILs (Lu et al., 2009). MiR-21 has also been labeled as an "oncomir" for reported antiapoptotic effects (Seike et al., 2009). This antiapoptotic effect has recently been extended to the nervous system with work showing miR-21 can protect against spinal cord damage (Hu et al., 2013) and traumatic brain injury (Ge et al., 2014). Thus, miR-21 upregulation in epilepsy models may regulate both inflammation and cell death. Functional studies of miR-21 may now be warranted in epilepsy models although with the caveat that the dual inflammatory and apoptosis functions may serve as functional antagonists.

MiR-146a is the most consistently upregulated miRNA in epilepsy (Table 6.2). Its primary targets are those involved in inflammatory signaling. Induction of miR-146a is driven by IL-1β and inflammation-associated transcription factors including nuclear factor κB (Taganov et al., 2006; Iyer et al., 2012). Increased miR-146a is thought to function by dampening immune responses, specifically by targeting

IRAK-1, IRAK-2, and TRAF-6 (Iyer et al., 2012). *In vivo* studies are now required to determine whether increasing miR-146a can protect against seizures or prevent or reverse epilepsy.

## 6.4.7   MicroRNAs Implicated in the Regulation of Dendritic Spines and Neuronal Activity: miR-132, miR-134, and miR-128

A number of miRNAs have been identified with important roles in the control of dendritic spines, the major contact points for excitatory communication in the central nervous system. These include miR-128, miR-132, miR-134, and miR-138 (Siegel et al., 2011). Among these, we now have functional evidence that miR-128, miR-132, and miR-134 influence brain excitability, seizures, and epilepsy-associated brain pathology.

The first miRNA for which functional studies were undertaken in epilepsy was miR-132 (Jimenez-Mateos et al., 2011). Prior to this, miR-132 had been described as an activity-regulated miRNA and overexpression of miR-132 increased dendritic length and branching via targeting p250GAP (Wayman et al., 2008). Consistent with this, overexpression of miR-132 was found to increase the frequency and amplitude of miniature excitatory currents in neurons (Edbauer et al., 2010). However, miR-132 may also serve an anti-inflammatory function via targeting of acetylcholinesterase (Shaked et al., 2009). Multiple studies have detected increased miR-132 after seizures in rodents and miR-132 is increased in some studies of human epilepsy (Nudelman et al., 2010; Jimenez-Mateos et al., 2011; Song et al., 2011; Peng et al., 2013). Upload to the RISC was also confirmed by Ago2 pull-downs after status epilepticus in mice (Jimenez-Mateos et al., 2011). Two studies have reported functional effects of miR-132 manipulation in seizure models. *In vivo* silencing of miR-132 using locked nucleic acid–modified antagomirs was shown to protect against hippocampal damage following intra-amygdala kainic acid (Jimenez-Mateos et al., 2011). Targeting was not, however, associated with significant effects on seizures. In a more recent study, silencing miR-132 was reported to reduce spontaneous seizures in rats. The protection was also associated with a reduction in neuronal death markers and reduced sprouting of mossy fibers (Huang et al., 2014).

Brain-enriched miR-134 was among the first miRNA for which functional data were obtained (Schratt et al., 2006). MiR-134 is a negative regulator of dendritic spines via targeting of LIM kinase 1, although other targets are now known. Expression of miR-134 is regulated by Mef2 (Fiore et al., 2009). Overexpression of miR-134 in neurons *in vitro* was shown to reduce dendritic spine volume, whereas silencing miR-134 resulted in a small increase (Schratt et al., 2006). *In vivo* effects of silencing miR-134 are slightly different and include a reduction in spine density as well as increased spine volume (Jimenez-Mateos et al., 2012, 2014). Recently, miR-134 has been implicated in the regulation of synaptic homeostasis in the presence of continuous low network activity (Fiore et al., 2014).

Increased miR-134 has been identified in a number of seizure models and in resected human brain tissue from TLE patients (Jimenez-Mateos et al., 2012; Peng et al., 2013; Wang et al., 2014). The first study to explore *in vivo* effects of miR-134 silencing found downregulation strongly reduced the severity of status epilepticus in the intra-amygdala kainic acid model (Jimenez-Mateos et al., 2012). More recently a similar effect was reported in the pilocarpine model, indicating silencing miR-134 is broadly effective against seizures triggered by different mechanisms (Jimenez-Mateos et al., 2014). *In vitro* antiseizure effects have also been reported (Wang et al., 2014). Remarkably, targeting miR-134 after status epilepticus resulted in up to 90% reduction in the later occurrence of spontaneous seizures in mice (Jimenez-Mateos et al., 2012). Even 2 months after a single injection of anti-miR-134, seizure rates were ~70% lower. These data suggest targeting miR-134 may have antiepileptogenic effects. Another finding in the same study was neuroprotective effects of miR-134 silencing (Jimenez-Mateos et al., 2012), which has since been corroborated by other groups using different models (Chi et al., 2014).

The last miRNA linked to neuronal activity and dendritic spines for which functional data are available in epilepsy is miR-128. Tan and colleagues (2013) found that mice with a conditional deletion of miR-128 developed fatal epilepsy which could be fully prevented by treatment with an AED. Mice lacking miR-128 also developed increased spine density. The authors confirmed that overexpression/rescue of miR-128 could suppress seizures and identified the extracellular signal regulating kinase pathway as a key source of miR-128 targets that might explain this potent effect (Tan et al., 2013). Altered expression of miR-128 has not, however, been a common finding in studies of miRNAs altered after status epilepticus or in established epilepsy.

### 6.4.8  MicroRNAs in Seizure-Induced Neuronal Death: miR-34a and miR-184

Finally, a number of miRNAs with roles in cell death have been explored in epilepsy. The first of these was miR-34a which was the first miRNA for which a role in apoptosis was demonstrated (Welch et al., 2007). Expression of miR-34a was later shown to be controlled by p53, a master controller of cell death which is known to be upregulated and contribute to seizure-induced neuronal death (Morrison et al., 1996; Engel et al., 2010a).

Regulation of miR-34a has been studied in two different rodent models. In the first model, miR-34a was found to be increased rapidly after status epilepticus in the CA3 and CA1 subfields and uploaded to the RISC (Sano et al., 2012). Silencing miR-34a in this model produced no effects on seizures or hippocampal damage. However, a small effect on apoptosis-related signaling was observed (Sano et al., 2012). Hu and colleagues also reported increased levels of miR-34a in a seizure model but found that inhibition of miR-34a reduced neuronal damage in their model. These differences may reflect model-specific contributions of the miRNA, timing of its induction relative to its targets, or technical issues around silencing.

McKiernan and colleagues (2012b) used a model of epileptic preconditioning to identify potentially protective miRNAs. Among those most upregulated after low-dose systemic kainic acid in mice, they identified miR-184. Although the targets of miR-184 were not fully explored, silencing miR-184 significantly increased seizure-induced neuronal death in two animal models. These data suggest miR-184 may protect against seizure-induced neuronal death (McKiernan et al., 2012b).

## 6.5 CONCLUSIONS AND FUTURE DIRECTIONS

MiRNAs represent a novel and emerging class of molecules with important roles in fine-tuning protein levels in cells. Because of their multi-targeting potential, they offer particular interest in epilepsy where there may be a need to impact on several targets within signaling pathways in order to disrupt disease pathogenesis. MiRNAs have proven targetable in preclinical animal models and a number of miRNAs have been identified with potent effects on seizure thresholds and epilepsy-associated hippocampal pathology. Future studies are required to identify more commonly regulated miRNAs by improving experimental design and collaboration between laboratories working with similar and different models. Alternative delivery routes for antagomirs should be tested that are more therapeutically relevant, such as intravenous injection. The BBB normally prevents passage of antagomirs into the brain (Krutzfeldt et al., 2005) but the BBB may open following epilepsy-precipitating injuries sufficient to allow passage of antagomirs. Other chemical modifications could allow directed delivery of miRNA inhibitors to specific cellular environments (Cheng et al., 2014). We need to learn more about the transcriptional control of miRNAs and identify more of their *in vivo* targets. Recent work by our group identified an important role for DNA methylation in the control of miRNA expression in human TLE (Miller-Delaney et al., 2014). Genetic variation may also exist in miRNAs or their biogenesis components which could explain disease risk. Finally, miRNAs in biofluids such as plasma may offer a rich source of biomarker in epilepsy. Preliminary work has already identified a number of miRNAs whose levels are altered in the blood at different phases of the epileptic process (Liu et al., 2010; Hu et al., 2011a; Gorter et al., 2014).

## ACKNOWLEDGMENTS

MicroRNA research in the authors' laboratory is supported by current grants from Science Foundation Ireland (12/RC/2272, 12/COEN/18, 13/IA/1891, 14/ADV/RC2721, 13/SIRG/2098, 13/SIRG/2114), Health Research Board (HRA-POR-2013-325), and Framework Programme 7 (No. 602130).

## REFERENCES

Allen A.S., Berkovic S.F., Cossette P. et al. 2013. *De novo* mutations in epileptic encephalopathies. *Nature* 501: 217–221.

Andre V., Ferrandon A., Marescaux C., Nehlig A. 2000. Electroshocks delay seizures and subsequent epileptogenesis but do not prevent neuronal damage in the lithium-pilocarpine model of epilepsy. *Epilepsy Research* 42: 7–22.

Aronica E., Fluiter K., Iyer A. et al. 2010. Expression pattern of miR-146a, an inflammation-associated microRNA, in experimental and human temporal lobe epilepsy. *European Journal of Neuroscience* 31: 1100–1107.

Babiarz J.E., Hsu R., Melton C. et al. 2011. A role for noncanonical microRNAs in the mammalian brain revealed by phenotypic differences in Dgcr8 versus Dicer1 knockouts and small RNA sequencing. *RNA* 17: 1489–1501.

Baram T.Z., Eghbal-Ahmadi M., Bender R.A. 2002. Is neuronal death required for seizure-induced epileptogenesis in the immature brain? *Progress in Brain Research* 135: 365–375.

Barik S. 2008. An intronic microRNA silences genes that are functionally antagonistic to its host gene. *Nucleic Acids Research* 36: 5232–5241.

Bartel D.P. 2009. MicroRNAs: Target recognition and regulatory functions. *Cell* 136: 215–233.

Baskerville S., Bartel D.P. 2005. Microarray profiling of microRNAs reveals frequent coexpression with neighboring miRNAs and host genes. *RNA* 11: 241–247.

Bernstein E., Caudy A.A., Hammond S.M., Hannon G.J. 2001. Role for a bidentate ribonuclease in the initiation step of RNA interference. *Nature* 409: 363–366.

Bialer M., White H.S. 2010. Key factors in the discovery and development of new antiepileptic drugs. *Nature Reviews Drug Discovery* 9: 68–82.

Bicker S., Khudayberdiev S., Weiss K. et al. 2013. The DEAH-box helicase DHX36 mediates dendritic localization of the neuronal precursor-microRNA-134. *Genes and Development* 27: 991–996.

Bicker S., Lackinger M., Weiss K., Schratt G. 2014. MicroRNA-132, -134, and -138: A microRNA troika rules in neuronal dendrites. *Cell and Molecular Life Sciences* 71: 3987–4005.

Bohnsack M.T., Czaplinski K., Gorlich D. 2004. Exportin 5 is a RanGTP-dependent dsRNA-binding protein that mediates nuclear export of pre-miRNAs. *RNA* 10: 185–191.

Boison D. 2008. The adenosine kinase hypothesis of epileptogenesis. *Progress in Neurobiology* 84: 249–262.

Bot A.M., Debski K.J., Lukasiuk K. 2013. Alterations in miRNA levels in the dentate gyrus in epileptic rats. *PLoS One* 8: e76051.

Buckmaster P.S., Lew F.H. 2011. Rapamycin suppresses mossy fiber sprouting but not seizure frequency in a mouse model of temporal lobe epilepsy. *Journal of Neuroscience* 31: 2337–2347.

Bumanglag A.V., Sloviter R.S. 2008. Minimal latency to hippocampal epileptogenesis and clinical epilepsy after perforant pathway stimulation-induced status epilepticus in awake rats. *Journal of Comparative Neurology* 510: 561–580.

Burroughs A.M., Ando Y., de Hoon M.J. et al. 2011. Deep-sequencing of human Argonaute-associated small RNAs provides insight into miRNA sorting and reveals Argonaute association with RNA fragments of diverse origin. *RNA Biology* 8: 158–177.

Cai X., Hagedorn C.H., Cullen B.R. 2004. Human microRNAs are processed from capped, polyadenylated transcripts that can also function as mRNAs. *RNA* 10: 1957–1966.

Cardinale F., Cossu M., Castana L. et al. 2013. Stereoelectroencephalography: Surgical methodology, safety, and stereotactic application accuracy in 500 procedures. *Neurosurgery* 72: 353–366.

Chang B.S., Lowenstein D.H. 2003. Epilepsy. *New England Journal of Medicine* 349: 1257–1266.

Cheng C.J., Bahal R., Babar I.A. et al. 2015. MicroRNA silencing for cancer therapy targeted to the tumour microenvironment. *Nature* 518: 107–110.

Cheung E.N., George S.R., Andrade D.M. et al. 2014. Neonatal hypocalcemia, neonatal seizures, and intellectual disability in 22q11.2 deletion syndrome. *Genetics in Medicine* 16: 40–44.

Chi W., Meng F., Li Y. et al. 2014. Downregulation of miRNA-134 protects neural cells against ischemic injury in N2A cells and mouse brain with ischemic stroke by targeting HSPA12B. *Neuroscience* 277: 111–122.

Clasadonte J., Haydon P.G. 2012 Astrocytes and epilepsy. In: *Jasper's Basic Mechanisms of the Epilepsies*, 4th Edition (Noebels JL, Avoli M, Rogawski MA, Olsen RW, Delgado-Escueta AV, eds). Oxford University Press, Bethesda, MD.

Cronin J., Dudek F.E. 1988. Chronic seizures and collateral sprouting of dentate mossy fibers after kainic acid treatment in rats. *Brain Research* 474: 181–184.

Czech B., Hannon G.J. 2011. Small RNA sorting: Matchmaking for Argonautes. *Nature Reviews Genetics* 12: 19–31.

Davis T.H., Cuellar T.L., Koch S.M. et al. 2008. Conditional loss of Dicer disrupts cellular and tissue morphogenesis in the cortex and hippocampus. *Journal of Neuroscience* 28: 4322–4330.

De Pietri Tonelli D., Pulvers J.N., Haffner C. et al. 2008. MiRNAs are essential for survival and differentiation of newborn neurons but not for expansion of neural progenitors during early neurogenesis in the mouse embryonic neocortex. *Development* 135: 3911–3921.

de Tisi J., Bell G.S., Peacock J.L. et al. 2011. The long-term outcome of adult epilepsy surgery, patterns of seizure remission, and relapse: A cohort study. *Lancet* 378: 1388–1395.

Devinsky O., Vezzani A., Najjar S., De Lanerolle N.C., Rogawski M.A. 2013. Glia and epilepsy: Excitability and inflammation. *Trends in Neuroscience* 36: 174–184.

Dingledine R., Varvel N.H., Dudek F.E. 2014. When and how do seizures kill neurons, and is cell death relevant to epileptogenesis? *Advances in Experimental Medicine and Biology* 813: 109–122.

Dorval V., Smith P.Y., Delay C. et al. 2012. Gene network and pathway analysis of mice with conditional ablation of Dicer in post-mitotic neurons. *PLoS One* 7: e44060.

Dube C.M., Brewster A.L., Richichi C., Zha Q., Baram T.Z. 2007. Fever, febrile seizures and epilepsy. *Trends in Neuroscience* 30: 490–496.

Dudek F.E., Obenaus A., Schweitzer J.S., Wuarin J.P. 1994. Functional significance of hippocampal plasticity in epileptic brain: Electrophysiological changes of the dentate granule cells associated with mossy fiber sprouting. *Hippocampus* 4: 259–265.

Ebert U., Brandt C., Loscher W. 2002. Delayed sclerosis, neuroprotection, and limbic epileptogenesis after status epilepticus in the rat. *Epilepsia* 43: 86–95.

Edbauer D., Neilson J.R., Foster K.A. et al. 2010. Regulation of synaptic structure and function by FMRP-associated microRNAs miR-125b and miR-132. *Neuron* 65: 373–384.

Eichhorn S.W., Guo H., McGeary S.E. et al. 2014. mRNA destabilization is the dominant effect of mammalian microRNAs by the time substantial repression ensues. *Molecular Cell* 56: 104–115.

Engel T., Murphy B.M., Hatazaki S. et al. 2010b. Reduced hippocampal damage and epileptic seizures after status epilepticus in mice lacking proapoptotic Puma. *FASEB Journal* 24: 853–861.

Engel T., Sanz-Rodgriguez A., Jimenez-Mateos E.M. et al. 2013. CHOP regulates the p53-MDM2 axis and is required for neuronal survival after seizures. *Brain* 136: 577–592.

Engel T., Tanaka K., Jimenez-Mateos E.M. et al. 2010a. Loss of p53 results in protracted electrographic seizures and development of an aggravated epileptic phenotype following status epilepticus. *Cell Death and Disease* 1: e79.

Fiore R., Khudayberdiev S., Christensen M. et al. 2009. Mef2-mediated transcription of the miR379-410 cluster regulates activity-dependent dendritogenesis by fine-tuning Pumilio2 protein levels. *EMBO Journal* 28: 697–710.

Fiore R., Rajman M., Schwale C. et al. 2014. MiR-134-dependent regulation of Pumilio-2 is necessary for homeostatic synaptic depression. *EMBO Journal* 33: 2231–2246.

Fisher R., Salanova V., Witt T. et al. 2010. Electrical stimulation of the anterior nucleus of thalamus for treatment of refractory epilepsy. *Epilepsia* 51: 899–908.

Fisher R.S. 2012. Therapeutic devices for epilepsy. *Annals of Neurology* 71: 157–168.

Fisher R.S., Acevedo C., Arzimanoglou A. et al. 2014. ILAE official report: A practical clinical definition of epilepsy. *Epilepsia* 55: 475–482.

Frotscher M., Jonas P., Sloviter R.S. 2006. Synapses formed by normal and abnormal hippocampal mossy fibers. *Cell Tissue Research* 326: 361–367.

Ge X.T., Lei P., Wang H.C. et al. 2014. MiR-21 improves the neurological outcome after traumatic brain injury in rats. *Scientific Reports* 4: 6718.

Goldberg E.M., Coulter D.A. 2013. Mechanisms of epileptogenesis: A convergence on neural circuit dysfunction. *Nature Reviews Neuroscience* 14: 337–349.

Gomez-Herreros F., Schuurs-Hoeijmakers J.H., McCormack M. et al. 2014. TDP2 protects transcription from abortive topoisomerase activity and is required for normal neural function. *Nature Genetics* 46: 516–521.

Gorter J.A., Iyer A., White I. et al. 2014. Hippocampal subregion-specific microRNA expression during epileptogenesis in experimental temporal lobe epilepsy. *Neurobiology of Disease* 62: 508–520.

Gorter J.A., van Vliet E.A., Aronica E. et al. 2006. Potential new antiepileptogenic targets indicated by microarray analysis in a rat model for temporal lobe epilepsy. *Journal of Neuroscience* 26: 11083–11110.

Graff J., Kim D., Dobbin M.M., Tsai L.H. 2011. Epigenetic regulation of gene expression in physiological and pathological brain processes. *Physiological Reviews* 91: 603–649.

Hanna M.G. 2006. Genetic neurological channelopathies. *Nature Clinical Practice Neurology* 2: 252–263.

He K., Xiao W., Lv W. 2014. Comprehensive identification of essential pathways and transcription factors related to epilepsy by gene set enrichment analysis on microarray datasets. *International Journal of Molecular Medicine* 34: 715–724.

Heng K., Haney M.M., Buckmaster P.S. 2013. High-dose rapamycin blocks mossy fiber sprouting but not seizures in a mouse model of temporal lobe epilepsy. *Epilepsia* 54: 1535–1541.

Henshall D.C. 2013. MicroRNAs in the pathophysiology and treatment of status epilepticus. *Frontiers in Molecular Neuroscience* 6: 1–11.

Henshall D.C., Engel T. 2013. Contribution of apoptosis-associated signaling pathways to epileptogenesis: Lessons from Bcl-2 family knockouts. *Frontiers in Cellular Neuroscience* 7: 110.

Hermeking H. 2010. The miR-34 family in cancer and apoptosis. *Cell Death and Differentiation* 17: 193–199.

Hoffmann K., Czapp M., Loscher W. 2008. Increase in antiepileptic efficacy during prolonged treatment with valproic acid: Role of inhibition of histone deacetylases? *Epilepsy Research* 81: 107–113.

Hoffmann K., Sivukhina E., Potschka H. et al. 2009. Retarded kindling progression in mice deficient in the extracellular matrix glycoprotein tenascin-R. *Epilepsia* 50: 859–869.

Houser C.R. 1990. Granule cell dispersion in the dentate gyrus of humans with temporal lobe epilepsy. *Brain Research* 535: 195–204.

Hsu R., Schofield C.M., Dela Cruz C.G. et al. 2012. Loss of microRNAs in pyramidal neurons leads to specific changes in inhibitory synaptic transmission in the prefrontal cortex. *Molecular and Cellular Neuroscience* 50: 283–292.

Hu J.Z., Huang J.H., Zeng L. et al. 2013. Anti-apoptotic effect of microRNA-21 after contusion spinal cord injury in rats. *Journal of Neurotrauma* 30: 1349–1360.

Hu K., Xie Y.Y., Zhang C. et al. 2012. MicroRNA expression profile of the hippocampus in a rat model of temporal lobe epilepsy and miR-34a-targeted neuroprotection against hippocampal neurone cell apoptosis post-status epilepticus. *BMC Neuroscience* 13: 115.

Hu K., Zhang C., Long L. et al. 2011a. Expression profile of microRNAs in rat hippocampus following lithium-pilocarpine-induced status epilepticus. *Neuroscience Letters* 488: 252–257.

Hu X.L., Cheng X., Cai L. et al. 2011b. Conditional deletion of NRSF in forebrain neurons accelerates epileptogenesis in the kindling model. *Cerebral Cortex* 21: 2158–2165.

Huang Y., Doherty J.J., Dingledine R. 2002. Altered histone acetylation at glutamate receptor 2 and brain-derived neurotrophic factor genes is an early event triggered by status epilepticus. *Journal of Neuroscience* 22: 8422–8428.

Huang Y., Guo J., Wang Q., Chen Y. 2014. MicroRNA-132 silencing decreases the spontaneous recurrent seizures. *International Journal of Clinical and Experimental Medicine* 7: 1639–1649.

Huang Y., Zhao F., Wang L. et al. 2012. Increased expression of histone deacetylases 2 in temporal lobe epilepsy: A study of epileptic patients and rat models. *Synapse* 66: 151–159.

Ivens S., Kaufer D., Flores L.P. et al. 2007. TGF-beta receptor-mediated albumin uptake into astrocytes is involved in neocortical epileptogenesis. *Brain* 130: 535–547.

Iyer A., Zurolo E., Prabowo A. et al. 2012. MicroRNA-146a: A key regulator of astrocyte-mediated inflammatory response. *PLoS One* 7: e44789.

Jakovcevski M., Akbarian S. 2012. Epigenetic mechanisms in neurological disease. *Nature Medicine* 18: 1194–1204.

Jeyaseelan K., Lim K.Y., Armugam A. 2008. MicroRNA expression in the blood and brain of rats subjected to transient focal ischemia by middle cerebral artery occlusion. *Stroke* 39: 959–966.

Jimenez-Mateos E.M., Bray I., Sanz-Rodriguez A. et al. 2011. MiRNA expression profile after status epilepticus and hippocampal neuroprotection by targeting miR-132. *American Journal of Pathology* 179: 2519–2532.

Jimenez-Mateos E.M., Engel T., Merino-Serrais P. et al. 2012. Silencing microRNA-134 produces neuroprotective and prolonged seizure-suppressive effects. *Nature Medicine* 18: 1087–1094.

Jimenez-Mateos E.M., Engel T., Merino-Serrais P. et al. 2015. Antagomirs targeting microRNA-134 increase hippocampal pyramidal neuron spine volume *in vivo* and protect against pilocarpine-induced status epilepticus. *Brain Structure and Function* 220: 2387–2399.

Jin P., Alisch R.S., Warren S.T. 2004. RNA and microRNAs in fragile X mental retardation. *Nature Cell Biology* 6: 1048–1053.

Kan A.A., van Erp S., Derijck A.A. et al. 2012. Genome-wide microRNA profiling of human temporal lobe epilepsy identifies modulators of the immune response. *Cell and Molecular Life Sciences* 69: 3127–3145.

Kharatishvili I., Pitkanen A. 2010. Association of the severity of cortical damage with the occurrence of spontaneous seizures and hyperexcitability in an animal model of post-traumatic epilepsy. *Epilepsy Research* 90: 47–59.

Klein M.E., Lioy D.T., Ma L. et al. 2007. Homeostatic regulation of MeCP2 expression by a CREB-induced microRNA. *Nature Neuroscience* 10: 1513–1514.

Knuckles P., Vogt M.A., Lugert S. et al. 2012. Drosha regulates neurogenesis by controlling neurogenin 2 expression independent of microRNAs. *Nature Neuroscience* 15: 962–969.

Kobow K., Jeske I., Hildebrandt M. et al. 2009. Increased reelin promoter methylation is associated with granule cell dispersion in human temporal lobe epilepsy. *Journal of Neuropathology and Experimental Neurology* 68: 356–364.

Kobow K., Kaspi A., Harikrishnan K.N. et al. 2013. Deep sequencing reveals increased DNA methylation in chronic rat epilepsy. *Acta Neuropathologica* 126: 741–756.

Konopka W., Kiryk A., Novak M. et al. 2010. MicroRNA loss enhances learning and memory in mice. *Journal of Neuroscience* 30: 14835–14842.

Kretschmann A., Danis B., Andonovic L. et al. 2014. Different microRNA profiles in chronic epilepsy versus acute seizure mouse models. *Journal of Molecular Neuroscience* 55: 466–479.

Krichevsky A.M., King K.S., Donahue C.P., Khrapko K., Kosik K.S. 2003. A microRNA array reveals extensive regulation of microRNAs during brain development. *RNA* 9: 1274–1281.

Krutzfeldt J., Rajewsky N., Braich R. et al. 2005. Silencing of microRNAs *in vivo* with 'antagomirs'. *Nature* 438: 685–689.

Kwak P.B., Tomari Y. 2012. The N domain of Argonaute drives duplex unwinding during RISC assembly. *Nature Structural and Molecular Biology* 19: 145–151.

Lee J.Y., Park A.K., Lee E.S. et al. 2014. MiRNA expression analysis in cortical dysplasia: Regulation of mTOR and LIS1 pathway. *Epilepsy Research* 108: 433–441.

Lee K., Kim J.H., Kwon O.B. et al. 2012. An activity-regulated microRNA, miR-188, controls dendritic plasticity and synaptic transmission by downregulating neuropilin-2. *Journal of Neuroscience* 32: 5678–5687.

Lee Y., Ahn C., Han J. et al. 2003. The nuclear RNAse III Drosha initiates microRNA processing. *Nature* 425: 415–419.

Lee Y., Kim M., Han J. et al. 2004. MicroRNA genes are transcribed by RNA polymerase II. *EMBO Journal* 23: 4051–4060.

Lei P., Li Y., Chen X., Yang S., Zhang J. 2009. Microarray based analysis of microRNA expression in rat cerebral cortex after traumatic brain injury. *Brain Research* 1284: 191–201.

Li M.M., Jiang T., Sun Z. et al. 2014. Genome-wide microRNA expression profiles in hippocampus of rats with chronic temporal lobe epilepsy. *Scientific Reports* 4: 4734.

Lin Q., Wei W., Coelho C.M. et al. 2011. The brain-specific microRNA miR-128b regulates the formation of fear-extinction memory. *Nature Neuroscience* 14: 1115–1117.

Liu D.Z., Tian Y., Ander B.P. et al. 2010. Brain and blood microRNA expression profiling of ischemic stroke, intracerebral hemorrhage, and kainate seizures. *Journal of Cerebral Blood Flow and Metabolism* 30: 92–101.

Liu J., Carmell M.A., Rivas F.V. et al. 2004. Argonaute2 is the catalytic engine of mammalian RNAi. *Science* 305: 1437–1441.

Lu T.X., Munitz A., Rothenberg M.E. 2009. MicroRNA-21 is up-regulated in allergic airway inflammation and regulates IL-12p35 expression. *Journal of Immunology* 182: 4994–5002.

Lugli G., Larson J., Martone M.E., Jones Y., Smalheiser N.R. 2005. Dicer and eIF2c are enriched at postsynaptic densities in adult mouse brain and are modified by neuronal activity in a calpain-dependent manner. *Journal of Neurochemistry* 94: 896–905.

Lugli G., Torvik V.I., Larson J., Smalheiser N.R. 2008. Expression of microRNAs and their precursors in synaptic fractions of adult mouse forebrain. *Journal of Neurochemistry* 106: 650–661.

Lukasiuk K., Pitkanen A. 2007. Gene and protein expression in experimental status epilepticus. *Epilepsia* 48(Suppl. 8): 28–32.

Lundberg J., Karimi M., von Gertten C. et al. 2009. Traumatic brain injury induces relocalization of DNA-methyltransferase 1. *Neuroscience Letters* 457: 8–11.

Maljevic S., Lerche H. 2014. Potassium channel genes and benign familial neonatal epilepsy. *Progress in Brain Research* 213: 17–53.

Maroso M., Balosso S., Ravizza T. et al. 2010. Toll-like receptor 4 and high-mobility group box-1 are involved in ictogenesis and can be targeted to reduce seizures. *Nature Medicine* 16: 413–419.

Matos G., Scorza F.A., Mazzotti D.R. et al. 2014. The effects of sleep deprivation on microRNA expression in rats submitted to pilocarpine-induced status epilepticus. *Progress in Neuropsychopharmacology and Biological Psychiatry* 51: 159–165.

McClelland S., Brennan G.P., Dube C. et al. 2014. The transcription factor NRSF contributes to epileptogenesis by selective repression of a subset of target genes. *Elife* 3: e01267.

McGonigal A., Bartolomei F., Regis J. et al. 2007. Stereoelectroencephalography in presurgical assessment of MRI-negative epilepsy. *Brain* 130: 3169–3183.

McKiernan R.C., Jimenez-Mateos E.M., Bray I. et al. 2012a. Reduced mature microRNA levels in association with dicer loss in human temporal lobe epilepsy with hippocampal sclerosis. *PLoS One* 7: e35921.

McKiernan R.C., Jimenez-Mateos E.M., Sano T. et al. 2012b. Expression profiling the microRNA response to epileptic preconditioning identifies miR-184 as a modulator of seizure-induced neuronal death. *Experimental Neurology* 237: 346–354.

McNamara J.O., Huang Y.Z., Leonard A.S. 2006. Molecular signaling mechanisms underlying epileptogenesis. *Science STKE* 2006: re12.

Merwick A., O'Brien M., Delanty N. 2012. Complex single gene disorders and epilepsy. *Epilepsia* 53(Suppl. 4): 81–91.

Miller-Delaney S.F., Bryan K., Das S. et al. 2015. Differential DNA methylation profiles of coding and non-coding genes define hippocampal sclerosis in human temporal lobe epilepsy. *Brain* 138: 616–631.

Miller-Delaney S.F., Das S., Sano T. et al. 2012. Differential DNA methylation patterns define status epilepticus and epileptic tolerance. *Journal of Neuroscience* 32: 1577–1588.

Mizoguchi H., Nakade J., Tachibana M. et al. 2011. Matrix metalloproteinase-9 contributes to kindled seizure development in pentylenetetrazole-treated mice by converting pro-BDNF to mature BDNF in the hippocampus. *Journal of Neuroscience* 31: 12963–12971.

Moon J., Lee S.T., Choi J. et al. 2014. Unique behavioral characteristics and microRNA signatures in a drug resistant epilepsy model. *PLoS One* 9: e85617.

Morita S., Horii T., Kimura M. et al. 2007. One Argonaute family member, Eif2c2 (Ago2), is essential for development and appears not to be involved in DNA methylation. *Genomics* 89: 687–696.

Morrison R.S., Wenzel H.J., Kinoshita Y. et al. 1996. Loss of the p53 tumor suppressor gene protects neurons from kainate-induced cell death. *Journal of Neuroscience* 16: 1337–1345.

Moschos S.A., Williams A.E., Perry M.M. et al. 2007. Expression profiling *in vivo* demonstrates rapid changes in lung microRNA levels following lipopolysaccharide-induced inflammation but not in the anti-inflammatory action of glucocorticoids. *BMC Genomics* 8: 240.

Moshe S.L., Perucca E., Ryvlin P., Tomson T. 2015. Epilepsy: New advances. *Lancet* 385: 884–898.

Mouri G., Jimenez-Mateos E., Engel T. et al. 2008. Unilateral hippocampal CA3-predominant damage and short latency epileptogenesis after intra-amygdala microinjection of kainic acid in mice. *Brain Research* 1213: 140–151.

Myer D.J., Gurkoff G.G., Lee S.M., Hovda D.A., Sofroniew M.V. 2006. Essential protective roles of reactive astrocytes in traumatic brain injury. *Brain* 129: 2761–2772.

Nakagawa A., Shi Y., Kage-Nakadai E., Mitani S., Xue D. 2010. Caspase-dependent conversion of Dicer ribonuclease into a death-promoting deoxyribonuclease. *Science* 328: 327–334.

Ndode-Ekane X.E., Pitkanen A. 2013. Urokinase-type plasminogen activator receptor modulates epileptogenesis in mouse model of temporal lobe epilepsy. *Molecular Neurobiology* 47: 914–937.

Nissinen J., Lukasiuk K., Pitkanen A. 2001. Is mossy fiber sprouting present at the time of the first spontaneous seizures in rat experimental temporal lobe epilepsy? *Hippocampus* 11: 299–310.

Nudelman A.S., DiRocco D.P., Lambert T.J. et al. 2010. Neuronal activity rapidly induces transcription of the CREB-regulated microRNA-132, *in vivo*. *Hippocampus* 20: 492–498.

O'Carroll D., Mecklenbrauker I., Das P.P. et al. 2007. A slicer-independent role for Argonaute 2 in hematopoiesis and the microRNA pathway. *Genes and Development* 21: 1999–2004.

O'Carroll D., Schaefer A. 2013. General principals of miRNA biogenesis and regulation in the brain. *Neuropsychopharmacology* 38: 39–54.

Ortinski P.I., Dong J., Mungenast A. et al. 2010. Selective induction of astrocytic gliosis generates deficits in neuronal inhibition. *Nature Neuroscience* 13: 584–591.

Park H.G., Yu H.S., Park S. et al. 2014. Repeated treatment with electroconvulsive sei-
    zures induces HDAC2 expression and down-regulation of NMDA receptor-related
    genes through histone deacetylation in the rat frontal cortex. *International Journal of
    Neuropsychopharmacology* 17: 1487–1500.
Peng J., Omran A., Ashhab M.U. et al. 2013. Expression patterns of miR-124, miR-134, miR-
    132, and miR-21 in an immature rat model and children with mesial temporal lobe
    epilepsy. *Journal of Molecular Neuroscience* 50: 291–297.
Piccione M., Vecchio D., Cavani S. et al. 2011. The first case of myoclonic epilepsy in a child
    with a *de novo* 22q11.2 microduplication. *American Journal of Medical Genetics A*
    155A: 3054–3059.
Pitkanen A., Lukasiuk K. 2011. Mechanisms of epileptogenesis and potential treatment tar-
    gets. *Lancet Neurology* 10: 173–186.
Pitkanen A., Sutula T.P. 2002. Is epilepsy a progressive disorder? Prospects for new therapeu-
    tic approaches in temporal-lobe epilepsy. *Lancet Neurology* 1: 173–181.
Ravizza T., Lucas S.M., Balosso S. et al. 2006. Inactivation of caspase-1 in rodent brain: A
    novel anticonvulsive strategy. *Epilepsia* 47: 1160–1168.
Risbud R.M., Porter B.E. 2013. Changes in microRNA expression in the whole hippocampus
    and hippocampal synaptoneurosome fraction following pilocarpine induced status epi-
    lepticus. *PLoS One* 8: e53464.
Rodriguez A., Griffiths-Jones S., Ashurst J.L., Bradley A. 2004. Identification of mammalian
    microRNA host genes and transcription units. *Genome Research* 14: 1902–1910.
Rojas A., Jiang J., Ganesh T. et al. 2014. Cyclooxygenase-2 in epilepsy. *Epilepsia* 55:
    17–25.
Sano T., Reynolds J.P., Jimenez-Mateos E.M. et al. 2012. MicroRNA-34a upregulation during
    seizure-induced neuronal death. *Cell Death and Disease* 3: e287.
Schindler C.K., Pearson E.G., Bonner H.P. et al. 2006. Caspase-3 cleavage and nuclear
    localization of caspase-activated DNase in human temporal lobe epilepsy. *Journal of
    Cerebral Blood Flow and Metabolism* 26: 583–589.
Schratt G.M., Tuebing F., Nigh E.A. et al. 2006. A brain-specific microRNA regulates den-
    dritic spine development. *Nature* 439: 283–289.
Seike M., Goto A., Okano T. et al. 2009. MiR-21 is an EGFR-regulated anti-apoptotic factor
    in lung cancer in never-smokers. *Proceedings of the National Academy of Sciences
    United States of America* 106: 12085–12090.
Shaked I., Meerson A., Wolf Y. et al. 2009. MicroRNA-132 potentiates cholinergic anti-
    inflammatory signaling by targeting acetylcholinesterase. *Immunity* 31: 965–973.
Sheedy F.J., O'Neill L.A. 2008. Adding fuel to fire: MicroRNAs as a new class of mediators
    of inflammation. *Annals of the Rheumatic Diseases* 67(Suppl. 3): iii50–iii55.
Siegel G., Saba R., Schratt G. 2011. MicroRNAs in neurons: Manifold regulatory roles at the
    synapse. *Current Opinion in Genetics and Development* 21: 491–497.
Singh S., Singh P.K., Bhadauriya P., Ganesh S. 2012. Lafora disease E3 ubiquitin ligase
    malin is recruited to the processing bodies and regulates the microRNA-mediated gene
    silencing process via the decapping enzyme Dcp1a. *RNA Biology* 9: 1440–1449.
Sloviter R.S., Bumanglag A.V. 2013. Defining "epileptogenesis" and identifying "antiepilep-
    togenic targets" in animal models of acquired temporal lobe epilepsy is not as simple
    as it might seem. *Neuropharmacology* 69: 3–15.
Sng J.C., Taniura H., Yoneda Y. 2006. Histone modifications in kainate-induced status epilep-
    ticus. *European Journal of Neuroscience* 23: 1269–1282.
Song Y.J., Tian X.B., Zhang S. et al. 2011. Temporal lobe epilepsy induces differential expres-
    sion of hippocampal miRNAs including let-7e and miR-23a/b. *Brain Research* 1387:
    134–140.
Sutula T., Cascino G., Cavazos J., Parada I., Ramirez L. 1989. Mossy fiber synaptic reorgani-
    zation in the epileptic human temporal lobe. *Annals of Neurology* 26: 321–330.

Taganov K.D., Boldin M.P., Chang K.J., Baltimore D. 2006. NF-kappaB-dependent induction of microRNA miR-146, an inhibitor targeted to signaling proteins of innate immune responses. *Proceedings of the National Academy of Sciences United States of America* 103: 12481–12486.

Tan C.L., Plotkin J.L., Veno M.T. et al. 2013. MicroRNA-128 governs neuronal excitability and motor behavior in mice. *Science* 342: 1254–1258.

Tao J., Wu H., Lin Q. et al. 2011. Deletion of astroglial Dicer causes non-cell-autonomous neuronal dysfunction and degeneration. *Journal of Neuroscience* 31: 8306–8319.

Thom M., Mathern G.W., Cross J.H., Bertram E.H. 2010. Mesial temporal lobe epilepsy: How do we improve surgical outcome? *Annals of Neurology* 68: 424–434.

Thomas R.H., Berkovic S.F. 2014. The hidden genetics of epilepsy—A clinically important new paradigm. *Nature Reviews Neurology* 10: 283–292.

Tremblay M.E., Stevens B., Sierra A. et al. 2011. The role of microglia in the healthy brain. *Journal of Neuroscience* 31: 16064–16069.

Vezzani A. 2013. Fetal brain inflammation may prime hyperexcitability and behavioral dysfunction later in life. *Annals of Neurology* 74: 1–3.

Vezzani A., French J., Bartfai T., Baram T.Z. 2011. The role of inflammation in epilepsy. *Nature Reviews Neurology* 7: 31–40.

Wang X.M., Jia R.H., Wei D., Cui W.Y., Jiang W. 2014. MiR-134 blockade prevents status epilepticus like-activity and is neuroprotective in cultured hippocampal neurons. *Neuroscience Letters* 572: 20–25.

Wayman G.A., Davare M., Ando H. et al. 2008. An activity-regulated microRNA controls dendritic plasticity by down-regulating p250GAP. *Proceedings of the National Academy of Sciences United States of America* 105: 9093–9098.

Welch C., Chen Y., Stallings R.L. 2007. MicroRNA-34a functions as a potential tumor suppressor by inducing apoptosis in neuroblastoma cells. *Oncogene* 26: 5017–5022.

Wetherington J., Serrano G., Dingledine R. 2008. Astrocytes in the epileptic brain. *Neuron* 58: 168–178.

Williams-Karnesky R.L., Sandau U.S., Lusardi T.A. et al. 2013. Epigenetic changes induced by adenosine augmentation therapy prevent epileptogenesis. *Journal of Clinical Investigation* 123: 3552–3563.

Zeng L.H., Rensing N.R., Wong M. 2009. The mammalian target of rapamycin signaling pathway mediates epileptogenesis in a model of temporal lobe epilepsy. *Journal of Neuroscience* 29: 6964–6972.

Zeng L.H., Xu L., Rensing N.R. et al. 2007. Kainate seizures cause acute dendritic injury and actin depolymerization *in vivo*. *Journal of Neuroscience* 27: 11604–11613.

Zhang Z., Qin Y.W., Brewer G., Jing Q. 2012. MicroRNA degradation and turnover: Regulating the regulators. *Wiley Interdisciplinary Reviews on RNA* 3: 593–600.

Zhang Z.Y., Zhang Z., Fauser U., Schluesener H.J. 2007. Global hypomethylation defines a sub-population of reactive microglia/macrophages in experimental traumatic brain injury. *Neuroscience Letters* 429: 1–6.

Zhu X., Han X., Blendy J.A., Porter B.E. 2012. Decreased CREB levels suppress epilepsy. *Neurobiology of Disease* 45: 253–263.

Zucchini S., Marucci G., Paradiso B. et al. 2014. Identification of miRNAs differentially expressed in human epilepsy with or without granule cell pathology. *PLoS One* 9: e105521.

# 7 MicroRNAs and Pain

*Hjalte H. Andersen and Parisa Gazerani*

## CONTENTS

7.1 Burden of Pain: A Brief Overview ................................................................. 183
    7.1.1 Pain Processing Pathways ................................................................. 184
    7.1.2 Nonneuronal Components in Pain ..................................................... 185
7.2 Emerging Role of MicroRNAs as Modulators in Pain Processing .............. 185
    7.2.1 Transcriptional Changes of MicroRNA in Neurons
           in Response to Pain ........................................................................... 187
    7.2.2 Involvement of MicroRNA in Pain Processing ................................ 188
           7.2.2.1 MicroRNA Alterations in Inflammatory Pain .................. 188
           7.2.2.2 MicroRNA Alterations in Neuropathic Pain .................... 192
7.3 Role of NeurimmiRs as Modulators of Inflammation ................................. 195
7.4 MicroRNAs as Potential Biomarkers in Patients with Pain Conditions ....... 197
7.5 Conclusion and Future Perspectives ........................................................... 198
References ............................................................................................................ 199

## 7.1 BURDEN OF PAIN: A BRIEF OVERVIEW

Pain is a global health problem and one of the most common conditions for which people seek medical attention (Goldberg and McGee, 2011). Pain management has been considered a human right (Lohman et al., 2010). The International Association for the Study of Pain defines pain as an unpleasant sensory and emotional experience associated with actual or potential tissue damage or described in terms of such damage (Merskey, 1979; Merskey and Bogduk, 1986). Acute pain can be brief and short lasting or persistent until the disease or injury heals. However, chronic pain is pain without biological value that persists beyond the normal anticipated healing time. Chronic pain is defined as pain lasting more than 3 months and can be recurrent or persistent with a high negative impact on the quality of life and daily functioning (Breivik et al., 2006). About 20% of adults suffer from pain and 10% are newly diagnosed with chronic pain each year (Breivik et al., 2013). Recent studies in Europe have consistently estimated that 25%–35% of adults report chronic pain, which is considered a burden to health-care systems with estimated costs of billions of Euros annually (Reid et al., 2011). Pain affects all populations, regardless of age, sex, income, race/ethnicity, or geographical borders; although the percentage and format of its effect might vary extensively. For instance, it is known that pain increases with age and is overrepresented in women (Gerdle et al., 2004; Leresche, 2011). With the demographic changes of an ageing population, a challenge is expected globally (Tsang et al., 2008). Chronic pain is often seen in the neck, back, knee, hip, and other joints; but headaches and neuropathic pain are also common and to an equal extent disabling (Tsang et al., 2008). Several different risk factors including genetic, environmental, and personality traits have been associated with chronic pain; however, it is not yet completely understood how these factors can lead to abnormal processing of painful signals at molecular, cellular, and network levels (Denk et al., 2014). Not only is the etiology of pain complex but it also coexists with several other conditions such as depression, that makes it even more complicated to approach for a proper management (Dahan et al., 2014). A better understanding of pain processing is a key factor to assist identifying individuals at risk and eventually will lead to new therapeutic developments. Early identification and adequate management of chronic pain minimize long-term personal suffering and disability (Dworkin, 2012). Better diagnosis, prevention, or treatment of chronic pain may eventually lead to dramatic reduce of the global burden of this disorder.

## 7.1.1  Pain Processing Pathways

Pain processing is a complex phenomenon and consists of a series of signaling events that occurs at multiple levels of the peripheral and central nervous systems (CNS), (Basbaum et al., 2009). Acute pain signals are initiated by the peripheral stimulation of the nerve endings (Aδ and C-fiber nociceptors), propagated as action potentials through the cell soma residing in the sensory ganglia (e.g., dorsal root ganglion) to the dorsal horn of the spinal cord, and then sent through second-order neurons in the thalamus to the cortex for higher-order processing. Descending control of nociceptive signaling originates in the somatosensory cortex and travels through the midbrain to the spinal cord. Descending pathway serves to produce an analgesic effect through, for example, endogenous opioid receptors and inhibitory γ-aminobutyric acid (GABA)-ergic signaling (Latremoliere and Woolf, 2009; Woolf, 2011). While acute pain is the result of detecting noxious stimuli and serves as a protective mechanism for survival, chronic pain is associated with neuroplastic changes in the peripheral, and CNS in response to nociceptive input (Basbaum, 1999). In the development of disabling chronic pain, peripheral sensitization, central sensitization, and descending modulation are all involved to some extent (Latremoliere and Woolf, 2009; Woolf, 2011). Peripheral sensitization or peripheral nociceptor hyperexcitability is the consequence of series of events that occur following peripheral nerve stimulation (e.g., peripheral tissue damage, trauma, or injury). Increased spontaneous firing and alterations in the transduction, conduction, or sensitivity of nociceptive afferent fibers together with a phenomenon called neurogenic inflammation, whereby inflammatory products are released by activated nociceptors, lead to changes manifested as enhanced ion channel permeability or gene expression that collectively lead to occurrence of peripheral sensitization. Central sensitization, however, is persistent increased excitability of nociceptive neurons of the dorsal horn of the spinal cord or higher centers in the CNS. Enhanced activation of central synapses may result in long-term potentiation. Regardless of the cause, chronic pain can persist with or without an external stimulus manifested as spontaneous pain, hyperalgesia, and allodynia. Therefore, while peripheral sensitization results from an increase in responsiveness of the peripheral nociceptors, chronic or persistent pain is the result of pathologic increases in excitability, or sensitization, of one or more peripheral or central components of pain transduction pathways. A major component of persistent pain is now thought to arise from adaptive changes in structure and function of a number of central brain regions involved in processing of pain-related sensory information (Latremoliere and Woolf, 2009; Swieboda et al., 2013). Descending inhibitory pathways involving structures such as the periaqueductal gray, rostroventral medulla, and nucleus caudalis influence pain modulation. Opioidergic, serotonergic, and noradrenergic mechanisms have been known as key players (Staud, 2013). Supraspinal processing in the modulation of pain has also been implicated at higher centers, including thalamocortical connections, somatosensory cortices, anterior cingulate cortex, insula, motor cortex, dorsolateral prefrontal cortex, orbitofrontal cortex, and amygdala (Ossipov et al., 2010).

### 7.1.2 NONNEURONAL COMPONENTS IN PAIN

While it is well-known that neuronal processes are essential in pathological pain signaling, a growing body of evidence supports a key role of nonneuronal cells, glia, in pain processing. Both phenotypic and functional changes of glial cells located in the periphery (e.g., Schwann cells and satellite glial cells) or CNS (e.g., microglia and astrocytes) have been identified (Ji et al., 2013) in animal models of neuropathic or inflammatory pain. It has been presented that glial cells interact with neuronal signaling, become activated, and release substances (e.g., proinflammatory cytokines and chemokines) that directly or indirectly act on neurons and amplify pain under pathological pain conditions (Liu and Yuan, 2014). The concept of neuron–glia interactions in the genesis of pain or its maintenance (chronic pain) have giving rise to the idea of targeting of this abnormal cross talk for pain management. Inhibition of glial cell activation or blocking the released products of these cells have been the subject of extensive research recently as a potential therapeutic approach for neuropathic pain (Vallejo et al., 2010). Glial cells are also considered to play a critical role in contralateral spreading and widespread sensitization of pain neuronal networks (Jancalek, 2011).

## 7.2  EMERGING ROLE OF MicroRNAs AS MODULATORS IN PAIN PROCESSING

Since Bai et al. (2007) published the first paper specifically pertaining to the role of miRNAs in pain processing, accumulating evidence has proved that the small inhibitory class of noncoding RNAs play crucial role in fine-tuning how painful insults are processed (Bai et al., 2007; Andersen et al., 2014). This has been confirmed to be the case both for acute and chronic pain and for pain of neuropathic and inflammatory origin within a wide range of experimental models and more recently clinical pain disorders (Niederberger et al., 2011; Kynast et al., 2013a; Andersen et al., 2014; Lutz et al., 2014; Andersen et al., 2015). This chapter aims to present the advances made in understanding the role of miRNAs as modulators and biomarkers within in the field of pain. It must be stated that covering all miRNA alterations in diseases that can give rise to pain per se is not the objective. Many diseases of the nervous system for instance brain cancer or multiple sclerosis involve extensive well-investigated miRNA dysregulations, and are associated with excruciating pain

(Edvardsson and Persson, 2012; Willems et al., 2014); however, miRNA research within these areas is predominantly related to the miRNA-involvement in the specific pathogenesis and not pain processing (Osterberg and Boivie, 2010; Edvardsson and Persson, 2012; Møller et al., 2013; Willems et al., 2014). The studies used to compile this book chapter can broadly be subdivided into the following three categories, some of which cover multiple approaches (see Figure 7.1):

1. Studies of miRNA alterations induced by inflammatory and neuropathic pain models *in vivo*.
2. Biomarker discovery studies investigating miRNA alterations occurring in patients suffering from diseases related to significant recurrent or chronic pain.
3. Studies that explore the functional properties of one or more specific miRNA in relation to a particular pain models and explore opportunities for targeted therapy.

As highlighted in Section 7.1.1, neuronal adaptation, or more specifically peripheral and central sensitization, play an important role in the initiation and maintenance of a wide range of pain conditions. Dysfunctional inhibitory control and increased pain facilitation is recognized as crucial contributors to the clinical presentation of chronic pain of various origin including highly prevalent conditions such as fibromyalgia, knee osteoarthritis, and different neuropathic pain conditions.

Underlying these clinical states of increased pain sensitivity are alterations in protein expression driving long-term hyperexcitability of nociceptive neurons, thus contributing to the development or maintenance of chronic pain (Latremoliere and Woolf, 2009; Gold and Gebhart, 2010; Woolf, 2011). The process of sensitization can occur centrally, initiated and maintained by synaptic plasticity in neurons, changes in microglia, astrocytes, gap junctions, and neuronal membrane excitability (Scholz and Woolf, 2007). Similarly, sensitization can occur in the periphery affecting primary afferents and typically involving alterations in gene expression and expressional changes of signaling molecules, ion channels, or structural

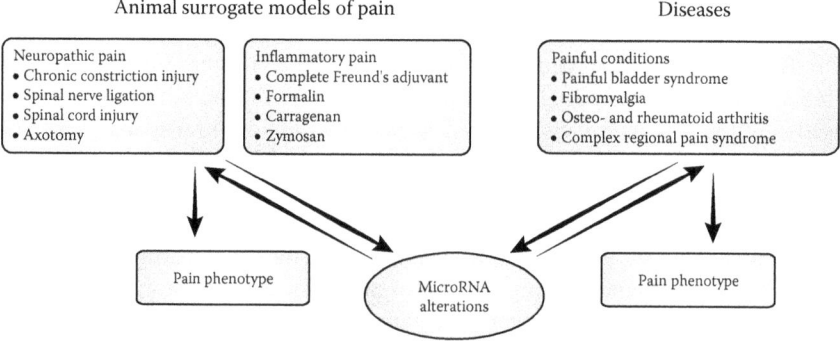

**FIGURE 7.1**    An overview of the inflammatory and neuropathic animal surrogate pain models and human diseases characterized by miRNA alterations.

proteins, leading to a more excitable afferent phenotype (Ji et al., 2003; Gold and Gebhart, 2010; Reichling et al., 2013). Being posttranscriptional mRNA inhibitors, miRNAs are capable of modulating many elements of this process centrally as well as peripherally. For instance, when considering that it is estimated that at least three out of five of the mammalian genes are miRNA targets, it is evident that miR-NAs exert posttranscriptional regulation of proteins involved in pain processing pathways (Friedman and Farh, 2009). In accordance, the dysregulation of miR-NAs suggested to target important molecular regulators of pain processing such as cyclooxygenase 2 (Akhtar and Haqqi, 2012; Park et al., 2013), $GABA_{\alpha 1}$ (Sengupta et al., 2013), transient receptor potential cation channel subfamily V member 1 (Li et al., 2011), and multiple $Na^+$ and $Ca^{2+}$ channels (Favereaux et al., 2011; Li et al., 2011; von Schack et al., 2011) has been observed in various surrogate models of pain. Similarly, it has been shown that let-7b and miR-23b regulate the expression of μ-opioid receptors (Wu et al., 2008; He and Wang, 2012; Ni et al., 2013) and that opioid-tolerance associated with prolonged morphine administration is potentially a consequence of miR-23b upregulation, thus extending the role of miRNA to the alteration of endogenous analgesic mechanisms (Wu et al., 2008). This highlights the importance of understanding the mechanisms behind the transition from acute physiological pain to pathological chronification, and in the context of this chapter, particularly mapping to which extend miRNA-induced transcriptional repression of genes is involved in the pathophysiology of pain.

### 7.2.1 TRANSCRIPTIONAL CHANGES OF MICRORNA IN NEURONS IN RESPONSE TO PAIN

A few studies implicate an activity-dependent dysregulation of certain miRNAs. For instance, the transcriptional factor, cAMP response element-binding protein (CREB), known to be important in mediating neuronal plasticity, induces rapid transcription of miRNA-132. In accordance with its mRNA transcription profile including, for example, c-Fos and brain-derived neurotrophic factor (BDNF), over-expression of miRNA-132 increases neuronal dendrite morphogenesis, while block-ing of miRNA-132 decreased dendrite outgrowth (Vo et al., 2005; Wayman et al., 2008). While limited evidence is available on the signaling processes that cause spe-cific transcriptional regulation of miRNAs in response to neuropathic and inflam-matory conditions, it is clear that the miRNA-processing machinery (described in detail in an earlier chapter) plays a role in the neuronal response leading to pain. An early study to demonstrate a significant contribution of miRNAs to processing of pain was based on a conditional Dicer deletion in Nav1.8+ nociceptors in dorsal root ganglia (Zhao et al., 2010). Surprisingly, through extensive miRNA alterations the Dicer knockout decreased the expression of pain-relevant mRNA transcripts, such as the Nav1.8 and CaMKII. In addition, knockouts exhibited a decreased pain behavior in response to the inflammatory pain and a decrease in c-FOS positive neurons in the spinal cord highlighting those alterations in primary afferents caused a decreased pain transmission in the CNS. Interestingly, the conditional knockout of Dicer did not affect Aδ- and C-fiber–mediated acute pain transmission. Additionally, the pain behavior of the spinal nerve ligation (SNL)-induced neuropathy, in which

Nav1.8⁺ neurons are believed to play a minor role, was not affected by the Dicer knockout (Zhao et al., 2010). This highlights that miRNA modulation could perhaps predominantly constitute a detrimental mechanism in the context of prolonged inflammation-induced pain. Another study used Dicer-deficient neurons in combination with the sciatic nerve crush (SNC) surrogate model of neuropathic injury *in vivo* and investigated regenerative axon growth *in vitro*. Upon SNC, the Dicer-deficient mutants exhibited a significantly decreased ability to reestablish normal mechanical sensitivity. In accordance, isolated Dicer knockout sciatic nerves showed an impaired repossession of conduction velocity and a decreased number of redeveloping nerve fibers compared to their wild-type counterparts, indicative of the crucial role of miRNAs in functional recovery of damaged nerves (Wu et al., 2012). Lastly, a study utilizing a rodent sciatic nerve injury model investigated both the differential expression of miRNA in dorsal root ganglion (DRG) neurons compared to sham injured mice, and the effect of the nerve injury on proteins involved in miRNA biogenesis. Interestingly, while subunits of the RNA-induced silencing complex were highly upregulated, levels of various p-body machinery proteins differed significantly between neuropathic and scam injured mice, indicating that not only miRNA derangements but also regulatory proteins in the miRNA processing are involved in the peripheral response to injury (Wu et al., 2011).

## 7.2.2 INVOLVEMENT OF MicroRNA IN PAIN PROCESSING

In the following sections, two distinctions will be made pertaining to the etiology of the pain model applied, which can be either inflammatory or neuropathic, and to the tissue analyzed which can be either peripheral, for example, DRG neurons or central, for example, neurons of the spinal dorsal horn (Table 7.1 provides an overview).

### 7.2.2.1 MicroRNA Alterations in Inflammatory Pain

Various *in vivo* models of inflammation have been used to assess the role of miRNA in response to induced inflammatory pain processing. It is clear that upon a peripheral inflammatory insult such as injection of complete Freund's adjuvant (CFA) or formalin; widespread miRNA alterations are observable at different levels of the nervous system. The present section highlights the results pertaining to the role of miRNA in inflammatory pain and hyperalgesia.

Five studies have been published using CFA to induce long-lasting inflammatory hyperalgesia (see Table 7.1). In a pioneering study in 2007 by Bai et al., it was shown that inflammation and subsequent hyperalgesia induced by injecting CFA into the masseter muscle of rats resulted in a trigeminal downregulation of miR-10a, -29a, -98, -99a, -124, -134, and -183, of which particularly miR-124 and miR-134 have since been reiterated as crucial miRNAs in the context of pain (Bai et al., 2007; Nakanishi et al., 2010; Ponomarev et al., 2011; Xiao et al., 2011; Willemen et al., 2012; Ni et al., 2013). Also applying CFA as a model of inflammation, Kusuda et al. (2011) quantified the expression of four brain-specific miRNAs of which miR-1 and miR-16 expression was markedly downregulated for 7 days in the DRG, but surprisingly upregulated in the SDH along with miR-206, 24 h after the induction.

**TABLE 7.1**

**An Overview of the Studies on MicroRNA (miRNA) Modulation in Animal Surrogate Models of Inflammatory and Neuropathic Pain**

| | Pain Model | Highlighted MiRNAs | Suspected Targets | References |
|---|---|---|---|---|
| **Animal Models of Inflammatory Pain** | | | | |
| Peripheral | Complete Freund's adjuvant-induced trigeminal pain | TG: miR-10a, miR-29a, miR-98, miR-99a, miR-124, miR-134, miR-183 ↓ | | Bai et al. (2007) |
| | Complete Freund's adjuvant-induced pain | DRG: miR-143 ↓ | VCAN1/2 | Tam Tam et al. (2011) |
| | Surgically induced joint injury | DRG: miR-146a, miR-183 ↓ | TLR-2/4, IL1β, IL-6, TNFα, NFκB, TRAF-6 | Li et al. (2013b) |
| | Complete Freund's adjuvant-induced pain | DRG: miR-134 ↓ | MOR1 | Ni et al. (2013) |
| | Complete Freund's adjuvant-induced pain | DRG: miR-219 ↓ | CAMK2G | Pan et al. (2014) |
| Central | Facial inflammation | DRG: miR-155 ↑ miR-233 ↑ | C/EBP-β | Poh et al. (2011) |
| | IL1β-induced inflammatory hyperalgesia | SCM: miR-124 ↓ | iNOS, C/EBP-α | Willemen et al. (2012) |
| Both | Formalin-induced inflammatory pain | SDH: miR-124a ↓ | MECP2 | Kynast et al. (2013b) |
| | Inflammatory joint pain | DRG and SDH: miR-146a ↓ | TNFα, COX-2, iNOS, IL-6, IL8, RANTS, TRPV1 | Li et al. (2011) |
| | Complete Freund's adjuvant-induced pain | DRG: miR-1 ↑ miR-16 ↑ miR-206 ↑ SDH: miR-1 ↓ miR-16 ↓ miR-206 ↓ | BDNF, MAPK3, CALM2, NGFR, PLA2G4A, IGF1, TRPC3, OPRD1 | Kusuda et al. (2011) |
| **Animal Models of Neuropathic Pain** | | | | |
| Peripheral | Spinal nerve ligation | DRG: miR-103 ↓ | CACNA1C, CACNA2D1, CACNB1 | Favereaux et al. (2011) |
| | Sciatic nerve crush | DRG: miR-124 ↓ miR-221 ↑ miR-142-5p ↑ miR-21 ↑ | | Wu et al. (2011) |

(Continued)

**TABLE 7.1 (Continued)**

**An Overview of the Studies on MicroRNA (miRNA) Modulation in Animal Surrogate Models of Inflammatory and Neuropathic Pain**

| | Pain Model | Highlighted MiRNAs | Suspected Targets | References |
|---|---|---|---|---|
| | Chronic constriction injury | DRG: miR-21, miR-221, miR-500 ↑ | GABAAR, CREB-1 | Genda et al. (2013) |
| | Chronic constriction injury | DRG: miR-96 ↓ | SCN3A | Chen et al. (2014) |
| | Spinal nerve ligation | DRG: miR-96 ↑ miR-182 ↑ miR-183 ↑ | SCN3A, SCN9A, BDNF TAC1, PAI-2, TREK-1 | Aldrich et al. (2009) |
| Central | Contusive spinal cord injury | SDH: miR-137, miR-181a, miR-219-2-3p, miR-7a ↓ miR-21 ↑ | TNFα, IL1β, ICAM1, PLA2, ANXA1/2 | Liu et al. (2009) |
| | Spinal cord injury | SC: miR-223 ↑, miR-124a ↓ | | Nakanishi et al. (2010) |
| | Sciatic nerve lesion | ACC: miR-34c ↓, miR-200b ↓ miR-429 ↓ | ZEB2, MECP2, DNMT3a, EPS8, GPM6α SNAP25 | Imai et al. (2011) |
| | Spinal cord injury | SDH: miR-23b ↓ | NOX4 | Im et al. (2012) |
| | Chronic constriction injury | HC: miR-125b ↓ miR-132 ↓ | BDNF, NR2A | Arai et al. (2013) |
| | Bilateral constriction injury | SDH: miR-203 ↓ | RAP1A | Li et al. (2014) |
| Both | Spinal nerve ligation (SNL) and axotomy | Axotomy, SDH: miR-1 ↑ miR-16 ↑ miR-206 ↑ DRG: miR-1 ↓ SNL, DRG: miR-1 ↓ miR-206 ↓ | BDNF, MAPK3, CALM2, NGFR, PLA2G4A, IGF1, TRPC3, OPRD1 | Kusuda et al. (2011) |
| | Chronic constriction injury | DRG: miR-341 ↑ SDH: miR-203 ↓ miR-541-5p ↓ miR-181a-1 ↓ | NMDAR | Li et al. (2013a) |

*Note:* The "suspected targets" column includes both *in silico* predicted, qPCR correlated and validated targets. Those validated with the use of luciferase-reporter assays are marked in bold. Refer to the individual papers to explore miRNA-mRNA target status. The abbreviation in from of the individual highlighted (not comprehensive) miRNAs denotes the origin of the sample: TG = trigeminal ganglion, SDH = spinal dorsal horn, DRG = dorsal root ganglion, SCM = spinal cord microglia, HC = hippocampus, SC = spinal cord and ACC = anterior cingulate cortex. Arrows denotes up- and downregulation upon painful insult.

In the same year, a comparative study showed that miR-143 was downregulated in the DRG upon CFA induction but notably not in a neuropathic pain model applying sciatic nerve transection (Kusuda et al., 2011), highlighting differentiated miRNA response patterns to neuropathic and inflammatory insults. The authors suggested the transcript Vcan1/2, known to modulate neurogenesis, as a putative target of miR-143 based on an ipsilateral increase in mRNA product, but did not perform a luciferase-based target validation (Kusuda et al., 2011). In a CFA-model study by Ni et al. (2013), it was confirmed that miR-134 is downregulated in the DRG upon inflammatory induction and that its expression was inversely correlated with MOR1 expression. The study subsequently applied a luciferase assay to validate MOR1 as a target of miR-134 elucidating a potential endogenous pain modulation mechanism and highlighting a novel therapeutic opportunity (Ni et al., 2013). Another notable study found that CFA-induced prolonged inflammatory pain significantly reduced miR-219 in the SDH of mice and demonstrated that overexpression of spinal miR-219 prevented thermal and mechanical hyperalgesia. Furthermore, CaMKIIγ was luciferase validated as an mRNA target of miR-219 and as a key player in establishing pain-responsive behavior upon CFA treatment (Pan et al., 2014). Very recently, a study using CFA-induced orofacial inflammation identified a dysregulated 7-miRNA-signature in the trigeminal ganglion, which was subsequently qPCR validated. Further studies of the significantly downregulated miR-125a-3p revealed that its expression was correlated with increased expression of mitogen-activated protein kinase-$\alpha$ and calcitonin gene-related peptide in ipsilateral TGs compared with control (Dong et al., 2014). Interestingly, migraine attacks have been intimately associated with CGRP surges and trigeminal sensitization (Cutrer, 2010; Pietrobon and Moskowitz, 2013). Furthermore, this study demonstrated that miR-125a-3p negatively regulates p38 alpha gene mRNA expression and exhibits positive correlation with the rat head withdrawal thresholds (Dong et al., 2014).

Several studies have investigated the potential role of miRNAs in osteoarthritis by using different surrogate models less generic than the CFA models to study the miRNA changes associated with the pathogenesis of this prevalent and painful condition (Li et al., 2011, 2013b). The few examples selected in this section focus on miRNA alterations of the nervous system as opposed to various peripheral nonneuronal cell types such as chondrocytes and synoviocytes, also known to be involved in the disease etiology. Li et al. (2011) used the monosodium iodoacetate OA model, subsequently analyzing miRNA expression in both DRG and SDH (Li et al., 2011). Surprisingly, in the rodent model of OA, a steep decrease in miR-146a expression in both DRG and SDH was seen contrasting the known upregulation in peripherally affected cells of the knee joint. Elaborating on this discrepancy, the influence of miR-146a was investigated by transient transfection of miR-146a in both astrocytes and microglia. This attempt resulted in a significant decrease in proinflammatory products such as COX2, TNF$\alpha$, nitric oxide synthase, and interleukin 6, indicating that miRNA alterations could also be of a protective nature (Li et al., 2011). In a surgical animal model of OA analyzing miRNA expression in both DRG and SDH, a significant reduction in the expression of miR-146a and miR-183 was found (Li et al., 2013b). This finding concurred with a steep increase in mRNA levels of several of proinflammatory molecules such as TNF$\alpha$, IL-1$\beta$, and NF$\kappa\beta$ but actual mRNA target validation was not conducted. The study conducted a series of loss/gain-of-function experiments on microglia and astrocytes supporting the previously presented notion of miR-146a to be a key regulator in inflammation (Li et al., 2011, 2013b). Along these lines, miR-124 has been found to be centrally downregulated in two studies applying intraplantar IL-1β injection and formalin application, to induce peripheral inflammatory hyperalgesia. Interestingly, miR-124 was validated to target the mRNA transcript of MeCP2, previously shown to play a role in inflammatory nociception (Willemen et al., 2012; Kynast et al., 2013b). Further experiments with miR-124 showed that it is capable of modulating microglia reactivity by promoting M1/M2 switch, whereby it could modulate prolonged central sensitization through an essentially nonneuronal mechanism (Willemen et al., 2012). Also, through intravenous administration of miR-124, Kynast et al. (2012) successfully decreased pain behavior for a short period of the assessment time, while the opposite results were true for the administration of anti-miR-124 (Kynast et al., 2013b). Other established centrally acting pain-related miRNAs include miR-155 and miR-233, both of which are upregulated in the prefrontal cortex of rats after carrageenan-induced inflammation (Poh et al., 2011).

Recently, visceral inflammation (a subcomponent of conditions affecting a currently increasing number of patients suffering from, e.g., painful bladder syndrome (Davis et al., 2014) and inflammatory bowel diseases (Lucendo and Hervías, 2014)) have been assessed for potential miRNA contribution (Gheinani et al., 2013; Sengupta et al., 2013). Most significantly, a study investigated alterations in miRNA expression in the developing spinal cord following induced neonatal cystitis in rodents by application of zymosan. Here, a steep increase in miR-181a transcript was revealed alongside suppression of $GABA_{A\alpha1}$ subunit expression. The investigators carried on to administer $GABA_A$ receptor agonist muscimol, which was unable to attenuate the visceromotor response to colon distension in rats with neonatal cystitis, but importantly remained effective in adult rats (Sengupta et al., 2013). The authors concluded that it is likely that miR-181a is directly involved in toning spinal cord inhibition, unmasking excitatory pathways to facilitate prolonged pain or even chronic pain and hyperalgesia (Sengupta et al., 2013). In a pioneering study by Park et al. (2014), extracellular miR-let-7b was demonstrated to induce inward currents in rat DRG through coupling between the nociceptive ion-channel TRPA1 and Toll-like receptor 7. The study elucidated this direct nociceptive effect of miR-let-7b *in vivo*, documenting that miR-let-7b is released from nociceptive neurons in response to stimulation and showed that administration of miR-let-7b inhibitor reduced in formalin-induced pain (Park et al., 2014). In summary, it is clear from numerous animal models that inflammatory insults cause peripheral as well as central alterations in miRNA expression leading to significant changes in pain processing. Reversal of specific inflammation-induced miRNA alterations could be therapeutically relevant from an analgesics' development perspective.

### 7.2.2.2   MicroRNA Alterations in Neuropathic Pain

Oppositely to inflammatory pain, neuropathic pain is characterized by being related to lesions of the somatosensory nervous system. As for sensitization, it is generally classified according to its origin as either being central or peripheral. It is estimated that around 8% of the Western population suffers from at least on type of pain-associated neuropathy with the largest patient subgroup being those with diabetic small-fiber neuropathy, a common complication of diabetes (Torrance et al., 2006). A number of studies have investigated changes in miRNA expression in different rodent models and have found a significant dysregulation of numerous miRNAs (see Table 7.1), many of which subsequently shown to have functional implications in pain behavior. Genda et al. (2013) found the expression of 111 miR-NAs to be altered in the SDH after chronic constriction injury (CCI) in mice, however, no functional characterization of the dysregulated miRNAs were performed (Genda et al., 2013). As a notable illustration of the lacking consolidation in this research field, a recent study, also using CCI, was unable to show any significant difference between CCI- and sham-operated mice (Brandenburger et al., 2012), in an experimental design similar to that of Genda et al. (2013). Utilizing a sciatic nerve injury model in mice, differential expression of miRNA in DRG neurons compared to scam injured mice was investigated (Wu et al., 2011). Among several significant miRNA dysregulations, a pronounced increase in miR-142-5p and miR-21 expression was found, in line with results obtained from the cerebral cortex of rats after traumatic brain injury, indicating that the alterations are likely reflective of neural regeneration and not pain per se (Lei et al., 2009). In an SNL model at L5, 63, miRNAs were found to exhibit altered expression, 59 of which were similarly dysregulated in the ipsilateral L4 DRG neurons, suggesting that miRNA changes in the uninjured primary afferents adjacent to the damaged neurons may also contribute to the development and continuation of neuropathic pain and highlighting the need for a separate control group. The study did not elaborate further on these discovered miRNA alterations, nor their implications for pain processing upon ligation injury (von Schack et al., 2011). A recent study also using nerve ligation model utilized a genome-wide miRNA-sequencing approach in two genetically modified rat strains displaying opposite responses to neuropathic injuries; a pain-susceptible and a pain-resistant strain. The study discovered three miRNAs that were highly downregulated in the pain-susceptible strain and subsequently applied *in silico* target prediction, but not luciferase validation, to associate the miRNAs: rno-miR-30d-5p and rno-miR-125b-5p with the pain-related genes TNFα and BDNF (Bali and Kuner, 2014). In addition to the peripheral miRNA expressional changes in response to neuropathic pain, several studies have assessed neuropathy-induced miRNA deregulation at the CNS level. In a study using CCI in mice, the hippocampus was analyzed for miRNA expression yielding 51 miRNAs that were found to be significantly altered compared with sham surgery on the contralateral side. With complementary qRT-PCR, the study investigated the downregulation of miRNAs: miR-125b and -132 at days 7 and 15 postinjury but conducted no miRNA-mRNA target validation to elaborate the findings (Arai et al., 2013). Similarly, Imai et al. (2011) found that sciatic nerve lesion in mice induced a significant downregulation of both miR-200b and -429 in the postsynaptic neurons of the nucleus accumbens and demonstrated that the downregulation corresponded with an increased levels of DNA(cytosine-5)-methyltransferase 3A (DNMT3a). The authors proposed that the increase in DNMT3a associated with diminished expression of miR-200b and -429 is a factor in the generation of dysfunctional mesolimbic motivation/valuation circuitry involved in linking prolonged nociceptive stimuli with comorbid conditions such as sleep disturbances and mood disorders (Imai et al., 2011). In a recent study, using a bilateral CCI model followed by miRNA microarray analysis on tissue from DRG, SDH, hippocampus, and anterior cingulate cortex, a significant upregulation of miR-341 expression was observed at the DRG level, while a downregulation of miR-181a-3-p, -203, and -541-3p was seen in the SDH, demonstrating that miRNA alterations are clearly dissimilar between the CNS and the PNS in response to the same neuropathic injury (Li et al., 2013a). In contrast, Arai et al. (2013) showed that there were no significant miRNA alterations in the hippocampus in response to injury, representing another considerable discrepancy in the literature (Li et al., 2013a). The study did not perform functional characterization of the differentially expressed miRNAs (Li et al., 2013a). Several studies have applied an interventional approach to explore the significance of miRNA-induced alterations in the DRG and SDH in response to neuropathic pain. This is generally achieved by establishing a miRNA of suspected importance in pain or hyperalgesia that exhibits altered expression upon injury and applying miR mimics or anti-miRs to reverse the dysregulation. This constitutes a form of targeted genetic therapy, which ultimately hinders the expression of specific proteins by posttranscriptional mRNA repression, thus decreasing activity in the nociceptive system (Bartel, 2009; Niederberger et al., 2011; Im et al., 2012; Kynast et al., 2013b; Andersen et al., 2014). In a model of painful small-fiber diabetic neuropathy using streptozotocin in rats, a miRNA-construct targeting $Na_v\alpha$-subunit was delivered by herpes simplex virus-based vector leading to a successful reversing of thermal and mechanical hyperalgesia (Chattopadhyay et al., 2012). In another comprehensive study by Im et al. (2013), intrathecal administration of miR-23b was shown to alleviate both mechanical and thermal hyperalgesia in a mice model of traumatic spinal injury. Interestingly, the study also demonstrated that miR-23b targets NADPH oxidase 4, a known inflammation-promoting enzyme that also decreases production of the main inhibitory neurotransmitter GABA (Im et al., 2012). Along these lines, miR-124 (which is highly expressed in quiescent microglia) has been shown to shift microglial activity toward the anti-inflammatory phenotype known as M2 and hereby decreases hyperalgesia and pain behavior in a spared nerve mice model, thus highlighting the potential applicability of this miRNA as a therapeutic opportunity in response to both inflammatory and neuropathic pain conditions (Willemen et al., 2012). Recently, Chen et al. (2014) demonstrated that CCI leads to a steep decrease of miR-96 expression in DRG neurons, coinciding with an increase of $Na_v1.3$. This ion-channel implicated in neuropathic hyperexcitability and was a predicted mRNA target of miR-96. Subsequently, the study showed that intrathecal administration of miR-96 was efficient in alleviating both mechanical and thermal hyperalgesia caused by the CCI (Chen et al., 2014). It is clear that miRNAs have a potential as future therapeutic entities within neuropathic and inflammatory pain conditions and further studies on the functional aspects of pain-associated miRNA in combination with exploration of delivery methods could hold promise for future targeted therapy of pain (Andersen et al., 2014; Bali et al., 2014; Lutz et al., 2014). Figure 7.2 highlights selected miRNA alterations and their associated targets occurring at various levels in the nervous system in response to inflammatory and neuropathic pain conditions.

(d)        MicroRNAs in the brain

| Condition | MicroRNA | Gene target |
|---|---|---|
| Inflammatory pain | MiR-200b, MiR-429↓ | DNMT3a |
| Neuropathic pain | MiR-155↓ | C/EBP*I |

(a)     MicroRNAs in circulation

| Condition | MicroRNA | Sample tissue |
|---|---|---|
| CRPS | 18-miRNA signature | Whole blood |
| Fibromyalgia | 9-miRNA signature | Cerebrospinal fluid |
| IBS | MiR-29a↑ | Plasma |

(b)        MicroRNAs in DRGs

| Condition | MicroRNA | Gene target |
|---|---|---|
| Inflammatory pain | MiR-134↑ | MORl |
| Neuropathic pain | MiR-183↓ | CACNAID* |

(c)        MicroRNAs in the SDH

| Condition | MicroRNA | Gene target |
|---|---|---|
| Inflammatory pain | MiR-181a↑ | GABRAl |
|  | MiR-124↓ | MECP2 |
| Neuropathic pain | MiR-29a↓ | CACNAlC |
|  | MiR-23b↓ | NOX4 |

**FIGURE 7.2** The levels at which miRNA alterations occur in response to induced pain and painful disorders. Tables list examples of well-investigated miRNAs and corresponding mRNA targets. MiRNA modulation occurs in: (a) the circulatory system, (b) DRG neurons, (c) SDH, glial cells, and in several brain areas (d). DRG = dorsal root ganglion and SDH = spinal dorsal horn. Asterisk (*) = not validated with a luciferase-reporter assay. The up/down arrows denote miRNA alterations. (From Andersen, H. H., Duroux, M., and Gazerani, P. 2014. *Neurobiol Dis*, 71, 159–168. With permission.)

## 7.3   ROLE OF NeurimmiRs AS MODULATORS OF INFLAMMATION

Cellular and molecular networks that regulate pain and inflammation have been studied extensively in the recent years (see Figure 7.3). Protein regulators of inflammation have been the major focus, however, as mentioned earlier, recent evidence also demonstrates an essential role of miRNAs, in managing certain features of the inflammatory process (O'Connell et al., 2012). Dysregulation of miRNA expression

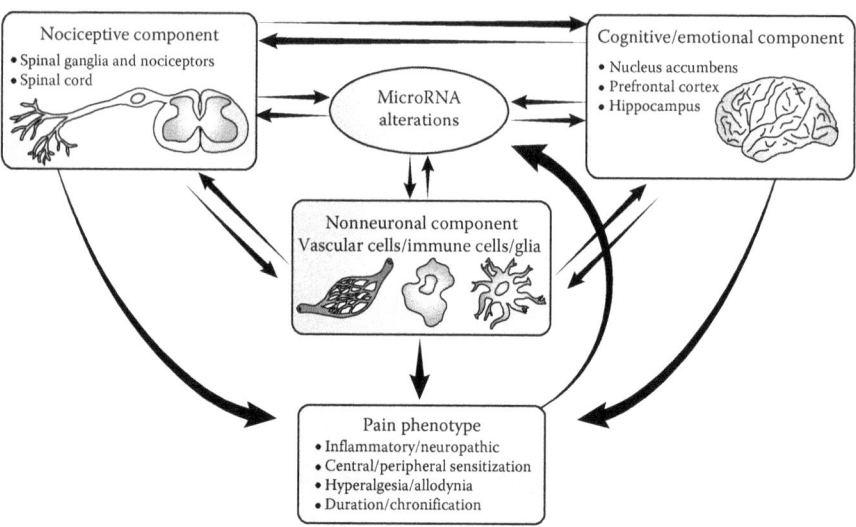

**FIGURE 7.3**  Alterations in miRNA expression in neuronal and nonneuronal compartments are inter-related and associated with pain phenotype and perception.

has been reported in variety of neuropathic and inflammatory pain models (Andersen et al., 2014). Specific patterns of circulating miRNA can be identified to improve diagnosis, patient stratification, mechanism-based treatment, and targeted prevention strategies for high-risk individuals in the response to pain or analgesics (Bali et al., 2014). Those miRNAs that modulate both neuronal and immune functions in pain processing are known as "neurimmiRs," which possibly target genes controlling both immune and neuronal components of pain processing (neuroinflammatory processes; Soreq and Wolf, 2011; Kress et al., 2013). Considering the fact that pain is a multidimensional phenomenon, the role of miRNAs in cognitive, emotional, and behavioral components of pain is also acknowledged (Kress et al., 2013). NeurimmiRs are found both in the CNS (e.g., brain) and peripheral tissues (e.g., immune system cells) and demonstrate that a crosstalk exists between these two compartments. Changes in neurimmiR levels following peripheral or central immune insults may influence or alter neuronal functions either directly by suppressing genes within neurons or indirectly by influencing the functioning of immune or glial cells. Similarly, neuronal insults may influence the immune cells responses. Accumulating evidence supports that neuroimmune alterations in the peripheral and CNS contribute in pathogenesis of neurogenic and neuropathic pain (Watkins et al., 2003; McMahon and Malcangio, 2009; Kuner, 2010). One example is neuroinflammation in complex regional pain syndrome (CRPS). Although psychological aspects are not neglectable, recent evidence has demonstrated that an autoimmune reaction in the peripheral and CNS may induce or exacerbate CRPS (Cooper and Clark, 2013). Elevated levels of proinflammatory cytokines (e.g., IL-1β and IL-6) have been found in cerebrospinal fluid of these patients, while anti-inflammatory cytokines (e.g., IL-4 and IL-10) are reduced (Alexander et al., 2005). In peripheral tissues (e.g., affected limbs in CPRS patients), proinflammatory cytokines have also been found (Schinkel et al., 2006). It has been shown that transcutaneous electrical stimulation exacerbates the condition proposing the presence of neurogenic inflammation (Birklein and Schmelz, 2008). Circulatory miRNA levels are also altered in CRPS patients—see the section below (Orlova et al., 2011).

## 7.4 MicroRNAs AS POTENTIAL BIOMARKERS IN PATIENTS WITH PAIN CONDITIONS

When addressing miRNA alterations in clinical pain conditions, it is important to state that the source of the miRNA dysregulation might not be neurons. So while features of chronic pain conditions certainly involves derangements of the nociceptive neurons these are not necessarily reflected on an extracellular or nonneuronal level. Since tissue samples cannot normally be drawn from the DRG, let alone the SDH of human patients, miRNA alterations manifesting in these cohorts are typically found in serum, whole blood or nonneuronal biopsies. This makes it difficult to interpret the alterations from a mechanistic perspective, since they may merely be unrelated epiphenomena to the specific pathogenesis and their cellular origin is often unknown. However, for example, serum miRNA alterations may serve as putative venues for discovery of biomarkers where a traditional protein-based search of such has failed. Moreover, in light of the recent study by Park et al. (2014), demonstrating a direct receptor-mediated role for let-7b, it is not unconceivable that extracellular miRNAs could have physiological relevance, just as their intracellular counterparts (Park et al., 2014). The concept of using miRNAs derived from biofluids as prognostic and/ or diagnostic biomarkers are well progressed in various fields, particularly neurodegeneration, oncology, and cardiology; however, the idea is fairly recent with respect to pain conditions (Ahmad et al., 2013; Kinet et al., 2013). Some advantages of using miRNAs as biomarkers are their presence in virtually all biofluids, their superior stability to mRNA and their apparent sensitivity (Jung et al., 2010; Kemppainen et al.; Tomaselli et al., 2012). As previously referred to, one of the first studies attempting to correlate the expression of circulating miRNA to a pain conditions was conducted by Orlova et al. (2011). Here, whole blood samples were taken from 41 patients suffering from complex regional pain syndrome (CRPS) and analyzed for miRNA expression, cytokines and numerous clinical parameters, upon which the data were compared to that of 20 healthy matched controls. As expected, cytokines such as CCL2 and VEGF were notably elevated in the CRPS group compared to control and a CRPS-miRNA signature was found. Interestingly, extensive correlational analyses revealed that four of the dysregulated miRNAs were significantly and positively correlated with CRPS-associated pain level, miR-150 was correlated with the frequency of migraine attacks within the CRPS patient cohort and an extensive array of miR-NAs was found to correlate with the levels of circulating cytokines (Orlova et al., 2011). A recent study with a similar design investigated the miRNA profile of the cerebrospinal fluid in fibromyalgia patients and identified 10 miRNAs differentially expressed between affected patients and healthy controls. Most notably, the study found that decreased levels of miR-145-5p in the CSF were associated with reported symptomatology, that is, pain intensity and fatigue (Bjersing et al., 2013). The same group recently published a similar study identifying a number of serum-extracted circulatory miRNAs significantly correlated to fibromyalgia-related pain intensity (miR-103a-3p and miR-320a) and psychophysically assessed pain threshold (miR-374b-5p), (Bjersing et al., 2014). Despite the positive prospects of applying miRNAs as biofluid-derived biomarkers in pain conditions numerous obstacles adhere: (1) a number of issues are associated with sampling, particularly hemolysis and low isolation yields. (2) it is still not clear why and to which extent miRNAs are selectively exported (Boon and Vickers, 2013). (3) circulatory miRNA expression is influenced by age, gender, hormonal, and metabolic factors (Tomaselli et al., 2012).

## 7.5   CONCLUSION AND FUTURE PERSPECTIVES

Two potential aspects are acknowledged in potential application of miRNAs in pain: (1) Identification and validation of miRNAs as translational biomarkers of pain, where these biomarkers can be used as unique signatures for different pain conditions. Those will not only assist in elucidating underlying mechanisms of pain but also pave a new way toward improved prognosis, diagnosis, and patient stratification, which help in identification of population at risk or personalized medicine for prevention strategies, or follow up the curse of treatment or responsiveness to certain analgesics. (2) Application of miRNA as potential therapeutic targets, which can imply a mechanism-based treatment strategy or disease modifying approach instead of symptomatic management of pain. Despite a large beneficial value in these aspects, progress in miRNA in pain faces some challenges that call for scientific and practical solutions to overcome the current limitations in this field. Studies on the role of miRNA in pain processing have mainly been conducted in various animal pain models or in fewer cases in pain patient cohorts, which pose some inherent limitations as to the translatability between results. Using human surrogate pain models have been vastly beneficial in improving understanding of pain processing by temporarily mimicking concise pain symptoms characteristic for specific clinical conditions, such as mechanical and heat hyperalgesia or cold allodynia (Arendt-Nielsen and Yarnitsky, 2009; Andersen et al., 2013). Hence, studies mapping the human local and systemic miRNA alterations in surrogate translational models (e.g., capsaicin, NGF, and UVB models; Arendt-Nielsen and Yarnitsky, 2009) could be highly beneficial to substantiate the knowledge obtained from preclinical pain models. One should consider though the translatability of the findings from human models of pain into patient population as many of the current models are short lasting due to the ethical considerations and only some aspects of pain can be investigated. Identification and validation of the miRNA biomarkers depend on future comparable studies to yield a valid and reliable biomarker, which can be selective and specific to an acceptable level.

The idea behind development of miRNA and miR-mimics as potential therapeutic agents for pain management is highly valuable. Administration of apparently analgesic anti-miRs and miR-mimics has already been successfully carried out in animal models of pain, however, it remains necessary to increase the current knowledge on how pain-related miRNAs function during various normal and pathological pain conditions in humans before any clinically viable diagnostic or therapeutic utilization can be accomplished (Chattopadhyay et al., 2012; Im et al., 2012; Willemen et al., 2012; Chen et al., 2014). An additional problem is represented by the challenge of distributing miRNA-mimics or anti-miRNAs to specific target tissue, an objective that become particularly difficult when it comes to targeting tissue situated behind the blood–brain barrier. This would be a prerequisite to develop a centrally acting miRNA-based analgesic and this challenges the formulation and delivery of such compound. Novel results indicate that exosomes might be of great potential in this regard as they are already known to carry RNAs possibly mediating cell-to-cell communication (Chen et al., 2012; Johnsen et al., 2014). In a pioneering study by Alvarez-Erviti et al. (2011), modified exosomes expressing a neuron-specific targeting surface peptide was delivered to mice by systemic injection. This enabled cell-specific delivery of the siRNA-cargo to cross the blood–brain barrier (Alvarez-Erviti et al., 2011). There seems to be a long way ahead in terms of identifying the most optimal delivery, safety, and efficacy of novel miRNA-based therapeutic approaches. Based on the available data and the exponentially increase in new studies, the future in this field looks bright.

In conclusion, miRNAs are emerging as important players involved in the processing of inflammatory and neuropathic pain. The small noncoding RNAs molecules represent a promising venue for biomarker discovery and targeted therapy in a variety of human diseases, including pain conditions. However, being a field in its infancy, significant bench-to-bedside hindrances warrant additional studies before therapeutic or biomarker miRNAs can be added to the toolbox of monitoring and treating pain patients.

## REFERENCES

Ahmad, J., Hasnain, S. E., Siddiqui, M. A., Ahamed, M., Musarrat, J., and Al-Khedhairy, A. A. 2013. MicroRNA in carcinogenesis and cancer diagnostics: A new paradigm. *Indian J Med Res*, 137, 680–94.

Akhtar, N. and Haqqi, T. M. 2012. MicroRNA-199a* regulates the expression of cyclooxygenase-2 in human chondrocytes. *Ann Rheum Dis*, 71, 1073–80.

Aldrich, B. T., Frakes, E. P., Kasuya, J., Hammond, D. L., and Kitamoto, T. 2009. Changes in expression of sensory organ-specific microRNAs in rat dorsal root ganglia in association with mechanical hypersensitivity induced by spinal nerve ligation. *Neuroscience*, 164, 711–23.

Alexander, G. M., Van Rijn, M. A., Van Hilten, J. J., Perreault, M. J., and Schwartzman, R. J. 2005. Changes in cerebrospinal fluid levels of pro-inflammatory cytokines in CRPS. *Pain*, 116, 213–9.

Alvarez-Erviti, L., Seow, Y., Yin, H., Betts, C., Lakhal, S., and Wood, M. J. A. 2011. Delivery of siRNA to the mouse brain by systemic injection of targeted exosomes. *Nat Biotechnol*, 29, 341–5.

Andersen, H. H., Duroux, M., and Gazerani, P. 2014. MicroRNAs as modulators and biomarkers of inflammatory and neuropathic pain conditions. *Neurobiol Dis*, 71, 159–168.

Andersen, H. H., Olsen, R. V., Møller, H. G., Eskelund, P. W., Gazerani, P., and Arendt-Nielsen, L. 2013. A review of topical high-concentration L-menthol as a translational model of cold allodynia and hyperalgesia. *Eur J Pain*, 18, 315–325.

Andersen, H. H., Duroux, M., and Gazerani, P. 2015. Serum microRNA signatures in migraineurs during attacks and in pain-free periods. *Mol Neurobiol*, [Epub ahead of print].

Arai, M., Genda, Y., and Ishikawa, M. 2013. The miRNA and mRNA changes in rat hippocampi after chronic constriction injury. *Pain Med*, 14, 720–729.

Arendt-Nielsen, L. and Yarnitsky, D. 2009. Experimental and clinical applications of quantitative sensory testing applied to skin, muscles and viscera. *J Pain*, 10, 556–72.

Bai, G., Ambalavanar, R., Wei, D., and Dessem, D. 2007. Downregulation of selective microRNAs in trigeminal ganglion neurons following inflammatory muscle pain. *Mol Pain*, 3, 15.

Bali, K. K., Hackenberg, M., Lubin, A., Kuner, R., and Devor, M. 2014. Sources of individual variability: MiRNAs that predispose to neuropathic pain identified using genome-wide sequencing. *Mol Pain*, 10, 22.

Bali, K. K. and Kuner, R. 2014. Noncoding RNAs: Key molecules in understanding and treating pain. *Trends Mol Med*, 1–12.

Bartel, D. P. 2009. MicroRNAs: Target recognition and regulatory functions. *Cell*, 136, 215–233.

Basbaum, A. I. 1999. Spinal mechanisms of acute and persistent pain. *Reg Anesth Pain Med*, 24, 59–67.

Basbaum, A. I., Bautista, D. M., Scherrer, G., and Julius, D. 2009. Cellular and molecular mechanisms of pain. *Cell*, 139, 267–84.

Birklein, F. and Schmelz, M. 2008. Neuropeptides, neurogenic inflammation and complex regional pain syndrome (CRPS). *Neurosci Lett*, 437, 199–202.

Bjersing, J. L., Bokarewa, M. I., and Mannerkorpi, K. 2014. Profile of circulating microRNAs in fibromyalgia and their relation to symptom severity: An exploratory study. *Rheumatol Int*.

Bjersing, J. L., Lundborg, C., Bokarewa, M. I., and Mannerkorpi, K. 2013. Profile of cerebrospinal microRNAs in fibromyalgia. *PLoS One*, 8, e78762.

Boon, R. A. and Vickers, K. C. 2013. Intercellular transport of microRNAs. *Arterioscler Thromb Vasc Biol*, 33, 186–92.

Brandenburger, T., Castoldi, M., Brendel, M. et al. 2012. Expression of spinal cord microRNAs in a rat model of chronic neuropathic pain. *Neurosci Lett*, 506, 281–286.

Breivik, H., Collett, B., Ventafridda, V., Cohen, R., and Gallacher, D. 2006. Survey of chronic pain in Europe: Prevalence, impact on daily life, and treatment. *Eur J Pain*, 10, 287–333.

Breivik, H., Eisenberg, E., and O'brien, T. 2013. The individual and societal burden of chronic pain in Europe: The case for strategic prioritisation and action to improve knowledge and availability of appropriate care. *BMC Public Health*, 13, 1229.

Chattopadhyay, M., Zhou, Z., Hao, S., Mata, M., and Fink, D. J. 2012. Reduction of voltage gated sodium channel protein in DRG by vector mediated miRNA reduces pain in rats with painful diabetic neuropathy. *Mol Pain*, 8, 17.

Chen, H.-P., Zhou, W., Kang, L.-M. et al. 2014. Intrathecal miR-96 Inhibits Nav1.3 expression and alleviates neuropathic pain in rat following chronic construction injury. *Neurochem Res*, 76–83.

Chen, X., Liang, H., Zhang, J., Zen, K., and Zhang, C.-Y. 2012. Secreted microRNAs: A new form of intercellular communication. *Trends Cell Biol*, 22, 125–132.

Cooper, M. S. and Clark, V. P. 2013. Neuroinflammation, neuroautoimmunity, and the comorbidities of complex regional pain syndrome. *J Neuroimmune Pharmacol*, 8, 452–69.

Cutrer, F. M. 2010. Pathophysiology of migraine. *Semin Neurol*, 30, 120–30.

Dahan, A., Van Velzen, M., and Niesters, M. 2014. Comorbidities and the complexities of chronic pain. *Anesthesiology*, 121, 675–7.

Davis, N. F., Brady, C. M., and Creagh, T. 2014. Interstitial cystitis/painful bladder syndrome: Epidemiology, pathophysiology and evidence-based treatment options. *Eur J Obstet Gynecol Reprod Biol*, 175, 30–7.

Denk, F., Mcmahon, S. B., and Tracey, I. 2014. Pain vulnerability: A neurobiological perspective. *Nat Neurosci*, 17, 192–200.

Dong, Y., Li, P., Ni, Y., Zhao, J., and Liu, Z. 2014. Decreased MicroRNA-125a-3p contributes to upregulation of p38 MAPK in rat trigeminal ganglions with orofacial inflammatory pain. *PLoS One*, 9, e111594.

Dworkin, R. H. 2012. Mechanism-based treatment of pain. *Pain*, 153, 2300.

Edvardsson, B. and Persson, S. 2012. Cluster headache and parietal glioblastoma multiforme. *The Neurologist*, 18, 206–7.

Favereaux, A., Thoumine, O., Bouali-Benazzouz, R. et al. 2011. Bidirectional integrative regulation of Cav1.2 calcium channel by microRNA miR-103: Role in pain. *EMBO J*, 30, 3830–41.

Friedman, R. C. and Farh, K. K. H. 2009. Most mammalian mRNAs are conserved targets of microRNAs. *Genome Res*, 19, 92–105.

Genda, Y., Arai, M., Ishikawa, M., Tanaka, S., Okabe, T., and Sakamoto, A. 2013. MicroRNA changes in the dorsal horn of the spinal cord of rats with chronic constriction injury: A TaqMan® low density array study. *Int J Mol Med*, 31, 129–37.

Gerdle, B., Björk, J., Henriksson, C., and Bengtsson, A. 2004. Prevalence of current and chronic pain and their influences upon work and healthcare-seeking: A population study. *J Rheumatol*, 31, 1399–406.

Gheinani, A. H., Burkhard, F. C., and Monastyrskaya, K. 2013. Deciphering microRNA code in pain and inflammation: Lessons from bladder pain syndrome. *Cell Mol Life Sci*, 70, 3773–89.

Gold, M. S. and Gebhart, G. F. 2010. Nociceptor sensitization in pain pathogenesis. *Nat Med*, 16, 1248–57.

Goldberg, D. S. and Mcgee, S. J. 2011. Pain as a global public health priority. *BMC Public Health*, 11, 770.

He, Y. and Wang, Z. J. 2012. Let-7 microRNAs and opioid tolerance. *Frontiers in Genetics*, 3, 110.

Im, Y. B., Jee, M. K., Choi, J. I., Cho, H. T., Kwon, O. H., and Kang, S. K. 2012. Molecular targeting of NOX4 for neuropathic pain after traumatic injury of the spinal cord. *Cell Death Dis*, 3, e426.

Imai, S., Saeki, M., Yanase, M. et al. 2011. Change in microRNAs associated with neuronal adaptive responses in the nucleus accumbens under neuropathic pain. *J Neurosci*, 31, 15294–9.

Jancalek, R. 2011. Signaling mechanisms in mirror image pain pathogenesis. *Ann Neurosci*, 18, 123–127.

Ji, R.-R., Berta, T. and Nedergaard, M. 2013. Glia and pain: Is chronic pain a gliopathy? *Pain*, 154 Suppl, S10–28.

Ji, R.-R., Kohno, T., Moore, K. A., and Woolf, C. J. 2003. Central sensitization and LTP: Do pain and memory share similar mechanisms? *Trends Neurosci*, 26, 696–705.

Johnsen, K. B., Gudbergsson, J. M., Skov, M. N., Pilgaard, L., Moos, T., and Duroux, M. 2014. A comprehensive overview of exosomes as drug delivery vehicles—Endogenous nanocarriers for targeted cancer therapy. *Biochim Biophys Acta*, 1846, 75–87.

Jung, M., Schaefer, A., Steiner, I. et al. 2010. Robust MicroRNA stability in degraded RNA preparations from human tissue and cell samples. *Clin Chem*, 56, 998–1006.

Kempppainen, J., Shelton, J., Kelnar, K. et al. 2006. MicroRNAs as biomarkers in blood and other biofluids—Commercial publication. *AsuraGen*, 1–1.

Kinet, V., Halkein, J., Dirkx, E., and Windt, L. J. D. 2013. Cardiovascular extracellular microRNAs: Emerging diagnostic markers and mechanisms of cell-to-cell RNA communication. *Front Genet*, 4, 214.

Kress, M., Hüttenhofer, A., Landry, M. et al. 2013. MicroRNAs in nociceptive circuits as predictors of future clinical applications. *Front Mol Neurosci*, 6, 33–33.

Kuner, R. 2010. Central mechanisms of pathological pain. *Nat Med*, 16, 1258–66.

Kusuda, R., Cadetti, F., Ravanelli, M. I. et al. 2011. Differential expression of microRNAs in mouse pain models. *Mol Pain*, 7, 17.

Kynast, K. L., Russe, O. Q., Geisslinger, G., and Niederberger, E. 2013a. Novel findings in pain processing pathways: Implications for miRNAs as future therapeutic targets. *Expert Rev Neurother*, 13, 515–25.

Kynast, K. L., Russe, O. Q., Möser, C. V., Geisslinger, G., and Niederberger, E. 2013b. Modulation of central nervous system–specific microRNA-124a alters the inflammatory response in the formalin test in mice. *Pain*, 154, 368–76.

Latremoliere, A. and Woolf, C. J. 2009. Central sensitization: A generator of pain hypersensitivity by central neural plasticity. *J Pain*, 10, 895–926.

Lei, P., Li, Y., Chen, X., Yang, S., and Zhang, J. 2009. Microarray based analysis of microRNA expression in rat cerebral cortex after traumatic brain injury. *Brain Res*, 1284, 191–201.

Leresche, L. 2011. Defining gender disparities in pain management. *Clin Orthop Relat Res*, 469, 1871–7.

Li, H., Huang, Y., Ma, C., Yu, X., Zhang, Z., and Shen, L. 2014. MiR-203 involves in neuropathic pain development and represses Rap1a expression in nerve growth factor differentiated neuronal PC12 cells. *Clin J Pain*, 31, 36–43.

Li, H., Shen, L., Ma, C., and Huang, Y. 2013a. Differential expression of miRNAs in the nervous system of a rat model of bilateral sciatic nerve chronic constriction injury. *Int J Mol Med*, 32, 219–26.

Li, X., Gibson, G., Kim, J.-S. et al. 2011. MicroRNA-146a is linked to pain-related pathophysiology of osteoarthritis. *Gene*, 480, 34–41.

Li, X., Kroin, J. S., Kc, R. et al. 2013b. Altered spinal microRNA-146a and the microRNA-183 cluster contribute to osteoarthritic pain in knee joints. *J Bone Miner Res*.

Liu, F. and Yuan, H. 2014. Role of glia in neuropathic pain. *Front Biosci*, 19, 798–807.

Liu, N.-K., Wang, X.-F., Lu, Q.-B., and Xu, X.-M. 2009. Altered microRNA expression following traumatic spinal cord injury. *Exp Neurol*, 219, 424–9.

Lohman, D., Schleifer, R., and Amon, J. J. 2010. Access to pain treatment as a human right. *BMC Med*, 8, 8.

Lucendo, A. J. and Hervías, D. 2014. Epidemiology and temporal trends (2000–2012) of inflammatory bowel disease in adult patients in a central region of Spain. *Eur J Gastroenterol Hepatol*, 26, 1399–407.

Lutz, B. M., Bekker, A., and Tao, Y.-X. 2014. Noncoding RNAs: New players in chronic pain. *Anesthesiology*, 121, 409–417.

Mcmahon, S. B. and Malcangio, M. 2009. Current challenges in glia-pain biology. *Neuron*, 64, 46–54.

Merskey, H. 1979. Pain terms: A list of definitions and notes on usage. *Pain*, 6, 247–252.

Merskey, H. and Bogduk, N. 1986. Classification of chronic pain. Descriptions of chronic pain syndromes and definitions of pain terms. Prepared by the International Association for the Study of Pain, Subcommittee on Taxonomy. *Pain Suppl*, 3, S1–226.

Møller, H. G., Rasmussen, A. P., Andersen, H. H., Johnsen, K. B., Henriksen, M., and Duroux, M. 2013. A systematic review of microRNA in glioblastoma multiforme: Micromodulators in the mesenchymal mode of migration and invasion. *Mol Neurobiol*, 47, 131–44.

Nakanishi, K., Nakasa, T., Tanaka, N. et al. 2010. Responses of microRNAs 124a and 223 following spinal cord injury in mice. *Spinal Cord*, 48, 192–6.

Ni, J., Gao, Y., Gong, S., Guo, S., Hisamitsu, T., and Jiang, X. 2013. Regulation of μ-opioid type 1 receptors by microRNA134 in dorsal root ganglion neurons following peripheral inflammation. *Eur J Pain*, 17, 313–23.

Niederberger, E., Kynast, K., Lötsch, J., and Geisslinger, G. 2011. MicroRNAs as new players in the pain game. *Pain*, 152, 1455–8.

O'connell, R. M., Rao, D. S., and Baltimore, D. 2012. MicroRNA regulation of inflammatory responses. *Annu Rev Immunol*, 30, 295–312.

Orlova, I. A., Alexander, G. M., Qureshi, R. A. et al. 2011. MicroRNA modulation in complex regional pain syndrome. *J Transl Med*, 9, 195.

Ossipov, M. H., Dussor, G. O., and Porreca, F. 2010. Central modulation of pain. 120. *J Clin Invest*, 120, 3779–87.

Osterberg, A. and Boivie, J. 2010. Central pain in multiple sclerosis—Sensory abnormalities. *Eur J Pain*, 14, 104–10.

Pan, Z., Zhu, L. J., Li, Y. Q. et al. 2014. Epigenetic modification of spinal miR-219 expression regulates chronic inflammation pain by targeting CaMKII. *J Neurosci*, 34, 9476–9483.

Park, C.-K., Xu, Z.-Z., Berta, T. et al. 2014. Extracellular microRNAs activate nociceptor neurons to elicit pain via TLR7 and TRPA1. *Neuron*, 82, 47–54.

Park, S. J., Cheon, E. J., and Kim, H. A. 2013. MicroRNA-558 regulates the expression of cyclooxygenase-2 and IL-1β-induced catabolic effects in human articular chondrocytes. *Osteoarthritis Cartilage*, 21, 981–9.

Pietrobon, D. and Moskowitz, M. A. 2013. Pathophysiology of migraine. *Annu Rev Physiol*, 75, 365–91.

Poh, K.-W., Yeo, J.-F., and Ong, W.-Y. 2011. MicroRNA changes in the mouse prefrontal cortex after inflammatory pain. *Eur J Pain*, 15, 801.e1–12.

Ponomarev, E. D., Veremeyko, T., Barteneva, N., Krichevsky, A. M., and Weiner, H. L. 2011. MicroRNA-124 promotes microglia quiescence and suppresses EAE by deactivating macrophages via the C/EBP-α-PU.1 pathway. *Nat Med*, 17, 64–70.

Reichling, D. B., Green, P. G., and Levine, J. D. 2013. The fundamental unit of pain is the cell. *Pain*, 154, S2–S9.

Reid, K. J., Harker, J., Bala, M. M. et al. 2011. Epidemiology of chronic non-cancer pain in Europe: Narrative review of prevalence, pain treatments and pain impact. *Curr Med Res Opin*, 27, 449–62.

Schinkel, C., Gaertner, A., Zaspel, J., Zedler, S., Faist, E., and Schuermann, M. 2006. Inflammatory mediators are altered in the acute phase of posttraumatic complex regional pain syndrome. *Clin J Pain*, 22, 235–239.

Scholz, J. and Woolf, C. J. 2007. The neuropathic pain triad: Neurons, immune cells and glia. *Nat Neurosci*, 10, 1361–8.

Sengupta, J. N., Pochiraju, S., Pochiraju, S. et al. 2013. MicroRNA-mediated GABA Aα-1 receptor subunit down-regulation in adult spinal cord following neonatal cystitis-induced chronic visceral pain in rats. *Pain*, 154, 59–70.

Soreq, H. and Wolf, Y. 2011. NeurimmiRs: MicroRNAs in the neuroimmune interface. *Trends Mol Med*, 17, 548–55.

Staud, R. 2013. The important role of CNS facilitation and inhibition for chronic pain. *Int J Clin Rheumtol*, 8, 639–46.

Swieboda, P., Filip, R., Prystupa, A., and Drozd, M. 2013. Assessment of pain: Types, mechanism and treatment. *Ann Agric Environ Med*, 1, 2–7.

Tam Tam, S., Bastian, I., Zhou, X. F. et al. 2011. MicroRNA-143 expression in dorsal root ganglion neurons. *Cell Tissue Res*, 346, 163–73.

Tomaselli, S., Panera, N., Gallo, A., and Alisi, A. 2012. Circulating miRNA profiling to identify biomarkers of dysmetabolism. *Biomark Med*, 6, 729–42.

Torrance, N., Smith, B. H., Bennett, M. I., and Lee, A. J. 2006. The epidemiology of chronic pain of predominantly neuropathic origin. Results from a general population survey. *J Pain*, 7, 281–9.

Tsang, A., Von Korff, M., Lee, S. et al. 2008. Common chronic pain conditions in developed and developing countries: Gender and age differences and comorbidity with depression-anxiety disorders. *J Pain*, 9, 883–91.

Vallejo, R., Tilley, D. M., Vogel, L., and Benyamin, R. 2010. The role of glia and the immune system in the development and maintenance of neuropathic pain. *Pain Practice*, 10, 167–84.

Vo, N., Klein, M. E., Varlamova, O. et al. 2005. A cAMP-response element binding protein-induced microRNA regulates neuronal morphogenesis. *Proc Natl Acad Sci USA*, 102, 16426–31.

Von Schack, D., Agostino, M. J., Murray, B. S. et al. 2011. Dynamic changes in the microRNA expression profile reveal multiple regulatory mechanisms in the spinal nerve ligation model of neuropathic pain. *PLoS One*, 6, e17670.

Watkins, L. R., Milligan, E. D., and Maier, S. F. 2003. Glial proinflammatory cytokines mediate exaggerated pain states: Implications for clinical pain. *Adv Exp Med Biol*, 521, 1–21.

Wayman, G. A., Davare, M., Ando, H. et al. 2008. An activity-regulated microRNA controls dendritic plasticity by down-regulating p250GAP. *Proc Natl Acad Sci USA*, 105, 9093–8.

Willemen, H. L. D. M., Huo, X.-J., Mao-Ying, Q.-L., Zijlstra, J., Heijnen, C. J., and Kavelaars, A. 2012. MicroRNA-124 as a novel treatment for persistent hyperalgesia. *J Neuroinflammation*, 9, 143.

Willems, L. M., Kwakkenbos, L., Leite, C. C. et al. 2014. Frequency and impact of disease symptoms experienced by patients with systemic sclerosis from five European countries. *Clin Exp Rheumatol*, 32 Suppl 8, 88–93.

Woolf, C. J. 2011. Central sensitization: Implications for the diagnosis and treatment of pain. *Pain*, 152, S2–15.

Wu, D., Raafat, A., Pak, E., Clemens, S., and Murashov, A. K. 2012. Dicer-microRNA pathway is critical for peripheral nerve regeneration and functional recovery *in vivo* and regenerative axonogenesis in vitro. *Exp Neurol*, 233, 555–65.

Wu, D., Raafat, M., Pak, E., Hammond, S., and Murashov, A. K. 2011. MicroRNA machinery responds to peripheral nerve lesion in an injury-regulated pattern. *Neuroscience*, 190, 386–97.

Wu, Q., Law, P.-Y., Wei, L.-N., and Loh, H. H. 2008. Post-transcriptional regulation of mouse mu opioid receptor (MOR1) via its 3' untranslated region: A role for microRNA23b. *FASEB J*, 22, 4085–95.

Xiao, J., Jing, Z.-C., Ellinor, P. T. et al. 2011. MicroRNA-134 as a potential plasma biomarker for the diagnosis of acute pulmonary embolism. *J Transl Med*, 9, 159.

Zhao, J., Lee, M.-C., Momin, A. et al. 2010. Small RNAs control sodium channel expression, nociceptor excitability, and pain thresholds. *J Neurosci*, 30, 10860–71.

# Section IV

MicroRNA Biomarkers
in Neurological Diseases
(Applications)

# 8 Circulating Cell-Free MicroRNAs as Biomarkers for Neurodegenerative Diseases

*Margherita Grasso, Francesca Fontana, and Michela A. Denti*

## CONTENTS

8.1 Introduction ...................................................................................................207
8.2 Biomarkers in Neurodegenerative Diseases .................................................210
8.3 MiRNA Biogenesis and Function.................................................................211
8.4 Circulating MiRNA.......................................................................................212
    8.4.1 Alzheimer's Disease and Circulating miRNAs................................214
    8.4.2 Parkinson's Disease and Circulating miRNAs.................................219
    8.4.3 Amyotrophic Lateral Sclerosis and Circulating miRNAs...............223
    8.4.4 Huntington's Disease and Circulating miRNAs...............................225
    8.4.5 Other Diseases .................................................................................225
8.5 Therapy..........................................................................................................226
References..............................................................................................................227

## 8.1 INTRODUCTION

Neurodegenerative diseases include a range of debilitating conditions of the central nervous system characterized by the progressive loss of neural tissues. More than 600 disorders afflict the nervous system, as hereditary or sporadic conditions that progressively cause neurodegeneration.

The main neurodegenerative diseases are

- Alzheimer's disease (AD) and other dementias (frontotemporal dementia, FTD)
- Parkinson's disease (PD) and PD-related disorders
- Huntington's disease (HD)
- Amyotrophic lateral sclerosis (ALS)

These diseases are multifactorial debilitating disorders of the nervous system accounting for a significant and increasing proportion of mortality in the world. All these neurodegenerative diseases are characterized by loss and death of neurons in specific brain areas and by different clinical manifestations.

Alzheimer disease is the most common cause of dementia. It is characterized by cortical atrophy and loss of neurons in the parietal and temporal lobes. It is a progressive neurodegenerative disorder characterized by memory impairment with executive dysfunction, motor problems, and/or language difficulties. AD is caused by mutations in three highly penetrant genes: amyloid precursor protein (*APP*), presenilin 1 (*PSEN1*), and presenilin 2 (*PSEN2*; Levy-Lahad et al., 1995), and one susceptibility gene (*APOE*; Strittmatter et al., 1993).

Frontotemporal dementia is characterized by atrophy of the brain that predominantly affects the frontal and temporal lobes, and is often used as a synonym for frontotemporal lobar degeneration (FTLD). It results in progressive behavioral changes and language dysfunction, but motor and cognitive impairment maybe present as well. Mutations in five genes have been linked to FTLD: microtubule-associated protein tau (*MAPT*; Hutton et al., 1998), progranulin (*GRN*; Baker et al., 2006), chromosome 9 open reading frame 72 (*C9orf72*; Renton et al., 2011), valosin-containing protein (*VCP*; Watts et al., 2004), and charged multivesicular body protein 2B (*CHMP2B*; Skibinski et al., 2005). Moreover, there are other two subtypes of FTLD called FTLD-TDP43, characterized by tau-negative inclusions containing aggregates of TDP-43 (TAR DNA-binding protein 43; Neumann et al., 2006) and FTLD-FUS, showing inclusions of FUS (fused in sarcoma) protein co-localized with ubiquitin-immunoreactive inclusions (Neumann et al., 2009).

Parkinson's disease is the second most common neurodegenerative disorder leading to loss of dopamine-producing brain cells and a progressive deterioration of motor function. It is characterized by symptoms such as tremor, stiffness, slowness, impaired balance, anxiety, depression, and dementia. With regard to PD, late-onset forms of disease show mutations in α-synuclein (*SNCA*; Polymeropoulos et al., 1996, 1997) and leucine-rich repeat kinase 2 (*LRRK2*; Funayama et al., 2002) genes, whereas early-onset forms of disease show Parkin (*PARK2*; Kitada et al., 1998), PTEN-induced putative kinase 1 (*PINK1*; Valente et al., 2001), oncogene DJ1 (*DJ1*; Van Duijn et al., 2001; Bonifati et al., 2003) mutations.

Huntington's disease is an autosomal dominant disorder associated with degeneration of the striatum. It is a genetic, progressive disorder characterized by the gradual development of involuntary muscle movements (chorea) and deterioration of cognitive processes and memory (dementia). Mutations in the Huntingtin (*HTT*) gene cause HD (MacDonald, 1993).

Amyotrophic lateral sclerosis is a progressive degenerative disorder affecting upper and lower motor neurons. Loss of motor neurons results in progressive loss of voluntary muscle movement, which in turn leads to muscle atrophy. Motor impairment may eventually affect respiratory systems but cognitive functions usually remain intact. ALS is caused by different mutations, such as mutations in superoxide dismutase 1 (*SOD1*; Rosen et al., 1993), in the genes coding for TDP-43 (*TARDBP*; Sreedharan et al., 2008), *FUS/TLS* (Sapp et al., 2003), and ubiquilin 2 (*UBQLN2*; Kaye and Shows, 2000). Moreover, mutations in *CHMP2B* (Parkinson et al., 2006),

*VCP* (Johnson et al., 2010), and *C9ORF72* (DeJesus-Hernandez et al., 2011; Renton et al., 2011) have also been implicated in ALS.

Although these are distinct pathologies in which neurodegeneration predominantly affects specific neuronal population (dopaminergic neurons in PD, striatal medium spiny neurons in HD, motor neurons in ALS, and cortical and hippocampal neurons in AD), there are converging lines of investigation showing a clinical, pathological, and genetic overlap. Some examples are the aggregation and deposition of misfolded proteins, dysfunction of RNA metabolism and processing, and protein homeostasis.

The accumulation of insoluble aggregates and deposition of misfolded proteins vary from disease to disease, such as amyloid plaques mainly constituted by β-amyloid protein (Aβ; Glenner and Wong, 1984) and neurofibrillary tangles composed of aggregates of hyperphosphorylated tau protein (Grundke-Iqbal et al., 1986) in AD, Lewy bodies formed by α-synuclein (Spillantini et al., 1997) in PD, intranuclear deposits of a polyglutamine-rich version of Huntingtin protein (DiFiglia et al., 1997) in HD, and aggregates mainly composed of superoxide dismutase (Bruijn et al., 1998) in ALS.

RNA processing is a very important step in cellular physiology in general and in neuronal function in particular. RNA-binding proteins (RBPs) and microRNAs (miRNAs), acting on mRNAs, add a level of gene expression regulation in the cell. Dysregulation of RBPs and miRNAs in neurons can affect RNA metabolism and, as a consequence, can lead to neuronal dysfunction and neurodegeneration.

One example of this complex regulation network comes from mislocalization of TDP-43 and FUS/TLS, observed in a large number of disorders (mainly in ALS and FTLD, but also in AD, HD, and PD). Altogether these specific kinds of neurodegenerative diseases are called TDP-43 and FUS/TLS proteinopathies. TDP-43 and FUS/TLS are both structurally close to the family of heterogeneous ribonucleoproteins (hnRNPs) and act as regulators of multiple levels of RNA processing including transcription, splicing, transport, translation, nucleocytoplasmic shuttling, formation of stress granules, and miRNA processing.

TDP-43 inclusions are found in ALS patients with TARDBP mutation but not in forms caused by *SOD1* mutations (Mackenzie et al., 2007), in sporadic FTLD as well as familial FTLD cases with mutations in *GRN* gene, in *VCP* gene, and in rare cases with *TARDP* mutations (Cairns et al., 2007; Neumann et al., 2007). TDP-43 inclusions have also been reported in Alzheimer (Uryu et al., 2008) and Parkinson patients (Hasegawa et al., 2007) and, in some instances, these inclusions coexist with tau or α-synuclein aggregates.

FUS/TLS inclusions are less well defined but they were found in ALS and FTLD patients and in association with another kind of inclusions found in postmortem analysis of patients with different polyQ diseases (such as HD; Doi et al., 2008).

Another example of defects in RNA metabolism comes from hexanucleotide (GGGCCC) expansion in *C9ORF72* gene (between alternative exons 1a and 1b) accounting for some ALS and FTD patients. There are at least three mechanisms of action of expanded hexanucleotide repeat in the *C9ORF72* gene: haploinsufficiency due to reduced expression of the allele containing the repeat expansion; RNA toxicity due to the production of a mutant RNA containing the repeat that may sequester

RBP; protein toxicity due to the expression of a mutant protein containing the repeat expansion (La Spada and Taylor, 2010).

Other factors implicated in neurodegeneration are mitochondrial dysfunctions (shape, size, fission–fusion, distribution, and movement), oxidative stress, and/or environmental factors.

Oxidative stress can be caused by an imbalance in the redox state of the cell, or by overproduction of reactive oxygen species, or by dysfunction of the antioxidant systems. A link has been demonstrated also between several environmental factors strongly associated with age, including pesticides, metals, head injuries, lifestyles, and dietary habits and an increased disease risk.

Overall, these findings show that several common multifactorial processes contributing to neuronal death and leading to functional impairments are shared by many neurodegenerative diseases, making a diagnosis extremely complex.

Neurons of the central nervous system cannot regenerate after cell death and, as a consequence, once a neurodegenerative disease has manifested, significant neuronal loss and damage are already present. Therefore, it is clear the importance to establish an early diagnosis to maximize the effectiveness of disease-modifying therapies. In recent years, large efforts have been made to identify disease neuropathological, biochemical, and genetic biomarkers to try to establish a diagnosis in the earlier stages. The finding that some related neurodegenerative diseases share common mechanisms suggests that these disorders may have similar targets for the development of diagnostic and therapeutic agents. Moreover, the fact that the current diagnosis is based on the patient's cognitive functions increases the need to investigate for common or similar noninvasive diagnostic methods, able to identify the disease at an early stage. Robust biomarkers would be valuable not only for the initial diagnosis, but also to classify various subtypes of the disease, to monitor responses to therapeutic agents, and to track disease progression.

## 8.2    BIOMARKERS IN NEURODEGENERATIVE DISEASES

Neurodegenerative disorders are different nervous system diseases characterized by the progressive loss of neuronal cells and tissues, without a possible regeneration process after the damage occurs (Rachakonda et al., 2004). The group is formed by hundreds of neurological diseases that show different symptoms, such as cognitive dysfunction or altered behavior (Krystal and State, 2014). The standard methodologies failed in the attempt to clarify and investigate on the pathological causes, due to the complex nature of neurodegenerative diseases that involve many different pathways and targets (Han et al., 2014). In the recent years, there was an increasing application of sequencing and genomic approaches for the study of neurodegenerative disorders (McCarroll et al., 2014) that lead to the discovery of novel risk genes and peripheral biomarkers useful for the investigation disease mechanisms. Although genetic studies represent a key method to identify genetic risk factors through DNA sequencing and investigate the heritability process of a disease state, their function of risk assessment is clearly different from the role of biomarkers (Gonzalez-Cuyar et al., 2011). Biomarkers are biological substances measured *in vivo* and used to indicate the onset or the presence of a specific disease (Rachakonda et al., 2004).

The hypothetical biomarker changes are thought to begin 10–20 years before the clinical onset of a disease event. However, this timing could be influenced by the type of disorder considered, the analysis applied, and the age of the subjects (Langbaum et al., 2013; Knickmeyer et al., 2014). According to this perspective, the possibility to identify disease-specific biomarkers at an early stage would lead to an early and maybe more effective treatment for the patients. Moreover, specific and reliable biomarkers could be applied to follow and quantify disease progression and investigate on the possible reactions to treatments, in order to achieve a better therapy (Rachakonda et al., 2004; Gonzalez-Cuyar et al., 2011). Biomarkers could also help in specific diagnosis, in discriminating similar diseases or in identifying combinations of diseases, which is enormously important for the assembling of subjects for clinical trials (Gonzalez-Cuyar et al., 2011). Body fluids, such as plasma, serum, urine, and cerebrospinal fluid (CSF), represent a useful reservoir of information of what is happening in the body. In particular, CSF that is closely linked with the brain could reflect neuropathological features of brain disorders (Ghidoni et al., 2011). Biomarkers that can be measured in different biological fluids can fall in two categories: proteins and miRNAs. Compared to proteins, miRNAs show many advantages such as a tissue- or cell type–specific expression, lower cost and shorter time required for the development of an assays, and the presence of an amplifiable signal (Chevillet et al., 2014).

## 8.3   MiRNA BIOGENESIS AND FUNCTION

The miRNAs are a group of small noncoding RNAs with important regulatory roles on the posttranscriptional expression of target mRNAs (Bartel, 2009; Ghildiyal and Zamore, 2009), found in animals, plants, green algae, and viruses (Griffiths-Jones et al., 2008). Specifically miRNAs are 21- to 22-nucleotides long, single-stranded RNA molecules, firstly discovered by Lee et al. (1993), generating from longer transcripts of different lengths (pri-miRNA), usually transcribed by RNA polymerase II, from intragenic or intergenic DNA regions (Lee et al., 2004; Garzon et al., 2010). MiRNAs can perform their function when they are loaded on the RNA-induced silencing complex (RISC) and associates with Argonaute-2 (Ago2; Meister and Tuschl, 2004; Meister et al., 2004). The miRNA–RISC complex interacts with the target mRNAs, through a binding between the seed region of the miRNA, localized on its 5′ end between nucleotides 2 and 8 (Bartel, 2009). However, a significant fraction of noncanonical interactions, that involve nonseed base pairing can occur between the miRNA–RISC complex and the target mRNA (Helwak et al., 2013). The RISC complex induces mRNA degradation if there is a perfect complementarity between the sequences of the miRNA and its target mRNA, while the interaction leads to translation inhibition in case of imperfect binding (Bartel, 2009). Frequently miRNAs recognized bind to sequences in the 3′ untranslated region (UTR) of target mRNAs, but also the coding region and the 5′-UTR of a mRNA could be involved in the interaction with miRNAs (Ørom et al., 2008; Rigoutsos, 2009).

MiRNAs modulate through their binding the transcriptome of cells (Guo et al., 2010). This process of posttranscriptional regulation seems to be complex and difficult to be fully understood, since one single miRNA could potentially target hundreds of different mRNAs and one single mRNA could be controlled by many different miRNAs. Moreover, miRNAs themselves can be regulated posttranscriptionally as suggested by Farajollahi and Maas (2010), creating an increasingly intricate network of regulation.

The assumption that miRNA dysregulation has the potential to lead to neurodegeneration derived from different experiments in which knockout of Dicer, with consequent disruption of miRNAs biogenesis causes neurodegenerative phenotypes. Specifically, deletion of Dicer was performed in mouse cerebellar neurons (Schaefer et al., 2007), in midbrain dopamine neurons (Kim et al., 2007), in striatal, retinal, spinal, and cortical neurons (Cuellar et al., 2008; Damiani et al., 2008; Davis et al., 2008; Haramati et al., 2010) and in glial cells (Shin et al., 2009; Tao et al., 2011; Wu et al., 2012). Several studies showed that the DGCR8 haploinsufficiency leads to a decreased production of miRNAs, with neuronal alteration as result (Stark et al., 2008; Fénelon et al., 2011; Schofield et al., 2011). Hébert and colleagues found an altered phosphorylation pattern of tau after the deletion of Dicer, suggesting the presence of miRNAs' control on specific aspects of the neurodegeneration process (Hébert et al., 2010).

Microarray analyses demonstrated a specific brain expression of different miRNAs (Lim et al., 2005; Manakov et al., 2009), which is particularly relevant during brain development (Miska et al., 2004; Kapsimali et al., 2007), whereas sequencing data leads to the development of a mammalian miRNAs expression atlas in different cell types (Landgraf et al., 2007). Recent studies showed an interesting correlation between the expression pattern of miRNAs in brains of different primates and human development (Somel et al., 2010, 2011; Hu et al., 2011), suggesting that changes in miRNAs profiles induced significant differences at mRNA and protein levels between human and other primates' brain. Therefore, not only the global loss of miRNAs can cause neurodegeneration, but also in some cases a specific alteration of a single miRNA pattern in the brain could be linked to a particular disease.

## 8.4   CIRCULATING MiRNA

RNA was considered for a long time to be unstable in the blood, due to the nuclease activity observed in human plasma (Kamm and Smith, 1972). However, this concept changed rapidly with the observation of circulating and stable cell-free miRNAs in healthy individuals and different types of cancer patients (Mitchell et al., 2008; Hu et al., 2010; Asaga et al., 2011; Schwarzenbach et al., 2011). Subsequently many circulating miRNAs that can be used as possible biomarkers were also described for different neurodegenerative disorders. However, it is not clear yet if all the reported miRNAs are directly released from cells linked with a disease or a product of a specific secondary response.

The observation of the rapid degradation of purified or synthetic miRNAs compared with the ones found in the plasma (Mitchell et al., 2008) lead to the hypothesis of a packaging system to protect them from RNase degradation. Exosomes, which are membrane vesicles of 50–100 nm diameter found in many body fluids (Weber et al., 2010), were shown to contain proteins, mRNAs, and miRNAs derived from the originating cell (Valadi et al., 2007). Moreover, microvesicles (secreted vesicles of 1–10 μm diameter) were found to contain miRNAs associated with invasion and migration in prostate cancer (Morello et al., 2013). Several studies observed *in vitro* the cellular uptake of exosomes and other vesicles, indicating the miRNA-containing exosomes as a possible way of communication between cells (Skog et al., 2008; Kharaziha et al., 2012; Montecalvo et al., 2012; Stoorvogel, 2012). In case of RNA transfer, the target cells' gene expression and protein translation would be modified, as a result (Pant et al., 2012). The origin of extracellular vesicles from multivesicular bodies indicates that they could have an important role for the clearance of toxins or altered proteins by the lysosomal pathway (Candelario and Steindler, 2014). Many neurodegenerative disorders, such as AD and PD, present lysosomal dysfunctions. Moreover, different proteins that lead to aggregation and accumulation in neurodegenerative disorders have been found secreted through extracellular vesicles, such as TDP-43 (Nonaka et al., 2013), Aβ (Rajendran et al., 2006), tau (Saman et al., 2012), and α-synuclein (Emmanouilidou et al., 2010). Different studies found a large part of miRNAs which is not associated with lipid vesicles but can be affected by protease digestion (Arroyo et al., 2011; Turchinovich et al., 2011). Therefore, miRNAs could circulate in biofluids also in association with different proteins, such as Ago2 (Arroyo et al., 2011) or nucleophosmin (Wang et al., 2010) and the high-density lipoprotein (Vickers et al., 2011).

Different methods are used for the isolation, quantification, and profiling of miRNAs. Total RNA extraction and isolation, guaranteeing recovery of miRNAs, are possible through commercially available column filtration protocols, or by using "tri-reagents" (acid phenol in combination with guanidinium thiocyanate and chloroform). MiRNAs quantification and profiling are obtained with different techniques: next generation sequencing (NGS), microarray, and real-time polymerase chain reaction (RT-PCR). NGS is rapidly evolving for its multiplexing capacities and provides accurate and sensitive miRNAs measurements. NGS gives the possibility to discover novel miRNAs, in contrast to microarray and quantitative RT-PCR (qRT-PCR) methods, detecting only already known miRNAs. However, NGS is labor consuming in sample preparation and data analysis and very expensive. Another technique used for miRNA profiling is represented by microarray, showing often problems of cross-hybridization between members of miRNA families and discrepancies in comparing results obtained with different microarray platforms. A common strategy is to validate the microarray data by qRT-PCR, to warrantee high sensitivity and specificity. Moreover, multiwell plate-based qRT-PCR assays could substitute microarrays in the high-throughput profiling of miRNAs. qRT-PCR is presently the most easily performed and cost-effective method when there is a need to measure the levels of a restricted number of miRNA as biomarkers. An important step during qRT-PCR analysis is the choice of a suitable normalization method, constituted by one or more stably expressed genes (called housekeeping or endogenous genes). This step is required to remove variations and increase the accuracy of miRNAs quantification. Ribosomal RNAs (rRNAs), U6 snRNA, or a combination of different miRNAs have been used as reference genes in miRNA profiling. Finally, two approaches are frequently used for the selection of promising circulating miRNA as disease biomarkers. The first is based on an initial screening of miRNAs with subsequent validation of potential biomarkers by qRT-PCR. This kind of approach gives the possibility to study a new mechanism not previously associated with the disease.

The limitations are linked to sensitivity and variability depending on the platforms used to profile miRNAs. The second approach is based on the analysis of miRNAs already associated with the disease, but it has the limitations of the potential involvement of the same miRNA in nonrelated diseases and of a lack of correlation between the expression of the specific miRNA in the affected organ and the relative expression of the same miRNA in plasma or CSF. This latter issue is specifically important in neurodegenerative diseases, in which the isolation of the nervous system from the rest of the body, by the blood–brain barrier, makes it difficult for the dysregulation of an miRNA in the brain to be reflected in body fluids.

## 8.4.1 Alzheimer's Disease and Circulating miRNAs

Alzheimer's disease is the most prevalent cause of dementia in the elderly (Blennow et al., 2006) and shows the typical accumulation of two modified proteins: the amyloid β (Aβ) peptide, derived from the amyloid precursor protein (APP), can accumulate in diffuse or neuritic plaques (O'Brien and Wong, 2011) whereas the microtubule associated protein tau accumulates in structures called neurofibrillary tangles (NFTs). The accumulation of these proteins leads to toxic effects and inflammatory responses, contributing to the disruption of the neuronal network important for the cognitive functions (Gascon and Gao, 2012).

Dosage of proteins associated with AD in CSF is already reported as a neurochemical way for the diagnosis of AD (Marksteiner et al., 2007; Blennow and Zetterberg, 2009). Specifically one of the most common peptides derived from the cleavage of APP, the 42-aminoacids-long $A\beta_{42}$, was found decreased in CSF of AD patients (Galasko and Montine, 2010). Similarly, elevated levels of total and phosphorylated tau in the CSF are AD biomarkers (Fagan et al., 2007; Marksteiner et al., 2007; Blennow and Zetterberg, 2009). Regarding the altered expression of miRNAs in AD tissues and their pathogenic role, a study identified miR-107 to be decreased during AD (Wang et al., 2008), through an miRNA expression profiling in cortex from control and AD patients. The mRNA of BACE1, the protease involved in the cleavage of APP was predicted as target of miR-107 and the miRNA's role was confirmed by expression studies. mRNA profiling showed that BACE1 mRNA levels tented to increase as miR-107 decreased during the progression of AD, indicating how this miRNA could be involved in the pathological process of AD (Wang et al., 2008). Regulation and control of BACE1 by another cluster of miRNAs (miR-29a/b-1) was further investigated by Hébert et al. (2008). Moreover, a study demonstrated upregulated levels of miR-9 and miR-128 in AD hippocampus (Lukiw et al., 2007) and another work found increased levels of miR-9, miR-125b, and miR-146a in the temporal lobe neocortex of affected AD individuals (Sethi and Lukiw, 2009), suggesting that miRNAs pattern can be altered in sporadic AD.

The recent research demonstrated that miRNAs are not only aberrantly expressed in AD or involved in the regulation of the main pathological processes, but also that their altered presence in different body fluids could be diagnostic of AD (Table 8.1). The first study that investigated peripheral miRNAs expression in Alzheimer condition was performed in 2007 (Schipper et al., 2007). The authors identified an increased level of miRNAs in blood mononuclear cells (BMC) of patients with

**TABLE 8.1**

**Alzheimer's Disease and Circulating MiRNAs**

| Sample | MiRNA | Trend | Experimental Approach | Pilot Study | Validation | References |
|---|---|---|---|---|---|---|
| PBMC | MiR-34a, miR-181b | Upregulation | MiRNA profiling | Microarray AD:16 NEC:16 | Taqman® miRNA qRT-PCR | Schipper et al. (2007) |
| Blood serum | MiR-137, miR-181c, miR-9, miR-29a/b | Downregulation | Candidate miRNAs | SYBR qRT-PCR AD:7 MCI/Early AD:7 CT:7 | | Geekiyanage and Chan (2012) |
| Plasma | MiR-132 family, miR-134 family, miR-491-5p, miR-370 | Ratio of miRNAs paired | Candidate miRNAs | Taqman® miRNA qRT-PCR MCI:10 CT:10 | TaqMan® qRT-PCR MCI: 20 AD: 20 CT: 20 CY: 20 | Sheinerman et al. (2012) |
| Blood | 12-miRNA signature: miR-112, miR-161, let-7d-3p, miR-5010-3p, hsa-miR-26a-5p, hsa-miR-1285-5p, and hsa-miR-151a-3p | Upregulation | MiRNA profiling | Next generation sequencing AD: 48 CT: 22 | SYBR qRT-PCR AD: 94 MCI: 18 MS: 16 PD: 9 DEP: 15 BD: 15 Schiz: 14 CT: 21 | Leidinger et al. (2013) |
| Blood | 12-MiRNA signature:miR-103a-3p, miR-107, miR-532-5p, miR-26b-5p, let-7f-5p | Downregulation | MiRNA profiling | Next generation sequencing AD: 48 CT: 22 | SYBR qRT-PCR AD: 94 MCI: 18 MS: 16 PD: 9 DEP: 15 BD: 15 Schiz: 14 CT: 21 | Leidinger et al. (2013) |
| Plasma | 7-MiRNA signature: hsa-let-7d-5p, hsa-let-7g-5p, hsa-miR-15b-5p, hsa-miR-142-3p, hsamiR-191-5p, hsa-miR-301a-3p, and hsa-miR-545-3p | | MiRNA profiling | Nanostring AD:11 MCI:9 CT:20 | Taqman® miRNA qRT-PCR AD:20 CT:17 | Kumar et al. (2013) |

*(Continued)*

**TABLE 8.1 (Continued)**
**Alzheimer's Disease and Circulating MiRNAs**

| Sample | MiRNA | Trend | Experimental Approach | Pilot Study | Validation | References |
|---|---|---|---|---|---|---|
| Plasma | MiR-15a associated with amyloid plaque score in hippocampus region | Downregulation | Candidate miRNAs | Microarray | qRT-PCR | Bekris et al. (2013) |
| Cell-free plasma or PBMC | MiR-34c | Upregulation | Candidate miRNAs | Taqman® miRNA qRT-PCR AD:110 CT:123 | | Bhatnagar et al. (2014) |
| Plasma | MiR-34a, miR-146a | Downregulation | Candidate miRNAs | qRT-PCR | | Kiko et al. (2014) |
| Serum | MiR-98-5p, miR-885-5p, miR-483-3p, miR-342-3p, miR-191-5p, miR.191-5p, miR-let-7d-5p | Upregulation and downregulation | miRNA profiling | Illumina HiSeq 2000 sequencing AD:50 CT:50 | qRT-PCR AD:158 CT:155 | Tan et al. (2014) |
| Serum | MiR-125b, miR-23a, miR-26b | Downregulation | miRNA profiling | Array AD:22 NINDCs:18 INDCs:18 FTD:10 | qRT-PCR | Galimberti et al. (2014) |
| CSF | 60 miRNAs differently express between Braak stage V and Braak stage I | Upregulation and downregulation | miRNA profiling | Taqman® miRNA qRT-PCR Braak stage V AD:10 Braak stage I CT:10 | | Cogswell et al. (2008) |

*(Continued)*

**TABLE 8.1 (Continued)**
**Alzheimer's Disease and Circulating MiRNAs**

| Sample | MiRNA | Trend | Experimental Approach | Pilot Study | Validation | References |
|---|---|---|---|---|---|---|
| CSF ECF | MiRNA-9, miRNA-125b, miR-146a, miR-155 | Upregulation | Candidate miRNAs | Microarray AD:5 CT:5 | LED-Northern dot-blot assay using primary human neuronal-glial cell-cultures and AD ECF | Alexandrov et al. (2012) |
| CSF | MiR-27a-3p | Downregulation | MiRNA profiling | qRT-PCR AD CT | qRT-PCR AD:35 CT:37 | Sala Frigerio et al. (2013) |
| CSF ECF | MiR-146a, miR-155 | Upregulation | Candidate miRNAs | Microarray AD:5 CT:5 | LED-Northern dot-blot assay using primary human neuronal-glial cell-cultures and AD ECF | Lukiw et al. (2012) |
| CSF | MiR-let-7b | Upregulation | Candidate miRNAs | Taqman® miRNA qRT-PCR AD:13 CT:11 | Mice experiments | Lehmann et al. (2012) |
| CSF | MiR-34a, miR-125b, miR-146a | Downregulation | Candidate miRNAs | qRT-PCR | | Kiko et al. (2014) |
| CSF | MiR-29a, miR-29b | Upregulation | Candidate miRNAs | qRT-PCR | | Kiko et al. (2014) |
| CSF | MiR-125b, miR-26b | Downregulation | MiRNA profiling | MiRNA PCR array AD:22 FTD:10 CT:18 | Taqman qRT-PCR | Galimberti et al. (2014) |

*Note:* PBMC: peripheral blood mononuclear cells; NEC: normal elderly controls; AD: Alzheimer disease; CT: control; NINDCs: noninflammatory neurological controls; INDCs: inflammatory neurological controls; MCI: mild cognitive impairment; MS: multiple sclerosis; PD: Parkinson disease; DEP: major depression; BD: bipolar disorder; Schiz: schizophrenia; CSF: cerebrospinal fluid; ECF: brain tissue–derived extracellular fluid; CY: young control; FTD: frontotemporal dementia.

sporadic AD using a microarray containing 462 human miRNAs. In particular, miR-34a and miR-181b were significantly increased in AD individuals, as validated through qPCR (Schipper et al., 2007). In 2012, a decreased level of some circulating miRs measured through SYBR green qRT-PCR in the blood serum of AD patients was suggested as a noninvasive diagnostic method for AD (Geekiyanage and Chan, 2012). Around the same year, another work used plasma miRNAs quantified through TaqMan miRNA qRT-PCR assay to detect mild cognitive impairment (MCI) that represents an intermediate state between normal aging and AD or other types of dementia (Sheinerman et al., 2012). This group demonstrated that ratio of plasma levels of miR-132 and miR-134 families paired with miR-491-5p and miR-370 differentiate MCI subjects from controls, with high sensitivity and specificity (Sheinerman et al., 2012). In particular, they identified two sets of paired biomarkers ratio: the miR-132 family (miR-128/miR-491-5p, miR-132/miR-491-5p, and miR-874/miR-491-5p) and the miR-134 family (miR-134/miR-370, miR-323-3p/miR-370, and miR-382/miR-370). Subsequently, Leidinger and coworkers identified in the blood of AD patients a signature of 12 miRNAs that can be useful for the diagnosis. They applied NGS to select a panel of 12 miRNAs that were further used through RT-qPCR in larger cohort of samples, differentiating with high accuracy and specificity between AD and controls or between AD patients and individuals suffering of other neurological diseases (Leidinger et al., 2013). The signature contains seven upregulated miRNAs in AD (brain-miR-112, brain-miR-161, hsa-let-7d-3p, hsa-miR-5010-3p, hsa-miR-26a-5p, hsa-miR-1285-5p, and hsa-miR-151a-3p) and five downregulated miRNAs in AD patients (hsa-miR-103a, hsa-miR-107, hsa-miR532-5p, hsa-miR-26b-5p, and hsa-miR-let-7f-5p; Leidinger et al., 2013). A different signature of seven miRNAs (miR-let-7d-5p, miR-let-7g-5p, miR-15b-5p, miR-142-3p, miR-191-5p, miR-301a-3p, and miR-545-3p) was identified in the plasma to distinguish AD patients from normal subjects with a good accuracy (Kumar et al., 2013). Another work that evaluated the possibility to use miRNAs expressed in human brain and biofluids as biomarkers of AD was performed by Bekris and collaborators, in which miR-15a level was shown to positively correlate with neuritic plaque score (Bekris et al., 2013). In particular, a higher plasma miR-15a level was suggested to be a feasible marker of high neuritic plaque in hippocampus (Bekris et al., 2013). Bhatnagar and collaborators studied through TaqMan array and validated with qRT-PCR, miRNAs circulating in plasma and peripheral BMCs (PBMCs) of AD patients, identifying miR-34c as a biomarker to individuate sporadic AD subjects, since its level is significantly increased in plasma samples of patients (Bhatnagar et al., 2014). During the last year, one work showed an altered expression of 6 miRNAs (miR-98-5p, miR-885-5p, miR-483-3p, miR-342-3p, miR-191-5p, and miR-let-7d-5p) in serum of AD patients compare to controls, with high specificity and sensitivity for miR-342-3p (Tan et al., 2014). More recently, Galimberti and colleagues correlated altered miRNAs expression between serum and CSF of AD patients compared to healthy individuals, finding downregulation of three miRNAs (miR-125b, miR-23a, and miR-26b) in serum derived from 22 AD patients. Downregulation of miR-125b and miR-26b was also observed and confirmed in CSF from AD individuals (Galimberti et al., 2014). The function of miR-125b was also recently validated *in vivo*, indeed injection of miR-125b into hippocampus in mice induces tau phosphorylation and learning impairment (Banzhaf-Strathmann et al., 2014).

The first time in which CSF was used to identify possible biomarkers for the diagnosis of AD was in 2008, Cogswell and coworkers discovered that miRNAs can be detected in this fluid and their expression is altered in the presence of AD pathology. In particular, they found altered miRNAs related to immune cell differentiation and innate immunity (Cogswell et al., 2008). Significant increases in the levels of miR-9, miR-125b, miR-146a, and miR-155 were found in CSF and in extracellular fluid (ECF) derived from short postmortem interval brain tissue of AD patients (Alexandrov et al., 2012). Some of these miRNAs are known to be overexpressed in the brain of patients and associated with the spreading of inflammatory neurodegeneration (Alexandrov et al., 2012). Another candidate biomarker for AD is miR-27a-3p, which was found reduced in the CSF of patients compare to control subjects in a pilot study (Sala Frigerio et al., 2013). The decrease in the level of miR-27a-3p is combined with high CSF tau levels and low CSF β-amyloid levels (Sala Frigerio et al., 2013).

As already reported (Cogswell et al., 2008), CSF and ECF derived from AD patients contain abundant levels of proinflammatory miR, in particular miR-146a and miR-155 (Lukiw et al., 2012). These two miRNAs were suggested to be involved in the spreading process of AD, since human neuronal-glial cocultures secrete these miRNAs upon cytokine tumor necrosis factor-$\alpha$ (TNF-$\alpha$) and A$\beta$42-peptide stress; in addition, it was observed that a conditioned medium containing miR-146a and miR-155 induces inflammatory gene expression and downregulation of complement factor H (CFH). This regulator is involved in inflammatory degeneration in AD and other disorders (Lukiw et al., 2012). A similar function was proposed for miR-let-7 that was found upregulated in the CSF of AD subjects. They demonstrated that extracellular introduction of let-7b in the CSF of mice induces neurodegeneration through the RNA-sensing Toll-like receptor (TLR) 7 (Lehmann et al., 2012). miR-146a was also identified in plasma and CSF of AD patients in another recent study with other candidate miRNAs (Kiko et al., 2014). However, its level was analyzed through qRT-PCR and found significantly downregulated in AD patients compared to controls in apparently contrast with previous reports (Alexandrov et al., 2012; Lukiw et al. 2012).

### 8.4.2 PARKINSON'S DISEASE AND CIRCULATING miRNAs

Approximately 1% of the population over the age of 55 is affected by PD. This neurodegenerative disorder is characterized by the degeneration of dopaminergic neurons of the substantia nigra that leads to rigidity, tremors, and slowed movements. Another pathological feature is the presence of inclusions primarily composed of $\alpha$-synuclein, called Lewy bodies. These inclusions are cytoplasmic and show a characteristic pattern in the brain. One of the biochemical markers used to recognize the onset of PD is the loss of the dopamine transporter or the identification of $\alpha$-synuclein protein in the Lewy bodies (Duyckaerts and Hauw, 2003). Ruling out the analysis of protein involved in the pathology in the brain, an increased level of oxidative stress markers in blood of PD patients, such as superoxide radicals and the coenzyme Q10 redox ratio were suggested as potential biomarker for PD (Michell et al., 2004). It was also shown that the proinflammatory factor TNF-α is three to four fold higher in CSF of PD patients compared to controls (Le et al., 1999). Decreased levels of α-synuclein concentration in CSF of PD patients have been found by different laboratories (Mollenhauer et al., 2011; Shi et al., 2011). In addition, some groups investigated as biomarker of PD, the CSF level of a multifunctional redox-sensitive protein, important for mitochondrial function, called DJ-1 (Waragai et al., 2006; Hong et al., 2010). However, the studies highlighted the necessity of controlling blood contamination of CSF, age during the analysis (Hong et al., 2010) and the clear need of an improved laboratory test with higher performance (Shi et al., 2011). So far none of these potential biomarker seems to be sufficiently robust and specific to be useful as a real diagnostic biomarker in clinical practice.

Regarding the presence of circulating miRNAs in the body fluids of PD patients (Table 8.2), the first study that investigate blood samples was performed by Margis and collaborators in the 2011 (Margis and Rieder, 2011). They found through qRT-PCR analysis three differential expressed miRNAs: miR-1, miR-22-5p, and miR-29 that could be used to distinguish between not-treated PD patients and normal control individuals; whereas miR-16-2-3p, miR-26a-2-3p, and miR-30a are differentially expressed between treated and untreated PD patients (Margis and Rieder, 2011). In the same year, another study found using miRCURY locked nucleic acid (LNA) microarrays, 18 miRNAs with an altered expression in the PBMCs of PD subjects, and predicted that target genes of these miRNAs were involved in PD's pathological pathways (Martins et al., 2011). More recently a study identified through RNA-Seq, 16 miRNAs differentially expressed in blood leukocytes of PD patients compared to healthy control volunteers (Soreq et al., 2013). Eleven miRNAs were modified after deep-brain stimulation (DBS) treatment, whereas five were changed inversely to the disease-induced changes (Soreq et al., 2013). Investigation and analysis of miRNA expression in plasma of PD patients were also performed by Khoo and coworkers (2012). Using microarrays and TaqMan qRT-PCR validation, they identified four miRNAs: miR-1826, miR-450b-3p, miR626, and miR-505, whose levels can be used in combination to obtain the highest predictive biomarker performance to individuate the presence of the disorder (Khoo et al., 2012). Another work that analyzed plasma from PD patients and normal controls through TaqMan miRNA qRT-PCR, identified miR-331-5p as a possible PD biomarker (Cardo et al., 2013). Recently a study investigated circulating miRNAs in serum of PD or multiple system atrophy (MSA) patients compared to healthy controls to distinguish individuals who are affected by these two different pathologies with overlapping features (Vallelunga et al., 2014). Specifically, through array analysis and qRT-PCR validation, the authors identified four miRNAs that are downregulated and five miRNAs that are upregulated in PD serum versus control subjects (Vallelunga et al., 2014). Profiling of miRNAs in serum derived from idiopathic PD, PD patients carriers the LRRK2 G2019S mutation and controls was also performed through RT-PCR-based TaqMan miRNA arrays, finding downregulation of miR-29a, miR-29c, miR-19a, and miR-19b in patients compare to healthy individuals (Botta-Orfila et al., 2014). So far there are no studies reporting miRNA biomarkers for PD in patients' CSF.

## TABLE 8.2
## Parkinson's Disease and Circulating MiRNAs

| Sample | MiRNA | Trend | Experimental Approach | Pilot Study | Validation | References |
|---|---|---|---|---|---|---|
| Total peripheral blood | MiR-1, miR-22-5p, miR-29 distinguish nontreated PD from healthy subjects | Downregulation | MiRNA profiling | qRT-PCR untreated PD: 8 treated PD: 4 early-onset PD: 7 CT:8 | | Margis and Rieder (2011) |
| Total peripheral blood | miR-16-2-3p, miR-26a-2-3p, miR-30a distinguish treated from untreated PD | Upregulation | MiRNA profiling | qRT-PCR untreated PD: 8 treated PD: 4 early-onset PD: 7 CT:8 | | Margis and Rieder (2011) |
| PBMC | 18 miRNAs | Downregulation | MiRNA profiling | Microarray PD:19 CT:13 | | Martins et al. (2011) |
| Leukocytes | 16 miRNAs differentially expressed: 6 miRNAs | Downregulation | MiRNA profiling | Next generation sequencing PD pre-DBS: 3 PD post-DBS: 3 CT:6 | | Soreq et al. (2013) |
| Leukocytes | 16 miRNAs differentially expressed: 10 miRNAs | Upregulation | MiRNA profiling | Next generation sequencing PD pre-DBS: 3 PD post-DBS: 3 CT:6 | | Soreq et al. (2013) |
| Plasma | MiR-1826, miR-450b-3p, miR-626, miR-505 | Upregulation of k-top scoring pairs (k-TSP1) (miR-1826/miR-450b-3p), miR-626, miR505 | MiRNA profiling | Agilent microarray PD:32 CT:32 | Taqman® miRNA qRT-PCR PD:30 MSA:4 PSP:5 CT:8 | Khoo et al. (2012) |

*(Continued)*

**TABLE 8.2 (*Continued*)**
**Parkinson's Disease and Circulating MiRNAs**

| Sample | MiRNA | Trend | Experimental Approach | Pilot Study | Validation | References |
|---|---|---|---|---|---|---|
| Plasma | MiR-331-5p | Upregulation | MiRNA profiling | Taqman® miRNA qRT-PCR PD:31 CT:25 | Taqman® miRNA qRT-PCR PD:25 MSA:25 CT:25 | Cardo et al. (2013) |
| Serum | MiR-339-5p, miR-652, miR-1274A, miR-34b | Downregulation | MiRNA profiling | Taqman® human miRNA Array PD:6 MSA:9 CT:5 | Taqman® miRNA qRT-PCR PD:25 MSA:25 CT:25 | Vallelunga et al. (2014) |
| Serum | MiR-223*, miR-324-3p, miR-24, miR-148b, miR-30c | Upregulation | MiRNA profiling | Taqman® human microRNA Array PD:6 MSA:9 CT:5 | Taqman® miRNA qRT-PCR PD:25 MSA:25 CT:25 | Vallelunga et al. (2014) |
| Serum | MiR-29a, miR-29c, miR-19a, miR-19b | Downregulation | MiRNA profiling | Taqman® human miRNA array IPD:10 PD:10 CT:10 | Taqman® human miRNA array IPD:20 PD:20 CT:20 | Botta-Orfila et al. (2014) |

*Note:*  PD: Parkinson disease; PSP: progressive supranuclear palsy; MSA: multiple system atrophy; DBS: deep brain stimulation; IPD: idiopathic Parkinson's disease.

### 8.4.3 Amyotrophic Lateral Sclerosis and Circulating miRNAs

As already mentioned, ALS is caused by mutations in SOD1; in RBPs as TDP-43 and FUS/TLS; and in CHMP2B, VCP, and C9ORF72. Some miRNAs are emerging as important contributors to ALS pathogenesis.

The involvement of TDP-43 and FUS/TLS in miRNA biogenesis has been uncovered, as these proteins directly bind key components of the miRNA processing pathway. Drosha is able to form two distinct protein complexes, a "more canonical" complex with DGCR8 and a larger complex, including TDP-43 and FUS, with limited pri-miRNA processing activity (Gregory et al., 2004). In addition, TDP-43 was shown to directly bind Dicer, Ago2, subsets of pri-miRNAs in the nucleus and pre-miRNAs in the cytoplasm (Kawahara and Mieda-Sato, 2012). *In vitro* depletion of TDP-43 and FUS proteins leads to a reduction of specific subsets of miRNAs implicated in neuromuscular development, neuronal function, and survival (Buratti et al., 2010; Kawahara and Mieda-Sato, 2012; Morlando et al., 2012).

To determine whether miRNAs are essential to motor neuron survival, Haramati's group in 2010 used Dicer knockdown mice. They demonstrated that the heavy neurofilament subunit is a target of miR-9, already reported to be downregulated in a genetic model of spinal muscular atrophy (SMA; Haramati et al., 2010). It has been shown that SMA and ALS are motor neuron diseases linked by a common molecular pathway: FUS, mutated in ALS, interacts with SMN, deficient in SMA (Yamazaki et al., 2012). Moreover, in a recent study, a miR-9 reduction was found in human neurons derived from induced pluripotent stem cells from patients with the pathogenic TARDBP M337V mutation, suggesting that miR-9 downregulation could be a common pathogenic event in FTD/ALS (Zhang et al., 2013). In a recent study, alteration in ALS of some miRNAs directly targeting neurofilament light chain mRNA (NEFL) has been shown. Among these dysregulated miRNAs, there are miR-146a* (upregulated) and miR-524-5p and miR-582-3p (downregulated) in spinal cord (SC) from sporadic ALS (sALS) patients (Campos-Melo et al., 2013). In addition, a group of 80 putative novel miRNAs from control and sALS SCs has been characterized. Among them, 24 have miRNA response elements (MREs) within the NEFL mRNA 3′-UTR and 2 of them, miR-b1336 and miR-b2403, are downregulated in ALS SC (Ishtiaq et al., 2014).

Changes in miRNAs have also been seen in peripheral ALS tissues. The muscle-specific miR-206 is upregulated in lower limbs of SOD1-G93A mice (Williams et al., 2009) and in a mouse model with miR-206 deletion, acceleration of disease progression was observed, suggesting that the high amount of miR-206 in SOD1-G93A mice may have a compensatory effect to reduce degeneration in ALS. A similar increase in miR-206 was also observed in ALS patients' muscle tissue (Russell et al., 2012). In the same work, the authors found that skeletal muscle mitochondrial dysfunction in ALS patients is associated with an increase in some miRNAs (miR-23a, miR-29b, miR-206, and miR-455) and with a reduction in peroxisome proliferator-activated receptor γ coactivator-1α (PGC-1α) signaling pathways (Russell et al., 2012).

In another work, high expression of miR-29a was observed in brain and SC of SOD1 (G93A) mice, a model for familial ALS. These results provide a first evidence for the possible therapeutic utility of modulation of miR-29a function in ALS (Nolan et al., 2014). In a very recent study, miR-141 and miR-200a are found to be linked with FUS by a feed-forward regulatory in which FUS upregulates miR-141/200a, which in turn regulate FUS protein synthesis. Moreover, Zeb1, a miR-141/200a target and at the same time a transcriptional repressor of these two miRNAs, is part of the circuitry and reinforces it (Dini Modigliani et al., 2014).

A small number of studies recently investigated miRNAs as ALS biomarkers in CSF and in blood (Table 8.3). A first study was performed on leukocytes from sALS patients with respect to healthy controls. The study reported a profile of eight miRNAs significantly up- or downregulated in sALS patients (De Felice et al., 2012). Another study performed the analysis on sorted CD14+ CD16− monocytes from ALS patients, which are monocytes activated and recruited to the SC in case of inflammation correlated with neuronal loss. This study showed an miRNAs profile constituting an inflammatory signature that could be useful as biomarker for disease stage or progression (Butovsky et al., 2012). The population of deregulated miRNAs found in leukocytes (De Felice et al., 2012) and monocytes (Butovsky et al., 2012) are not overlapping and comparable, although the two analyses were performed with the same technical approach, TaqMan miRNA assay-based qRT-PCR (qRT-PCR).

Recently, other two miRNAs emerged as possible candidates as ALS biomarkers in biofluids. In the first, miRNA alterations were studied in plasma of SOD1-G93A mice, and subsequently in the serum of human ALS patients (Toivonen et al., 2014). miR-206 was increased in plasma of symptomatic animals and in ALS patients,

## TABLE 8.3
## Amyotrophic Lateral Sclerosis and Circulating MiRNAs

| Sample | MiRNA | Trend | Experimental Approach | Pilot Study | Validation | References |
|---|---|---|---|---|---|---|
| Leukocytes | MiR-338-3p | Upregulation | MiRNA profiling | Microarray ALS:8 CT:12 | Taqman® miRNA qRT-PCR ALS:14 CT:14 | De Felice et al. (2012) |
| Leukocytes | MiR-451, miR-1275, miR-328, miR-638, miR-149, miR-665, miR-583 | Downregulation | MiRNA profiling | Microarray ALS:8 CT:12 | Taqman® miRNA qRT-PCR ALS:14 CT:14 | De Felice et al. (2012) |
| Monocytes | MiR-27a, miR-155, miR-146a, miR-532-3p | Upregulation | MiRNA profiling | Microarray ALS:8 MS:8 CT:8 | | Butovsky et al. (2012) |

*Note:* ALS: amyotrophic lateral sclerosis.

showing to be a promising candidate biomarker for this motor neuron disease (Toivonen et al., 2014). In the second study, upregulation of miR-338-3p in blood leukocytes as well in CSF, serum, and SC from sALS patients was detected (De Felice et al., 2014).

### 8.4.4 HUNTINGTON'S DISEASE AND CIRCULATING miRNAs

The causal mutation of HD is an expanded repetition of the CAG trinucleotide in the first exon of the gene encoding huntingtin (HTT). HTT associates with Ago2 in P bodies, and HTT knockdown has an effect on gene silencing mediated by miR-NAs as demonstrated by luciferase assay in which HTT kd abrogates let-7b silencing effect (Savas et al., 2008). MiRNAs were implicated in HD pathogenesis. The repressor element 1 silencing transcription (REST) factor, a transcriptional repressor acting to silence neuronal gene expression in nonneuronal cells, is elevated in HD neurons, and as a result, this upregulation gives a repression of key neuronal genes (Zuccato et al., 2007; Johnson et al., 2010). REST and its cofactor coREST have target sites for miR-9 and miR-9*, respectively (Packer et al., 2008), and these two miRNAs, together with miR-7, miR-124, miR-132, and other miRNAs result to be downregulated in HD patients (Johnson et al., 2008; Martí et al., 2010). There are also deregulated miRNAs not under REST control, suggesting that miRNA dysregulation is extensive in HD (Sinha et al., 2010; Jin et al., 2012). In cellular models of HD, miR-146a, miR-125b, and miR-150 are downregulated in the presence of mutant HTT protein (Sinha et al., 2010). Interestingly, miR-146a, miR-150, and miR-125b target HTT and are also predicted to interact with the TATA-binding protein mRNA. This protein is known to be recruited into mutant HTT aggregates (Sinha et al., 2010, 2011). In the cortex of mutant HTT mouse models at early stages of disease, miR-200 family is altered, compromising genes involved in neuronal plasticity and survival (Jin et al., 2012).

With regard to biofluids, it has been demonstrated that miR-34b is upregulated in response to mutant HTT in human plasma, suggesting a possible role for miR-34b as a biomarker for HD (Gaughwin et al., 2011).

### 8.4.5 OTHER DISEASES

Some miRNAs were linked to FTD through different mechanisms. For example, miR-29b and miR-107 regulate progranulin levels (Jiao et al., 2010; Wang et al., 2010). These miRNAs might decrease progranulin levels and, since it has been demonstrated that progranulin haploinsufficiency can cause FTD, might be a risk factor for this disease. Moreover, a genetic polymorphism (rs5848) in PGRN 3'-UTR, associated with a higher risk of FTD, affects miR-659–binding site resulting in a more efficient binding and, as a consequence, in decreased progranulin levels (Rademakers et al., 2008). Finally, several miRNAs were found deregulated in FTD with TDP-43 pathology (Kocerha et al., 2011). Unfortunately, no studies were performed about circulating miRNA in plasma or serum from patients with FTD, suggesting that the study of circulating miRNA in plasma/serum for diagnosis of this kind of disorder needs more detailed investigations.

## 8.5  THERAPY

The advantages in the use of miRNAs are that these very small RNAs have the ideal biomarker characteristics: ease of detection, extreme specificity, and remarkable stability. The use of miRNA as therapeutic agents for neurodegenerative diseases is an important opportunity but also a challenging topic for several reasons. Each miRNA can regulate numerous target of the same pathway and so they could be a promising tool in which low dose of the top miRNA would be enough to induce a focused change in the entire pathway. However, these same features also present great challenges. There are still many notions to be elucidated, such as all the functional targets for each miRNA in specific cell types or the hierarchic order of miRNA regulation in a specific pathway and, obviously, the delivery of therapeutics miRNAs into the brain due to the blood–brain barrier.

There are different miRNA-based therapeutic strategies *in vivo*. miRNA mimics are small RNA molecules very similar to miRNA precursors and could be used to potentiate the miRNA posttranscriptional regulation role in some disease conditions in which one miRNA is downregulated. On the other hand, it would be possible to block overexpressed miRNAs in other kind of diseases by injecting a complementary RNA sequence (anti-miRNA) that binds to and inactivates the target miRNAs. A different experimental strategy to inhibit miRNA function could be mediated by synthetic sponge mRNA, containing complementary binding sites for an miRNA of interest.

In the first strategy of miRNA-based therapeutics, some challenges are its potential off-target effects and the possibility to saturate the endogenous miRNA-processing machinery and to interfere with the normal functions of the cell. Another big challenge for this approach is the delivery of the miRNA to the right cells in the body. It has been shown that exosomes, endogenous nanovesicles transporting RNAs and proteins, can deliver siRNA to the brain of mice when injected intravenously (Alvarez-Erviti et al., 2011). This finding paves the way for a possible therapeutic approach mediated by exosomes opening the possibility to load miRNAs into the exosomes and to target them to the brain following systemic delivery. This approach has great potential but it will have to be explored further. In the second strategy LNAs, a class of bicyclic conformational analogs of RNA, exhibiting high-binding affinity to complementary RNA target molecules and high stability *in vivo* (Fluiter et al., 2003; Hutvágner et al., 2004) are used to enhance the specificity and to reduce the amount of anti-miRNA molecules. However, those "antagomirs" cannot cross the blood–brain barrier, but can enter into brain cells only if injected directly into the brain (Kuhn et al., 2010).

Finally, the sponge technology has some advantages in its experimental settings. For example, many miRNAs have the same seed sequence but are encoded by multiple distant loci and, therefore, sponges provide a way to sequestrate at the same time different miRNAs having the same seed sequence. However, also this technology needs further investigations to be a therapeutically viable strategy.

In conclusion, despite many scientific questions still open and a need of further investigations, the analysis of miRNA as biomarkers for neurodegenerative pathologies could be helpful to develop minimally invasive screening tests and to find diagnostic, prognostic, and therapeutic tool for these diseases.

# REFERENCES

Alexandrov P.N., Dua P., Hill J.M. et al. 2012. microRNA (miRNA) speciation in Alzheimer's disease (AD) cerebrospinal fluid (CSF) and extracellular fluid (ECF). *International Journal of Biochemistry and Molecular Biology* 3:365–373.

Alvarez-Erviti L., Seow Y., Yin H. et al. 2011. Delivery of siRNA to the mouse brain by systemic injection of targeted exosomes. *Nature Biotechnology* 29: 341–345.

Arroyo J.D., Chevillet J.R., Kroh E.M. et al. 2011. Argonaute2 complexes carry a population of circulating microRNAs independent of vesicles in human plasma. *Proceedings of the National Academy of Sciences of the United States of America* 108:5003–5008.

Asaga S., Kuo C., Nguyen T. et al. 2011. Direct serum assay for microRNA-21 concentrations in early and advanced breast cancer. *Clinical Chemistry* 57:84–91.

Banzhaf-Strathmann J., Benito E., May S. et al. 2014. MicroRNA-125b induces tau hyperphosphorylation and cognitive deficits in Alzheimer's disease. *EMBO Journal* 33:1667–1680.

Baker M., Mackenzie I.R., Pickering-Brown S.M. et al. 2006. Mutations in progranulin cause tau-negative frontotemporal dementia linked to chromosome 17. *Nature* 442:916–919.

Bartel D.P. 2009. MicroRNAs: Target recognition and regulatory functions. *Cell* 136:215–233.

Bekris L.M., Lutz F., Montine T.J. et al. 2013. MicroRNA in Alzheimer's disease: An exploratory study in brain, cerebrospinal fluid and plasma. *Biomarkers* 18:455–466.

Bhatnagar S., Cher Chertkow H., Schipper H.M. et al. 2014. Increased microRNA-34c abundance in Alzheimer's disease circulating blood plasma. *Frontiers in Molecular Neuroscience* 7:2.

Blennow K., de Leon M.J., Zetterberg H. 2006. Alzheimer's disease. *Lancet* 368:387–403.

Blennow K., Zetterberg H. 2009. Cerebrospinal fluid biomarkers for Alzheimer's disease. *Journal of Alzheimer's Disease* 18:413–417.

Bonifati V., Rizzu P., van Baren M.J. et al. 2003. Mutations in the DJ-1 gene associated with autosomal recessive early-onset parkinsonism. *Science* 299:256–269.

Botta-Orfila T., Morato X., Compta Y. et al. 2014. Identification of blood serum micro-RNAs associated with idiopathic and LRRK2 Parkinson's disease. *Journal of Neuroscience Research* 92:1071–1077.

Bruijn L.I., Houseweart M.K., Kato S. et al. 1998. Aggregation and motor neuron toxicity of an ALS-linked SOD1 mutant independent from wild-type SOD1. *Science* 281:1851–1854.

Buratti E., De Conti L., Stuani C. et al. 2010. Nuclear factor TDP-43 can affect selected microRNA levels. *FEBS Journal* 277:2268–2281.

Butovsky O., Siddiqui S., Gabriely G. et al. 2012. Modulating inflammatory monocytes with a unique microRNA gene signature ameliorates murine ALS. *Journal of Clinical Investigation* 122:3063–3087.

Candelario K.M., Steindler D. 2014. The role of extracellular vesicles in the progression of neurodegenerative disease and cancer. *Trends in Molecular Medicine* 20:368–374.

Cairns N.J., Neumann M., Bigio E.H. et al. 2007. TDP-43 in familial and sporadic frontotemporal lobar degeneration with ubiquitin inclusions. *The American Journal of Pathology* 171:227–240.

Campos-Melo D., Droppelmann C.A., He Z. et al. 2013. Altered microRNA expression profile in amyotrophic lateral sclerosis: A role in the regulation of NFL mRNA levels. *Molecular Brain* 6:26.

Cardo L.F., Coto E., de Mena L. et al. 2013. Profile of microRNAs in the plasma of Parkinson's disease patients and healthy controls. *Journal of Neurology* 260:1420–1422.

Chevillet J.R., Lee I., Briggs H.A. et al. 2014. Issues and prospects of microRNA-based biomarkers in blood and other body fluids. *Molecules* 19:6080–6105.

Cogswell J.P., Ward J., Taylor I.A. et al. 2008. Identification of miRNA changes in Alzheimer's disease brain and CSF yields putative biomarkers and insights into disease pathways. *Journal of Alzheimer's Disease* 14:27–41.

Cuellar T.L., Davis T.H., Nelson P.T. et al. 2008. Dicer loss in striatal neurons produces behavioral and neuroanatomical phenotypes in the absence of neurodegeneration. *Proceedings of the National Academy of Sciences of the United States of America* 105:5614–5619.

Damiani D., Alexander J.J., O'Rourke J.R. et al. 2008. Dicer inactivation leads to progressive functional and structural degeneration of the mouse retina. *Journal of Neuroscience* 28:4878–4887.

Davis T.H., Cuellar T.L., Koch S.M. et al. 2008. Conditional loss of Dicer disrupts cellular and tissue morphogenesis in the cortex and hippocampus. *Journal of Neuroscience* 28:4322–4330.

De Felice B., Guida M., Guida M. et al. 2012. A miRNA signature in leukocytes from sporadic amyotrophic lateral sclerosis. *Gene* 508:35–40.

De Felice B., Annunziata A., Fiorentino G. et al. 2014. miR-338-3p is over-expressed in blood, CFS, serum and spinal cord from sporadic amyotrophic lateral sclerosis patients. *Neurogenetics* 15:243–253.

DeJesus-Hernandez M., Mackenzie I.R., Boeve B.F. et al. 2011. Expanded GGGGCC hexanucleotide repeat in noncoding region of C9ORF72 causes chromosome 9p-linked FTD and ALS. *Neuron* 72:245–256.

DiFiglia M., Sapp E., Chase K.O. et al. 1997. Aggregation of huntingtin in neuronal intranuclear inclusions and dystrophic neurites in brain. *Science* 277:1990–1993.

Dini Modigliani S., Morlando M., Errichelli L. et al. 2014. An ALS-associated mutation in the FUS 3'-UTR disrupts a microRNA-FUS regulatory circuitry. *Nature Communications* 5:4335.

Doi H., Okamura K., Bauer P.O. et al. 2008. RNA-binding protein TLS is a major nuclear aggregate-interacting protein in huntingtin exon 1 with expanded polyglutamine-expressing cells. *Journal of Biological Chemistry* 283:6489–6500.

Duyckaerts C., Hauw J.J. 2003. Lewy bodies, a misleading marker for Parkinson's disease? *Bulletin de l'Académie Nationale de Médecine* 187:277–292.

Emmanouilidou E., Melachroinou K., Roumeliotis T. et al. 2010. Cell-produced alpha-synuclein is secreted in a calcium-dependent manner by exosomes and impacts neuronal survival. *Journal of Neuroscience* 30:6838–6851.

Fagan A.M., Roe C.M., Xiong C. et al. 2007. Cerebrospinal fluid tau/beta-amyloid(42) ratio as a prediction of cognitive decline in nondemented older adults. *Archives of Neurology* 64:343–349.

Farajollahi S., Maas S. 2010. Molecular diversity through RNA editing: A balancing act. *Trends in Genetics* 26:221–230.

Fénelon K., Mukai J., Xu B. et al. 2011. Deficiency of Dgcr8, a gene disrupted by the 22q11.2 microdeletion, results in altered short-term plasticity in the prefrontal cortex. *Proceedings of the National Academy of Sciences of the United States of America* 108:4447–4452.

Fluiter K., ten Asbroek A.L., de Wissel M.B. et al. 2003. *In vivo* tumor growth inhibition and biodistribution studies of locked nucleic acid (LNA) antisense oligonucleotides. *Nucleic Acids Research* 31:953–962.

Funayama M., Hasegawa K., Kowa H. et al. 2002. A new locus for Parkinson's disease (PARK8) maps to chromosome 12p11.2-q13.1. *Annals of Neurology* 51:296–301.

Galasko D., Montine T. 2010. Biomarkers of oxidative damage and inflammation in Alzheimer's disease. *Biomarkers in Medicine* 4:27–36.

Galimberti D., Villa C., Fenoglio C. et al. 2014. Circulating miRNAs as potential biomarkers in Alzheimer's disease. *Journal of Alzheimer Disease* 402(4):1261–1267.

Garzon R., Marcucci G., Croce C. 2010. Targeting microRNAs in cancer: Rationale, strategies and challenges. *Nature Reviews Drug Discovery* 9:775–789.

Gascon E., Gao F.-B. 2012. Cause or effect: Misregulation of microRNA pathways in neurodegeneration. *Frontiers in Neuroscience* 6:48.

Gaughwin P.M., Ciesla M., Lahiri N. et al. 2011. Hsa-miR-34b is a plasma-stable microRNA that is elevated in pre-manifest Huntington's disease. *Human Molecular Genetics* 20:2225–2237.

Geekiyanage H., Chan C. 2012. MicroRNA-137/181c regulates serine palmitoyltransferase and in turn amyloid beta, novel targets in sporadic Alzheimer's disease. *Journal of Neuroscience* 31:14820–14830.

Ghidoni R., Benussi L., Paterlini A. et al. 2011. Cerebrospinal fluid biomarkers for Alzheimer's disease: The present and the future. *Neurodegenerative Diseases* 8:413–420.

Ghildiyal M., Zamore P.D. 2009. Small silencing RNAs: An expanding universe. *Nature Review Genetics* 10:94–108.

Glenner G.G., Wong C.W. 1984. Alzheimer's disease: Initial report of the purification and characterization of a novel cerebrovascular amyloid protein. *Biochemical and Biophysical Research Communications* 120:885–90.

Gonzalez-Cuyar L.F., Sonnen J.A., Montine K.S. et al. 2011. Role of cerebrospinal fluid and plasma biomarkers in the diagnosis of neurodegenerative disorders and mild cognitive impairment. *Current Neurology and Neuroscience Reports* 11:455–463.

Gregory R.I., Yan K.P., Amuthan G. et al. 2004. The microprocessor complex mediates the genesis of microRNAs. *Nature* 432:235–240.

Griffiths-Jones S., Saini H.K., van Dongen S. et al. 2008. MiRBase: Tools for microRNA genomics. *Nucleic Acids Research* 36: D154–D158.

Grundke-Iqbal I., Iqbal K., Quinlan M. et al. 1986. Microtubule-associated protein tau. A component of Alzheimer paired helical filaments. *Journal of Biological Chemistry* 261:6084–6089.

Guo H., Ingolia N.T., Weissman J.S. et al. 2010. Mammalian microRNAs predominantly act to decrease target mRNA levels. *Nature* 466:835–840.

Han G., Sun J., Wang J. et al. 2014. Genomics in neurological disorders. *Genomics Proteomics Bioinformatics* 12:156–163.

Haramati S., Chapnik E., Sztainberg Y. et al. 2010. miRNA malfunction causes spinal motor neuron disease. *Proceedings of the National Academy of Sciences of the United States of America* 107:13111–13116.

Hasegawa M., Arai T., Akiyama H. et al. 2007. TDP-43 is deposited in the Guam parkinsonism-dementia complex brains. *Brain* 130:1386–1394.

Hébert S.S., Horre K., Nicolaï L. et al. 2008. Loss of microRNA cluster miR-29a/b-1 in sporadic Alzheimer's disease correlates with increased BACE1/β-secretase expression. *Proceedings of the National Academy of Sciences of the United States of America* 105:6415–6420.

Hébert S.S., Papadopoulou A.S., Smith P. et al. 2010. Genetic ablation of Dicer in adult forebrain neurons results in abnormal tau hyperphosphorylation and neurodegeneration. *Human Molecular Genetics* 19:3959–3969.

Helwak A., Kudla G., Dudnakova T. et al. 2013. Mapping the human miRNA interactome by CLASH reveals frequent noncanonical binding. *Cell* 153:654–665.

Hong Z., Shi M., Chung K.A. et al. 2010. DJ-1 and alpha-synuclein in human cerebrospinal fluid as biomarkers of Parkinson's disease. *Brain* 133:713–726.

Hu Z., Chen X., Zhao Y. et al. 2010. Serum microRNA signatures identified in a genome-wide serum microRNA expression profiling predict survival of non-small-cell lung cancer. *Journal of Clinical Oncology* 28:1721–1726.

Hu H.Y., Guo S., Xi J. et al. 2011. MicroRNA expression and regulation in human, chimpanzee, and macaque brains. *PLoS Genetics* 7:e1002327.

Hutton M., Lendon C.L., Rizzu P. et al. 1998. Association of missense and 5′-splice-site mutations in tau with the inherited dementia FTDP-17. *Nature* 393:702–705.

Hutvágner G., Simard M.J., Mello C.C. et al. 2004. Sequence-specific inhibition of small RNA function. *PLoS Biology* 2:E98.

Ishtiaq M., Campos-Melo D., Volkening K. et al. 2014. Analysis of novel NEFL mRNA targeting microRNAs in amyotrophic lateral sclerosis. *PLoS One* 9:e85653.

Jiao J., Herl L.D., Farese R.V. et al. 2010. MicroRNA-29b regulates the expression level of human progranulin, a secreted glycoprotein implicated in frontotemporal dementia. *PLoS One* 5:e10551.

Jin J., Cheng Y., Zhang Y. et al. 2012. Interrogation of brain miRNA and mRNA expression profiles reveals a molecular regulatory network that is perturbed by mutant huntingtin. *Journal of Neurochemistry* 123:477–490.

Johnson R., Zuccato C., Belyaev N.D. et al. 2008. A microRNA-based gene dysregulation pathway in Huntington's disease. *Neurobiology of Disease* 29:438–445.

Johnson R., Richter N., Jauch R. et al. 2010. Human accelerated region 1 noncoding RNA is repressed by REST in Huntington's disease. *Physiological Genomics* 41:269–274.

Johnson J.O., Mandrioli J., Benatar M. et al. 2010. Exome sequencing reveals VCP mutations as a cause of familial ALS. *Neuron* 68:857–864.

Kamm R., Smith A. 1972. Ribonuclease activity in human plasma. *Clinical Biochemistry* 5:198–200.

Kapsimali M., Kloosterman W.P., de Bruijn E. et al. 2007. MicroRNAs show a wide diversity of expression profiles in the developing and mature central nervous system. *Genome Biology* 8: R173.

Kawahara Y., Mieda-Sato A. 2012. TDP-43 promotes microRNA biogenesis as a component of the Drosha and Dicer complexes. *Proceedings of the National Academy of Sciences of the United States of America* 109:3347–3352.

Kaye F.J., Shows T.B. 2000. Assignment of ubiquilin2 (UBQLN2) to human chromosome xp11. 23—>p11.1 by GeneBridge radiation hybrids. *Cytogenetics and Cell Genetics* 89:116–117.

Kharaziha P., Ceder S., Li Q. et al. 2012. Tumor cell-derived exosomes: A message in a bottle. *Biochimica et Biophysica Acta* 1826:103–111.

Khoo S.K., Petillo D., Kang U.J. et al. 2012. Plasma-based circulating microRNA biomarkers for Parkinson's disease. *Journal of Parkinson's Disease* 2:321–331.

Kiko T., Nakagawa K., Tsuduki T. et al. 2014. MicroRNAs in plasma and cerebrospinal fluid as potential markers for Alzheimer's disease. *Journal of Alzheimer's Disease* 39:253–259.

Kim J., Inoue K., Ishii J. et al. 2007. A microRNA feedback circuit in midbrain dopamine neurons. *Science* 317:1220–1224.

Kitada T., Asakawa S., Hattori N. et al. 1998. Mutations in the parkin gene cause autosomal recessive juvenile parkinsonism. *Nature* 392:605–608.

Knickmeyer R.C., Wang J., Zhu H. et al. 2014. Common variants in psychiatric risk genes predict brain structure at birth. *Cerebral Cortex* 24:1230–1246.

Kocerha J., Kouri N., Baker M. et al. 2011. Altered microRNA expression in frontotemporal lobar degeneration with TDP-43 pathology caused by progranulin mutations. *BMC Genomics* 12:527.

Krystal J.H., State M.W. 2014. Psychiatric disorders: Diagnosis to therapy. *Cell* 157:201–214.

Kuhn D.E., Nuovo G.J., Terry A.V. et al. 2010. Chromosome 21-derived microRNAs provide an etiological basis for aberrant protein expression in human Down syndrome brains. *Journal of Biological Chemistry* 285:1529–1543.

Kumar P., Dezso Z., MacKenzie C. et al. 2013. Circulating miRNA biomarkers for Alzheimer's disease. *PLoS One* 8:e69807.

La Spada A.R., Taylor J.P. 2010. Repeat expansion disease: Progress and puzzles in disease pathogenesis. *Nature Reviews. Genetics* 11:247–258.

Landgraf P., Rusu M., Sheridan R. et al. 2007. A mammalian microRNA expression atlas based on small RNA library sequencing. *Cell* 129:1401–1414.

Langbaum J.B., Fleisher A.S., Chen K. et al. 2013. Ushering in the study and treatment of preclinical Alzheimer disease. *Nature Reviews Neurology* 9:371–381.

Le W.D., Rowe D.B., Jankovic J. et al. 1999. Effects of cerebrospinal fluid from patients with Parkinson disease on dopaminergic cells. *Archives of Neurology* 56:194–200.

Lee R.C., Feinbaum R.L., Ambros V. 1993. The *C. elegans* heterochronic gene lin-4 encodes small RNAs with antisense complementarity to lin-14. *Cell* 75:843–854.

Lee Y., Kim M., Han J. et al. 2004. MicroRNA genes are transcribed by RNA polymerase II. *EMBO Journal* 23:4051–4060.

Lehmann S.M., Krüger C., Park B. et al. 2012. An unconventional role for miRNA: Let-7 activates Toll-like receptor 7 and causes neurodegeneration. *Nature Neuroscience* 15:827–835.

Leidinger P., Backes C., Deutscher S. et al. 2013. A blood based 12-miRNA signature of Alzheimer disease patients. *Genome Biology* 14:R78.

Levy-Lahad E., Wasco W., Poorkaj P. et al. 1995. Candidate gene for the chromosome 1 familial Alzheimer's disease locus. *Science* 269:973–977.

Lim L.P., Lau N.C., Garrett-Engele P. et al. 2005. Microarray analysis shows that some microRNAs downregulate large numbers of target mRNAs. *Nature* 433:769–773.

Lukiw W.J., Zhao Y., Cui J.G. 2007. An NF-kappaB-sensitive micro RNA-146a-mediated inflammatory circuit in Alzheimer disease and in stressed human brain cells. *Journal of Biological Chemistry* 283:31315–31322.

Lukiw W.J., Alexandrov P.N., Zhao Y. et al. 2012. Spreading of Alzheimer's disease inflammatory signaling through soluble micro-RNA. *Neuroreport* 23:621–626.

MacDonald M. 1993. A novel gene containing a trinucleotide repeat that is expanded and unstable on Huntington's disease chromosomes. *Cell* 72:971–983.

Mackenzie I.R.A., Bigio E.H., Ince P.G. et al. 2007. Pathological TDP-43 distinguishes sporadic amyotrophic lateral sclerosis from amyotrophic lateral sclerosis with SOD1 mutations. *Annals of Neurology* 61:427–434.

Manakov S.A., Grant S.G.N., Enright A.J. 2009. Reciprocal regulation of microRNA and mRNA profiles in neuronal development and synapse formation. *BMC Genomics* 10:419.

Margis R., Rieder C.R. 2011. Identification of blood microRNAs associated to Parkinson's disease. *Journal of Biotechnology* 152:96–101.

Marksteiner J., Hinterhuber H., Humpel C. 2007. Cerebrospinal fluid biomarkers of Alzheimer's disease: Beta-amyloid (1–42), tau, phosphor-tau-181 and total protein. *Drugs of Today* 43:423–431.

Martí E., Pantano L., Bañez-Coronel M. et al. 2010. A myriad of miRNA variants in control and Huntington's disease brain regions detected by massively parallel sequencing. *Nucleic Acids Research* 38:7219–7235.

Martins M., Rosa A., Guedes L.C. et al. 2011 Convergence of miRNA expression profiling, α-synuclein interacton and GWAS in Parkinson's disease. *PLoS One* 6:e25443.

McCarroll S.A., Feng G., Hyman S.E. 2014. Genome-scale neurogenetics: Methodology and meaning. *Nature Neuroscience* 17:756–763.

Meister G., Landthaler M., Patkaniowska A. et al. 2004. Human Argonaute2 mediates RNA cleavage targeted by miRNAs and siRNAs. *Molecular Cell* 15:185–197.

Meister G., Tuschl T. 2004. Mechanisms of gene silencing by double-stranded RNA. *Nature* 431:343–349.

Miska E.A., Alvarez-Saavedra E., Townsend M. et al. 2004. Microarray analysis of microRNA expression in the developing mammalian brain. *Genome Biology* 5:R68.

Michell A.W., Lewis S.J., Foltynie T. et al. 2004. Biomarkers and Parkinson's disease. *Brain* 127:1693–1705.

Mitchell P.S., Parkin R.K., Kroh E.M. et al. 2008. Circulating microRNAs as stable blood-based markers for cancer detection. *Proceedings of the National Academy of Sciences of the United States of America* 105:10513–10518.

Mollenhauer B., Locascio J.J., Schulz-Schaeffer W. et al. 2011. α-Synuclein and tau concentrations in cerebrospinal fluid of patients presenting with parkinsonism: A cohort study. *The Lancet Neurology* 10:230–240.

Montecalvo A., Larregina A., Shufesky W.J. et al. 2012. Mechanism of transfer of functional microRNAs between mouse dendritic cells via exosomes. *Blood* 119:756–766.

Morello M., Minciacchi V.R., de Candia P. et al. 2013. Large oncosomes mediate intercellular transfer of functional microRNA. *Cell Cycle* 12:3526–3536.

Morlando M., Dini Modigliani S., Torrelli G. et al. 2012. FUS stimulates microRNA biogenesis by facilitating co-transcriptional Drosha recruitment. *EMBO Journal* 31:4502–4510.

Neumann M., Sampathu D.M., Kwong L.K. et al. 2006. Ubiquitinated TDP-43 in frontotemporal lobar degeneration and amyotrophic lateral sclerosis. *Science* 314:130–133.

Neumann M., Mackenzie I.R., Cairns N.J. et al. 2007. TDP-43 in the ubiquitin pathology of frontotemporal dementia with VCP gene mutations. *Journal of Neuropathology and Experimental Neurology* 66:152–157.

Neumann M., Rademakers R., Roeber S. et al. 2009. A new subtype of frontotemporal lobar degeneration with FUS pathology. *Brain* 132:2922–2931.

Nolan K., Mitchem M.R., Jimenez-Mateos E.M. et al. 2014. Increased expression of microRNA-29a in ALS mice: Functional analysis of its inhibition. *Journal of Molecular Neuroscience* 53:231–241.

Nonaka T., Masuda-Suzukake M., Arai T. et al. 2013. Prion-like properties of pathological TDP-43 aggregates from diseased brains. *Cell Reports* 4:124–134.

O'Brien R., Wong P. 2011. Amyloid precursor protein processing and Alzheimer's disease. *Annual Review of Neuroscience* 34:185–204.

Ørom U.A., Nielsen F.C., Lund A.H. 2008. MicroRNA-10a binds the 5'UTR of ribosomal protein mRNAs and enhances their translation. *Molecular Cell* 30:460–471.

Packer A.N., Xing Y., Harper S.Q. et al. 2008. The bifunctional microRNA miR-9/miR-9* regulates REST and CoREST and is downregulated in Huntington's disease. *Journal of Neuroscience* 28:14341–14346.

Pant S., Hilton H., Burczynski M.E. 2012. The multifaceted exosome: Biogenesis, role in normal and aberrant cellular function, and frontiers for pharmacological and biomarker opportunities. *Biochemical Pharmacology* 83:1484–1494.

Parkinson N., Ince P.G., Smith M.O. et al. 2006. ALS phenotypes with mutations in CHMP2B (charged multivesicular body protein 2B). *Neurology* 67:1074–1077.

Polymeropoulos M.H., Higgins J.J., Golbe L.I. et al. 1996. Mapping of a gene for Parkinson's disease to chromosome 4q21-q23. *Science* 274:1197–1199.

Polymeropoulos M.H., Lavedan C., Leroy E. et al. 1997. Mutation in the alpha-synuclein gene identified in families with Parkinson's disease. *Science* 276:2045–2047.

Rachakonda V., Hong P.T., Le W. 2004. Biomarkers of neurodegenerative disorders: How good are they? *Cell Research* 14:347–358.

Rademakers R., Eriksen J.L., Baker M. et al. 2008. Common variation in the miR-659 binding-site of GRN is a major risk factor for TDP43-positive frontotemporal dementia. *Human Molecular Genetics* 17:3631–3642.

Rajendran L., Honsho M., Zahn T.R. et al. 2006. Alzheimer's disease beta-amyloid peptides are released in association with exosomes. *Proceedings of the National Academy of Sciences of the United States of America* 103:11172–11177.

Renton A.E., Majounie E., Waite A. et al. 2011. A hexanucleotide repeat expansion in C9ORF72 is the cause of chromosome 9p21-linked ALS-FTD. *Neuron* 72:257–268.

Rigoutsos I. 2009. New tricks for animal microRNAS: Targeting of amino acid coding regions at conserved and nonconserved sites. *Cancer Research* 69:3245–3248.

Rosen D.R., Siddique T., Patterson D. et al. 1993. Mutations in Cu/Zn superoxide dismutase gene are associated with familial amyotrophic lateral sclerosis. *Nature* 362:59–62.

Russell A.P., Wada S., Vergani L. et al. 2012. Disruption of skeletal muscle mitochondrial network genes and miRNAs in amyotrophic lateral sclerosis. *Neurobiology of Disease* 49C:107–117.

Sala Frigerio C., Lau P., Salta E. et al. 2013. Reduced expression of hsa-miR-27a-3p in CSF of patients with Alzheimer disease. *Neurology* 81:2103–2106.

Saman S., Kim W., Raya M. et al. 2012. Exosome-associated tau is secreted in tauopathy models and is selectively phosphorylated in cerebrospinal fluid in early Alzheimer disease. *Journal of Biological Chemistry* 287:3842–3849.

Sapp P.C., Hosler B.A., McKenna-Yasek D. et al. 2003. Identification of two novel loci for dominantly inherited familial amyotrophic lateral sclerosis. *American Journal of Human Genetics* 73:397–403.

Savas J.N., Makusky A., Ottosen S. et al. 2008. Huntington's disease protein contributes to RNA-mediated gene silencing through association with Argonaute and P bodies. *Proceedings of the National Academy of Sciences of the United States of America* 105:10820–10825.

Schaefer A., O'Carroll D., Tan C.L. et al. 2007. Cerebellar neurodegeneration in the absence of microRNAs. *Journal of Experimental Medicine* 204:1553–1558.

Schipper H.M., Maes O.C., Chertkow H.M. et al. 2007. MicroRNA expression in Alzheimer blood mononuclear cells. *Gene Regulation and Systems Biology* 1:263–274.

Schofield C.M., Hsu R., Barker A.J. et al. 2011. Monoallelic deletion of the microRNA biogenesis gene Dgcr8 produces deficits in the development of excitatory synaptic transmission in the prefrontal cortex. *Neural Development* 6:11.

Schwarzenbach H., Hoon D.S.B., Pantel K. 2011. Cell-free nucleic acids as biomarkers in cancer patients. *Nature Reviews Cancer* 11:426–37.

Sethi P., Lukiw W.J. 2009. Micro-RNA abundance and stability in human brain: Specific alterations in Alzheimer's disease temporal lobe neocortex. *Neuroscience Letters* 459:100–104.

Sheinerman K.S., Tsivinsky V.G., Crawford F. et al. 2012. Plasma microRNA biomarkers for detection of mild cognitive impairment. *Aging* 4:590–605.

Shi M., Bradner J., Hancock A. et al. 2011. Cerebrospinal fluid biomarkers for Parkinson disease diagnosis and progression. *Annals of Neurology* 69:570–580.

Shin D., Shin J.Y., McManus M.T. et al. 2009. Dicer ablation in oligodendrocytes provokes neuronal impairment in mice. *Annals of Neurology* 66:843–857.

Sinha M., Ghose J., Das E. et al. 2010. Altered microRNAs in STHdh(Q111)/Hdh(Q111) cells: MiR-146a targets TBP. *Biochemical and Biophysical Research Communications* 396:742–747.

Sinha M., Ghose J., Bhattarcharyya N.P. 2011. Micro RNA -214,-150,-146a and -125b target Huntingtin gene. *RNA Biology* 8:1005–1021.

Skibinski G., Parkinson N.J., Brown J.M. et al. 2005. Mutations in the endosomal ESCRTIII-complex subunit CHMP2B in frontotemporal dementia. *Nature Genetics* 37:806–808.

Skog J., Würdinger T., van Rijn S. et al. 2008. Glioblastoma microvesicles transport RNA and proteins that promote tumour growth and provide diagnostic biomarkers. *Nature Cell Biology* 10:1470–1476.

Somel M., Guo S., Fu N. et al. 2010. MicroRNA, mRNA, and protein expression link development and aging in human and macaque brain. *Genome Research* 20:1207–1218.

Somel M., Liu X., Tang L. et al. 2011. MicroRNA-driven developmental remodeling in the brain distinguishes humans from other primates. *PLoS Biology* 9:e1001214.

Soreq L., Salomonis N., Bronstein M. et al. 2013. Small RNA sequencing-microarray analyses in Parkinson leukocytes reveal deep brainstimulation-induced splicing changes that classify brain region transcriptomes. *Frontiers in Molecular Neuroscience* 6:10.

Spillantini M.G., Schmidt M.L., Lee V.M. et al. 1997. Alpha-synuclein in Lewy bodies. *Nature* 388:839–840.

Sreedharan J., Blair I.P., Tripathi V.B. et al. 2008. TDP-43 mutations in familial and sporadic amyotrophic lateral sclerosis. *Science* 319:1668–1672.

Stark K.L., Xu B., Bagchi A. et al. 2008. Altered brain microRNA biogenesis contributes to phenotypic deficits in a 22q11-deletion mouse model. *Nature Genetics* 40:751–760.

Stoorvogel W. 2012. Functional transfer of microRNA by exosomes. *Blood* 119:646–648.

Strittmatter W.J., Saunders A.M., Schmechel D. et al. 1993. Apolipoprotein E: High-avidity binding to beta-amyloid and increased frequency of type 4 allele in late-onset familial Alzheimer disease. *Proceedings of the National Academy of Sciences of the United States of America* 90:1977–1981.

Tan L., Yu J.T., Tan M.S. et al. 2014. Genome-wide serum microRNA expression profiling identifies serum biomarkers for Alzheimer's disease. *Journal of Alzheimer Disease* 40(4):1017–1027.

Tao J., Wu H., Lin Q. et al. 2011. Deletion of astroglial Dicer causes non-cell autonomous neuronal dysfunction and degeneration. *Journal of Neuroscience* 31:8306–8319.

Toivonen J.M., Manzano R., Oliván S. et al. 2014. MicroRNA-206: A potential circulating biomarker candidate for amyotrophic lateral sclerosis. *PLoS One* 9:e89065.

Turchinovich A., Weiz L., Langheinz A. et al. 2011. Characterization of extracellular circulating microRNA. *Nucleic Acids Research* 39:7223–7233.

Uryu K., Nakashima-Yasuda H., Forman M.S. et al. 2008. Concomitant TAR-DNA-binding protein 43 pathology is present in Alzheimer disease and corticobasal degeneration but not in other tauopathies. *Journal of Neuropathology and Experimental Neurology* 67:555–564.

Valadi H., Ekström K., Bossios A. et al. 2007. Exosome-mediated transfer of mRNAs and microRNAs is a novel mechanism of genetic exchange between cells. *Nature Cell Biology* 9:654–659.

Valente E.M., Bentivoglio A.R., Dixon P.H. et al. 2001. Localization of a novel locus for autosomal recessive early-onset parkinsonism, PARK6, on human chromosome 1p35-p36. *American Journal of Human Genetics* 68:895–900.

Vallelunga A., Raqusa M., Di Mauro S. et al. 2014. Identification of circulating microRNAs for the differential diagnosis of Parkinson's disease and multiple system atrophy. *Frontiers in Cellular Neuroscience* 8:156.

Van Duijn C.M., Dekker M.C., Bonifati V. et al. 2001. Park7, a novel locus for autosomal recessive early-onset parkinsonism, on chromosome 1p36. *American Journal of Human Genetics* 69:629–634.

Vickers K., Palmisano B., Shoucri B.M. et al. 2011. MicroRNAs are transported in plasma and delivered to recipient cells by high-density lipoproteins. *Nature Cell Biology* 13:423–433.

Wang K., Zhang S., Weber J. et al. 2010. Export of microRNAs and microRNA-protective protein by mammalian cells. *Nucleic Acids Research* 38:7248–7259.

Wang W.X., Rajeev B.W., Stromberg A.J. et al. 2008. The expression of microRNA miR-107 decreases early in Alzheimer's disease and may accelerate disease progression through regulation of β-site amyloid precursor protein-cleaving enzyme 1. *Journal of Neuroscience* 28:1213–1223.

Wang W.X., Wilfred B.R., Madathil S.K. et al. 2010. MiR-107 regulates granulin/progranulin with implications for traumatic brain injury and neurodegenerative disease. *The American Journal of Pathology* 177:334–345.

Waragai M., Wei J., Fujita M. et al. 2006. Increased level of DJ-1 in the cerebrospinal fluids of sporadic Parkinson's disease. *Biochemical and Biophysical Research Communications* 345:967–972.

Watts G.D.J., Wymer J., Kovach M.J. et al. 2004. Inclusion body myopathy associated with Paget disease of bone and frontotemporal dementia is caused by mutant valosin-containing protein. *Nature Genetics* 36:377–381.

Weber J.A., Baxter D.H., Zhang S. et al. 2010. The microRNA spectrum in 12 body fluids. *Clinical Chemistry* 56:1733–1741.

Williams A.H., Valdez G., Moresi V. et al. 2009. MicroRNA-206 delays ALS progression and promotes regeneration of neuromuscular synapses in mice. *Science* 326:1549–1554.

Wu D., Raafat A., Pak E. et al. 2012. Dicer-microRNA pathway is critical for peripheral nerve regeneration and functional recovery *in vivo* and regenerative axonogenesis in vitro. *Experimental Neurology* 233:555–565.

Yamazaki T., Chen S., Yu Y. et al. 2012. FUS-SMN protein interactions link the motor neuron diseases ALS and SMA. *Cell Reports* 2:799–806.

Zhang Z., Almeida S., Lu Y. et al. 2013. Downregulation of microRNA-9 in iPSC-derived neurons of FTD/ALS patients with TDP-43 mutations. *PLoS One* 8:e76055.

Zuccato C., Belyaev N., Conforti P. et al. 2007. Widespread disruption of repressor element-1 silencing transcription factor/neuron-restrictive silencer factor occupancy at its target genes in Huntington's disease. *Journal of Neuroscience* 27:6972–6983.

# 9 Circulating Cell-Free MicroRNAs as Biomarkers for Neural Development and Their Importance in Fetal Programing for Postnatal Disease

*Mario Lamadrid-Romero,*
*Néstor Fabián Díaz Martínez,*
*and Anayansi Molina-Hernández*

## CONTENTS

Abstract .................................................................................................................. 237
9.1   MicroRNAs during Neural Development ................................................... 238
9.2   MiRNAs and Neurogenesis ....................................................................... 243
9.3   MicroRNAs and Its Potential as Central Nervous System
      Development Biomarkers ........................................................................... 245
9.4   Developmental Neurological Damage Can Be Related to Changes
      in MicroRNA Levels ................................................................................. 246
References ............................................................................................................. 249

## ABSTRACT

This chapter gives an overview of recent progress related to the role of microRNAs in the central nervous system development. One approach that highlights the role of these noncoding small RNAs during fetal central nervous system development is the use of transgenic animals in which the conditional knockout of Dicer has been accomplished during the development of the cerebral cortex, the hippocampus, the retina, or the midbrain. Another strategy has been to inhibit or overexpressed microRNAs that have been implicated in neurogenesis, one of the most studied process in the developmental biology. Although, little is known about how microRNAs are involved in the control of gene expression during the complex network formation of the central nervous system during development, the expression of several important transcriptional factors involved in central nervous system formation have been identify as microRNAs targets, suggesting an important participation of these molecules on this process. Finally, this chapter highlights the possibility of using microRNAs as biomarkers for central nervous system development and how altered conditions *in utero* might lead to the development of neurological diseases during postnatal life, an emerging field of research that certainly will impact on the etiology of human neural disorders.

## 9.1   MicroRNAs DURING NEURAL DEVELOPMENT

Diverse signals participate in central nervous system (CNS) patterning and organization synchronized on time and space. Among the molecules that are expressed in a temporal and spatial manner and considered essential during development are microRNAs (miRNAs; Lagos-Quintana et al., 2001; Lagos-Quintana et al., 2002; Gotz and Huttner, 2005; Fineberg et al., 2009; Li and Jin, 2010; Ji et al., 2013; Stappert et al., 2013). Interestingly, approximately 70% of the discovered miRNAs are expressed in the CNS (Krichevsky et al., 2003; Miska et al., 2004; Sempere et al., 2004; Ji et al., 2013).

MiRNAs are potent regulators of gene expression at posttranscriptional level through repression or translational inhibition of mRNAs targets (Lewis et al., 2003; Bartel, 2004; Lim et al., 2005; Barca-Mayo and De Pietri Tonelli, 2014). The participation of miRNAs in CNS development has been evidenced by diverse methodologies such as microarray, reverse transcription polymerase chain reaction (RT-qPCR), conditional null knockout of Dicer, northern blot, and gain–loss of function studies. Furthermore, miRNAs have been identified in several body fluids, an aspect that makes them potential molecules as biomarkers for neural development anomalies (Chen et al., 2008; Gilad et al., 2008; Cortez and Calin, 2009; Kroh et al., 2010; Gu et al., 2012; Blondal et al., 2013).

Microarray studies marked the birth of efforts to understand the participation of miRNAs in the developing CNS. Although new miRNAs are constantly discovered, the first effort to explore miRNA expression during CNS development was made by Krichevsky and collaborators in 2003. Using a 44 probes array, they detected changes on miRNA levels during corticogenesis from embryo day (E) 12 to E21 in murine. Changes on the levels of miRNAs were confirmed by northern blot analysis and an important finding was that several miRNAs showed up- or downregulation during brain development (Krichevsky et al., 2003). Further studies using animal models and stem cells have corroborated a dynamic miRNA expression in CNS development and in neural progenitor's proliferation and differentiation (Miska et al., 2004; Krichevsky et al., 2006).

Bioinformatics with qPCR validation is other tool that has been used to support the hypothesis that miRNAs are key molecule regulators of CNS development. In this sense, genome ontology analysis has revealed that downregulated miRNAs during neurogenesis have mRNA targets involved in the regulation of cellular migration and differentiation, while those which are upregulated have targets related to cell proliferation (Figure 9.1; Nielsen et al., 2009).

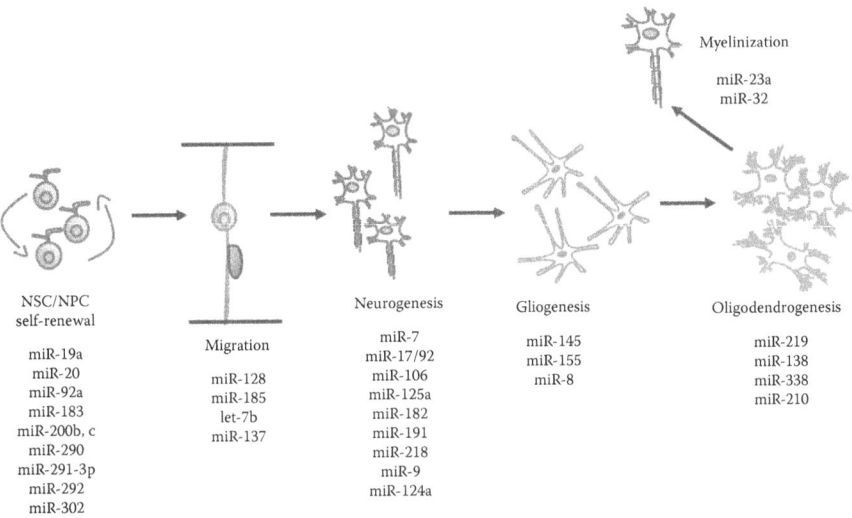

**FIGURE 9.1**    MicroRNAs (miRNAs-) expression. In neural stem/progenitor cells (NSC/NPC) renewal, neuron, glial, and oligodendrocyte differentiation and axon myelination.

Mice null for Dicer are not viable after E7.5 (Bernstein et al., 2003), indicating that endogenous miRNAs are critical for mammalian development. In order to investigate the role of miRNAs in CNS development, conditional deletions of Dicer have been used. Several studies regarding the ablation of miRNAs during fetal development using Dicer knockout (Dicer$^{-/-}$) *in vitro* and *in vivo* have revealed that miRNAs are important for neuron differentiation and maintenance (Bernstein et al., 2003; Cuellar et al., 2008; Davis et al., 2008; De Pietri Tonelli et al., 2008; Coolen and Bally-Cuif, 2009; Fineberg et al., 2009; Kawase-Koga et al., 2009; Huang et al., 2010; Peng et al., 2012; Saurat et al., 2013). The role for miRNAs in mammalian dopamine (DA) neuron differentiation, function, and survival has been assessed *in vitro* and *in vivo*, using a protocol to differentiate murine embryonic stem cells (mESC) in DA neurons (Kim et al., 2002). The mESC line expressing Dicer enzyme containing LoxP recombinase sites that flank both chromosomal copies of the Dicer gene (Kim et al., 2007; Murchison et al., 2005) were differentiated to a midbrain dopaminergic phenotype using the embryoid body protocol. The Cre-mediated deletion of the floxed Dicer alleles when the postmitotic DA neurons arise results in almost the complete loss of DA neurons and a reduction of GABAergic neurons due to an increased in apoptosis and a reduction in neurogenesis *in vitro*. The deletion of Dicer under the regulation of dopamine transporter resulted in a progressive loss of midbrain DA neurons and their nigrostriatal axonal processes due to apoptosis in mice (Figure 9.2; Chuhma et al., 2004; Kim et al., 2007).

Moreover, behavioral studies reported evident reduced locomotion in an open field assay. This study concludes that Dicer is essential for the terminal differentiation and maintenance of multiple neuron types, including midbrain dopaminergic neurons (Kim et al., 2007).

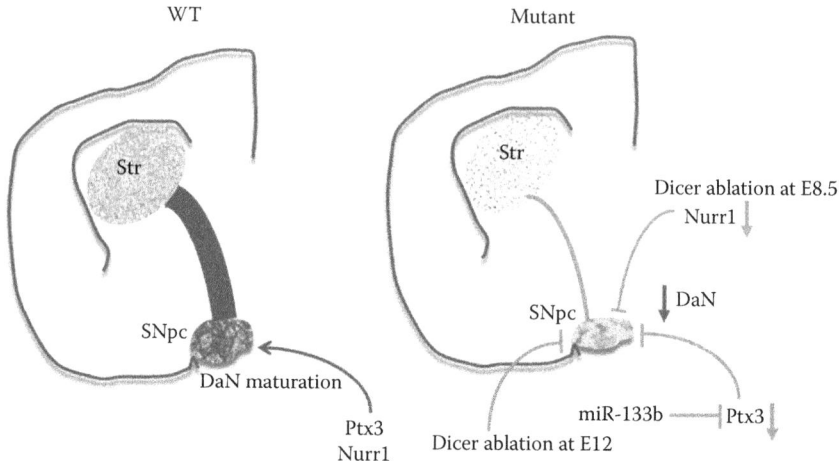

**FIGURE 9.2** Effect of mutant Dicer mice in dopaminergic differentiation and striatum innervation. In DATCRE/+: Dicer[flox/flox], Wnt1-Cre; Dicer[−/−] and miR-133 overexpression in Dopaminergic neuron differentiation (DaN) in the *sustancia nigra paris compacta* (SNpc) and its innervation to the striatum (Str). (Modified from Huang T. et al. 2010. *J Mol Cell Biol* 2:152–163; Nowakowski T.J., Mysiak K.S., O'Leary T. et al. 2013. *Dev Biol* 382:530–537; Nowakowski T.J. et al. 2011. *PLoS One* 6:e23013.)

In other series of experiment, the malformation of the midbrain was reported; in addition to the impairment of several neural crest derivate structures in Wnt1-Cre conditional null Dicer mice. This null mice presents a reduction of tyrosine hydroxylase in ventral midbrain and its ablation in the caudal region (E12.5–E14.5). These result suggested that miRNAs are involved in the development of the mid- and hindbrain regions (Figure 9.2; Huang et al., 2010).

The study of miRNAs in telencephalic mouse development has been addressed using the expression of Cre-recombinase driven by FoxG1 (expressed at E8), Emx1 and Nestin (expressed at E10.5) or calmodulin kinase II (CamkII, expressed at E15.5; Cuellar et al., 2008; Davis et al., 2008; Nowakowski et al., 2011).

The knockout of Dicer in the early developing forebrain using Foxg1-Cre Dicer1[fl/fl] mice demonstrates that when radial glia is normally generated, the expression of Sox9, ErbB2, and Nestin proteins is compromised (Nowakowski et al., 2011).

Moreover, these mice show that radial progenitors generate normal percentage of postmitotic neurons and an increased proportion of basal progenitor cells. However, these cell types are misplaced through the depth of the telencephalon, leading to an inappropriate specification of radial glia during early development due reduction of Nestin and Sox9 proteins (Nowakowski et al., 2011). Furthermore, the loss of Dicer in radial glia produced abnormally large numbers of Cux1-positive cells (cortical marker for layer II–IV neurons), a phenotype that is normally differentiated prenatally. This study showed that most of this Cux-1 neurons were abnormally generated after birth, suggesting that the loss of Dicer in radial glia lead to an increase on the time span of cortical neurogenesis (Figure 9.3d; Nowakowski et al., 2013).

Other null conditional for Dicer, the Emx1-Cre reveals that cortex-specific deletion of the enzyme results in a marked reduction in the phenotype cellular complexity of the cortex, due to a pronounced narrowing in the range of neuronal types generated by Dicer-null cortical stem and progenitor cells. Instead of generating different classes of lamina-specific neurons over the 6-day period of neurogenesis, Dicer-null cortical stem and progenitor cells continually produce only one class of deep layer projection neurons, the Tbr1 phenotype from cortical layers V–VI. However, gliogenesis in this Dicer-null cerebral cortex was not delayed, despite the loss of multipotency and the failure of neuronal lineage progression (Figure 9.3d; Saurat et al., 2013).

The function of miRNAs in the cortex and hippocampus have been studied by knocking down Dicer using a CaMKII-Cre$^{+/-}$ mouse, in which the expression of Cre recombinase is under the control of CaMKII promoter resulting in an inactivation of Dicer in excitatory forebrain neurons in the cortex and hippocampus of mice at E15 and forward (Dragatsis et al., 2000; Dragatsis and Zeitlin, 2000). The loss of Dicer in the cortex and hippocampus excitatory neurons led to dramatic effects on cellular and tissue morphology, axonal pathways, dendrite and spine morphology, and increased cortical apoptosis, which led to microcephaly and an increased lateral ventricle size, but has no effect on cortical lamination. Finally, these Dicer mutant mice are ataxic and present visible tremors and motor impairments after birth (Figure 9.3a–c) (Davis et al., 2008).

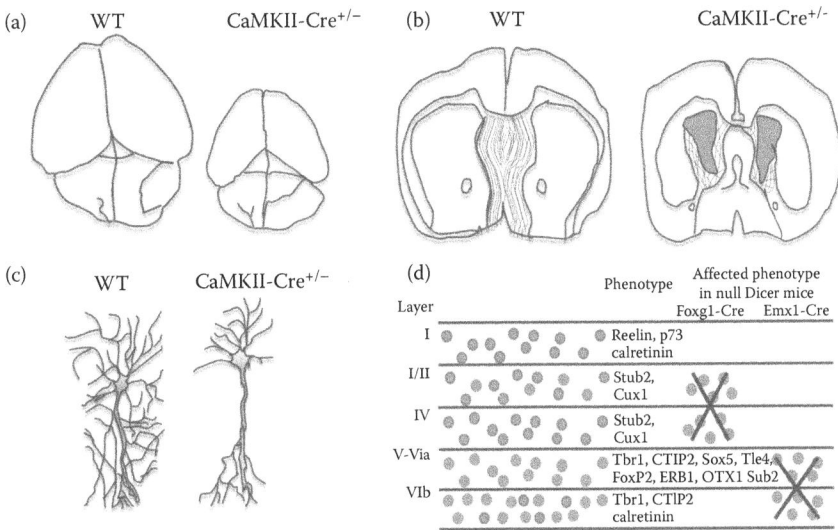

**FIGURE 9.3** Effect of the conditional loss of Dicer in cortex and hippocampus *in vivo*. A, B, and C are CaMKII-Cre and D the Foxg1-Cre Emx1-Cre Dicer-null mice. Decreases in brain size (a) and abnormally large ventricles (b) in brain mutant mice. A decreased branching of hippocampal CA1 neurons (c). And loss of deep neuron phenotypes (d). (Modified from Davis T.H. et al. 2008. *J Neurosci: Official J Soc Neurosci* 28:4322–4330; Nowakowski T.J. et al. 2013. *Dev Biol* 382:530–537; Nowakowski T.J. et al. 2011. *PLoS One* 6:e23013; Saurat N. et al. 2013. *Neural Dev* 8:14.)

The role of miRNAs in the control of axon pathfinder has been studied. The function of miRNAs in axon growth and guidance may be exerted through indirect mechanisms by regulation of factors important for tissue patterning, intracellular signaling or cell specification, or through direct regulation of axon guidance molecules critical for sensing the environment during axon extension (Pinter and Hindges, 2010). Interestingly, functional RISCs (RNA-induced silencing complex) have been found in developing axons (Hengst et al., 2006) and studies in zebrafish deficient for miRNAs have indicated general axon path finding defects during development (Giraldez et al., 2005). On the other hand, the mice retina has been used as a model of the participation of miRNAs in axon guidance. To study the conditional loss of Dicer in the retina, the Rx-Cre transgenic mouse has been used, where the Cre recombinase was expressed under the control of an Rx promoter element. The Cre-mediated recombination results in a complete ablation of Dicer activity, which resulted in alteration in eye size around E13. At E17.5, the vitreous humor was considerably reduced and the lens was not placed inside the retinal cup enclosed by the ciliary margin, but it was stuck out of the cup, with the ciliary margin about half way at the medial line of the lens. As in other conditional nulls for Dicer CNS structures, the reduction in the size of the retina is due to an increase in cell death, suggesting that mature miRNAs are essential for neural cell survival including the retina cells (Pinter and Hindges, 2010; Swindell et al., 2006).

The neurofilaments immunohistochemical retinal analysis showed that in Dicer mutant mice the retinal ganglion cells (RGC) axons extend correctly along the optic fiber layer, without growing aberrantly into other retinal layers, but with aberrant growth morphology of its axons in the optic fiber layer, since these appeared defasciculated and in a wave form while growing toward the optic disc, in comparison to the straight and fasciculated appearance of axons in the optic fiber layer of wild-type mice (Figure 9.4; Pinter and Hindges, 2010).

One of the major end points for RGC axons is the midline at the ventral hypothalamus, where in mice the majority of axons cross to the contralateral side and

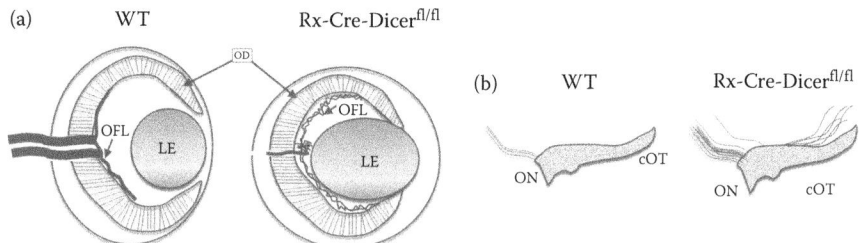

**FIGURE 9.4** Alternated retinal ganglion cell axons in the optic fiber layer in the Rx-Cre-Dicer[flox/flox] mice. A, is a representation of a coronal view at postnatal day 0 in native (a, left) and the mutant mice (b, right). Mutant animals shows a smaller optic disc, defasciculation and aberrant growth. LE, lens; OD, optic disc; OFL, optic fiber layer. In B, is a representation of the ventral view of the optic chiasm at E18.5 in a homozygous (right) Dicer mutant mice with aberrant fibers are formed. ON, optic nerve; OT: optic tract. (Modified from Pinter R., Hindges R. 2010. *PLoS One* 5:e10021.)

approximately 5% of the axons stay in an ipsilateral manner (Drager and Olsen, 1980). This process place between E13 and E17.5 in mice (Petros et al., 2008). Normally after birth the RGC axons are highly fasciculated across chiasm and in the optic tract. In contrast, in the Dicer mutant the axons are defasciculated which suggests that the lack of miRNAs disturbed the organization of the RGC axon trajectories at the midline and dispersed into the ventral hypothalamus bundles. This aspect was confirmed by retrograde staining (Pinter and Hindges, 2010).

## 9.2   MiRNAs AND NEUROGENESIS

The neurons are the first differentiated cell type during CNS development followed by astrocytes and oligodendrocytes (Figure 9.1). As mentioned above, miRNAs are known to play important roles during mammalian CNS development. These molecules present a wide diversity of expression in cerebral tissue during development and they are known to be essential for cerebral cortex morphogenesis (Krichevsky et al., 2003; Miska et al., 2004; Kapsimali et al., 2007; Nielsen et al., 2009).

The best studied process during CNS development is neurogenesis. During neuron differentiation the miRNAs that are consistently downregulated are miR-291-3p, miR-183, miR-200b, miR-200c, and miR-92a, whereas miR-9, miR-124a, miR-125a, miR-125b, and miR-7 are upregulated (Krichevsky et al., 2003; Nielsen et al., 2009). Predicted and experimental targets have been identified for some of these miRNAs. For example, some predicted targets for miR-183 encode proteins that promote neuronal differentiation, and the miR-92 targets mRNAs are involved in the negative regulation of the cortical progenitor cell cycle, reducing the inhibitor of the cell cycle cyclin-dependent kinase 1c translation (Nielsen et al., 2009).

The miR-183 is a member of a family that includes two other homologous miR-NAs (miR-96 and miR-182) that are transcribed from a single genetic locus in vertebrates. These miRNAs are expressed in cells of various tissues, including the olfactory epithelium, eye, neuromast, zebrafish ear, cranial and spinal ganglia sensory neurons, and the sensory cells of the eye and ear in chickens and mice, where it is involved in the development and maturation of sensory epithelia in the inner ear (Wienholds et al., 2005; Wienholds and Plasterk, 2005; Darnell et al., 2006; Kloosterman et al., 2006; Weston et al., 2006). This miRNA is considered an antiapoptotic and pro-oncogenic miRNA, as it inhibits early growth response protein 1 (Sarver et al., 2010) and promotes cerebellar granule neuron (CGN) progenitor proliferation in a cooperative manner with hedgehog signaling pathway (Zhang et al., 2013).

Another miRNA that is downregulated during neurogenesis is miR-92a, which is a member of the cluster miRNA-17/92. This cluster includes another five members; miR-17, miR-18a, miR-19a, miR-20a, and miR-19b-1. All of them are considered important in normal development as regulators of cell cycle, proliferation, and apoptosis (Mogilyansky and Rigoutsos, 2013).

Particularly, miR-92a has a direct participation in neuron maturation of postmitotic CGN from postnatal 8-day-old rats *in vitro*. Bioinformatics analysis showed that a putative mRNA target for this miRNA is the potassium-chloride co-transporter member 2 (KCC2). The interaction between miR-92 and KCC2 was demonstrated by a luciferase assay using a firefly reporter vector, supporting the role of miR-92 in the control of KCC2 expression during CGNs maturation (Barbato et al., 2010).

KCC2 is a neuron-specific chloride potassium symporter, responsible for the homeostasis of the chloride gradient in neurons, through the maintenance of low intracellular chloride concentrations that participate in determining the physiological response to the activation of ion-selective GABA and glycine receptors (Blaesse and Schmidt, 2015). Furthermore, KCC2 is a critical mediator of synaptic inhibition and cellular protection against excitotoxicity, and it has been involved in neuroplasticity (Takayama and Inoue, 2007; Barbato et al., 2010). The overexpression of miR-92 led to a change in the responsiveness to GABA inducing the shift in the reversal potential of GABA-induced chloride currents to a more positive voltage, an effect that is reversed by KCC2 overexpression, which implies an excitatory effect of this neurotransmitter before granular cell maturation (Barbato et al., 2010).

Interestingly, the miR-17/92 cluster is the first group of miRNAs to be implicated in a human disease, the Feingold syndrome, which is characterized by digital anomalies, microcephaly, facial dysmorphism, gastrointestinal atresia, and mild-to-moderate learning disability (Marcelis et al., 2008).

Regarding the miRNAs that are upregulated, the miR-9 is conserved among species, but its expression domains differ. In mice, gain and loss of function studies have shown that this miRNA is implicated in axonal growth and branch complexity, acting in a biphasic manner downstream of brain-derived neurotrophic factor (BDNF) signaling. First a reduction in miR-9 relieves the translational repression of microtubule-associated protein 1B (MAP1B, a protein that is required for dendritic spine development and synaptic maturation). The increased MAP1B level led to microtubule stabilization and axonal growth. In a second phase, the prolonged action of BDNF increases miR-9 expression inhibiting axon growth and promoting branching by decreasing MAP1b protein expression (Tortosa et al., 2011). *In vitro*, the ectopic expression of miR-9 leads to neurosphere formation during humanESC-derived neural progenitor differentiation. In contrast, the loss of miR-9 promoted migration of human neural progenitor cells *in vitro* and *in vivo* by suppressing stathmin (a protein that increases microtubule instability an event needed for migration) in mice cortical progenitors (Delaloy et al., 2010; Dajas-Bailador et al., 2012). One relevant target of miR-9 is Hes1, a transcriptional repressor of proneural genes such as Mash1 and Neurogenin2. During mice embryo development the telencephalon, diencephalon, and the spinal cord express high levels of miR-9, where this miRNA has a neurogenic effect by repressing Hes1 (Tan et al., 2012). Another important aspect of neurogenesis is the migration of neuron progenitor cells toward the ventricular zone in order to reach the site of final differentiation, a process where miR-9 is also involved (Barca-Mayo and De Pietri Tonelli, 2014).

MiR-124a is considered as a proneuronal miRNA. It has been established that this miRNA acting in a cooperative manner with mir-9 modulates the neuron-specific chromatin remodeling factor actin-related protein (BAF53a; Barca-Mayo and De Pietri Tonelli, 2014; Diaz et al., 2014). BAF53a regulates the transcription of RNA polymerase II promoter, which is involved in the establishment and/or maintenance of chromatin architecture, ATP-dependent chromatin remodeling, DNA repair, signal transduction, and DNA recombination.

Also, miR-124a is a key regulator of neuronal differentiation, brain development, and neuronal identity maintenance *in vitro* of ESCs, neural stem cells (NSCs), astrocytes, and cellular lines (P19 and HeLa). Lim and collaborators demonstrated that human (hsa)-miR-124 transfection into HeLa cells was sufficient to shift gene expression to a neuron-like type. MiR-124 is encoded in three different loci at chromosomes 2, 3, and 14; its transcription is controlled by the transcriptional repressor RE1-silencing transcription factor (REST), a restrictive neuronal factor. When REST is repressed alterations in radial migration in cerebral cortex can be detected (Visvanathan et al., 2007; Yu et al., 2008; Papagiannakopoulos and Kosik, 2009; Diaz et al., 2014).

An interesting finding is that miR-124a presents different profiles of expression between distinct areas of brain. For example, during the formation of the cerebral cortex the ventricular zone expresses miR-124 in low level, the subventricular zone presents intermediate expression, and the marginal zone has the higher expression. This gradient *in vivo* has been related with neural proliferation, migration, and neuron differentiation, respectively (Maiorano and Mallamaci, 2009, 2010).

Other targets for miR-124 with antineuronal function are: the small C-terminal domain phosphatase 1 (SCP1), PTBP1, PTBP2, LIM homeobox protein 2 (Lhx2), carboxy-terminal domain, RNA polymerase II, and polypeptide A (Ctdsp1; Makeyev et al., 2007; Visvanathan et al., 2007; Sanuki et al., 2011).

Finally, another important miRNA for CNS development is miR-125. This miRNA is homologous to the heterocronic miRNA lin4, the first miRNA described in the literature (Olsen and Ambros, 1999). The miR-125b is involved in neurogenesis and neuron maturation (Le et al., 2009) and reduces apoptosis. This miRNA is upregulated during mouse neurogenesis as well as in neural differentiation of mESCs, embryonic carcinoma cells (Krichevsky et al., 2006), neuroblastoma SK-N-BE cells (Laneve et al., 2007), SH-SY5Y cells, and cortical progenitor ReNcell cells. Process that is regulated by the translation inhibition of several targets, such as AP1M1, TK1IIP, PSMD8, ITCH, TBC1D1, TDG, MKNK2, DGAT1, GAB2, and SGPL1 (Le et al., 2009). Furthermore, the anti-apoptotic effects of miR-125b are due to the negative regulation of apoptotic genes from the p53 pathway (BAK1, TP53INP1, PPP1CA, and PRKRA; Le et al., 2011).

## 9.3  MicroRNAs AND ITS POTENTIAL AS CENTRAL NERVOUS SYSTEM DEVELOPMENT BIOMARKERS

As observed by the miRNA pattern of expression and dicer conditional null experiments, it is clear that mature miRNAs are essential during fetal CNS development. Besides, its presence in various body fluids such as urine, saliva, amniotic fluid, plasma, serum, and pleural fluid during pathological and physiological conditions (Cortez and Calin, 2009; Cortez et al., 2011), making them attractive as molecules that can be used as fetal CNS development biomarkers. In this sense, an important feature of miRNAs is that, unlike most RNAs, they are extremely stable in adverse conditions such as boiling, extreme pH, extended storage, and freeze-thaw cycles (Chen et al., 2008).

## 9.4 DEVELOPMENTAL NEUROLOGICAL DAMAGE CAN BE RELATED TO CHANGES IN MicroRNA LEVELS

The phenomenon whereby a stimulus occurring during a critical window during development which can cause lifelong changes in the structure and function of the body is defined as programming. Epidemiological studies have suggested intra-uterine environment as a leading cause of neurological pathologies such as schizo-phrenia, depression, obesity, Parkinson's, and Alzheimer's disease, among others in adult life (Rees and Inder, 2005; Alvarez-Buylla et al., 2008; Simeoni and Barker, 2009; Hanley et al., 2010; Glover, 2011). These have been supported by animal experiments in which alterations in fetal neurotransmitter balance, neural differ-entiation, migration, and death, as well as axon and dendritic abnormalities are observed in several maternal pathological conditions, such as diabetes, intrauter-ine growth restriction, maternal starvation, among others (Krichevsky et al., 2006; Loeken, 2006; Boren et al., 2008; Ha et al., 2008; Eixarch et al., 2009; Jawerbaum and White, 2010; Molina-Hernandez et al., 2012; Rodriguez-Martinez et al., 2012; van Vliet et al., 2013).

Routine fetal biopsies or analysis of amniotic liquid during development for miRNA profiling is not an option, therefore, researchers are turning toward less inva-sive procedures, involving circulating miRNA as biomarkers or "specific signatures" (Bernardo et al., 2012). Circulating miRNAs in humans have been proposed for a diagnostic role in neural tube defects and neurodegenerative processes (Gu et al., 2012). Nevertheless, several important aspects must be considered when selecting a specific or a group of miRNAs, as a biomarker for fetal CNS development, such as the moment during development in which the miRNAs of interest present changes in their levels of expression, the up- or downregulation of their expression during cer-tain stages of development and the magnitude of the change, the correlation between the tissue and the circulating levels at the same stage of CNS development, and the technique to be used for the measurement.

As mentioned above, a very important aspect to consider when proposing miRNAs as possible biomarkers, it is to choose the best technique to measure changes on the levels of circulating miRNAs in terms of time, sample processing, and cost. Taking together these characteristics, it has been suggested using the standard extraction of RNA based on the phenol–chloroform technique by using TRIZOL extraction, with an additional column purification and concentration, retrotranscription reaction using a stem-loop primers (or a miRNA polyadenilation and the retrotranscription with Oligo(dT) primers; Chen et al., 2005), a pre-amplification to enrich miRNAs (single or a pool of miRNAs) and a final qPCR amplification (Moldovan et al., 2014).

The changes in the expression of miRNAs in the blood have been assessed for acute myocardial infarction (Gilad et al., 2008; Ai et al., 2010; Widera et al., 2011), atherosclerosis (Menghini et al., 2013), nonsmall cell lung cancer (Huang et al., 2014), neurodegenerative diseases (Cheng et al., 2013; Jin et al., 2013; Sheinerman and Umansky, 2013a,b), skin fibrosis (Babalola et al., 2013), and osteoarthritis

(Gonzalez, 2013). However, there is no evidence in the literature regarding changes in maternal miRNAs levels related to human CNS development in health or disease.

For these reason, animal models are useful to determinate the relationship between tissue and circulating miRNA levels in the SNC. The most used animal models are mice and rats. These models permit researchers to assess in a short period during pregnancy (19–21 days) several stages (proliferative, neurogenic, and gliogenic) in both embryo and maternal plasma or serum (Figure 9.5).

For example, some CNS altered functions in offspring that are related with diabetes during pregnancy are: food intake regulation (Clausen et al., 2009; Steculorum and Bouret, 2011), language, vocabulary, and motor impairment (Rizzo et al., 1991; Ornoy et al., 1998, 1999, 2001).

It has been reported that embryos from diabetic-induced mice present early neuron differentiation and/or maturation in ventral and dorsal telencephalon, a structure that will give rise to the cerebral cortex, the hippocampus, the striatum, the hypothalamus among others, suggesting that the molecular programs that maintain stem-cell state or promotes neuron differentiation are impaired. Specific miRNAs have been involved in maintaining the undifferentiated pool of stem or progenitor cells, whereas others promote neuron differentiation, these functions are altered in fetuses from diabetic mice (Jawerbaum and White, 2010; Steculorum and Bouret, 2011). These evidence suggest that miRNAs could be upregulated during early development (before the neurogenic period) promoting early neurogenesis of the CNS.

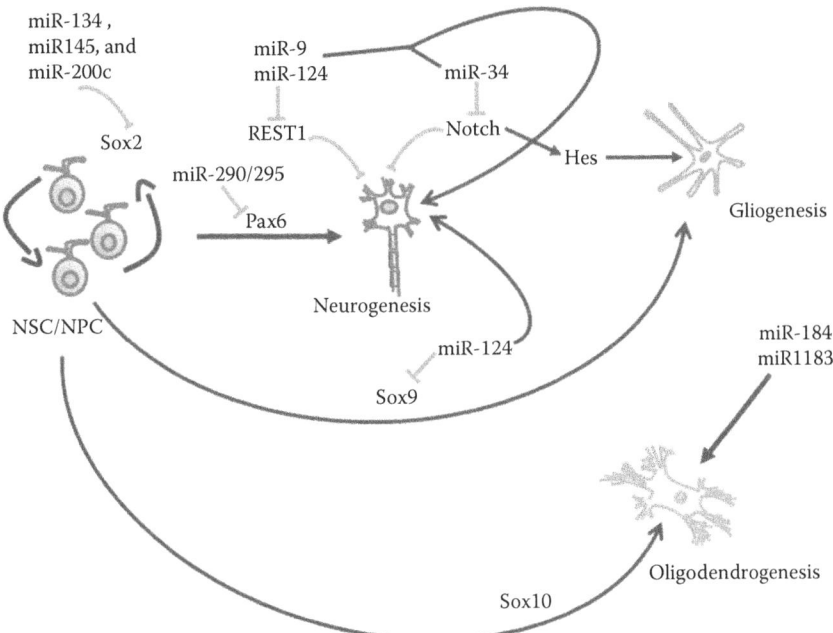

**FIGURE 9.5** Representative microRNAs and their targets involved in neural stem cell renewal and differentiation.

Some factors involved in CNS maintenance and differentiation are Notch signaling and the family of Sox transcription factors. Ligand activation of the Notch signal leads to cleavage of the transmembrane Notch receptor, which liberates the intracellular domain of the receptor that translocates to the nucleus, where it serves as a transcriptional activator (Louvi and Artavanis-Tsakonas, 2006). During early CNS development, Notch signaling blocks neuronal differentiation, via the activation of Hes transcription factors, which repress expression of factors that promotes neuron differentiation such as the neurogenin transcription factors. One of the targets by which miR-9 increases neuron differentiation is Notch (Figure 9.5).

The Sox protein family has opposite effects, it can promote stem-cell maintenance or differentiation. While Sox1, 2, and 3 maintain the stem-cell fate, Sox21 has the opposite role, and promotes neuronal differentiation. Sox9 is critical for promoting glial rather than neuronal fates, whereas Sox10 is required for oligodendrocyte differentiation (Bylund et al., 2003; Wegner, 2005). MiRNAs transcription regulation for all Sox factors have not been reported. Nevertheless, some miRNAs have been identified to target some of them.

It has been shown that there is a relation between the level of several miRNAs and the state in which the cell is. For example, miRNAs that are upregulated during the stage of self-renewal of stem/progenitor cells such as let-7, miR-134, miR-145, miR-200c, and miR-290/295 family target mRNA that promotes cell differentiation, apoptosis, and cell cycle inhibitors. Other examples are the miRNAs that promote neuron differentiation which targets mRNA related with glial phenotype or neural progenitor stage.

Particularly, the let-7/Lin-28 pathway has been linked to the pluripotency factor SOX2 in and this miRNA has been implicated in keeping the balance between self-renewal and commitment in NSCs (Cimadamore et al., 2013). Other examples are miRNA-124a which repress the gliogenic factor Sox9 translation and miR-9 promotes the translational repression of other glial factors such as Hes1 and Noch (Figure 9.5; Cheng et al., 2009).

Additional factors and signaling pathways act in parallel or cross-talk with Notch signaling and Sox factors to control CNS stem-cell proliferation and lineage decision. One factor that maintains CNS stem cells in an undifferentiated state is the zinc finger factor REST which was first shown to be a repressor of neuronal genes in non-neuronal cells. Also, some miRNAs that are known to regulate REST translation and promote neuron differentiation are miR-124 and miR-9 (Conaco et al., 2006; Visvanathan et al., 2007; Packer et al., 2008; Barca-Mayo and De Pietri Tonelli, 2014; Diaz et al., 2014). Other neurogenic factor Pax6 (Englund et al., 2005) is known to be regulated by miR-290/295. The downregulation of Pax6 protein by miR-290/295 family members decreases cell differentiation and increase cell proliferation (Figure 9.5; Kaspi et al., 2013).

All the data generated in the last 10 years suggests that miRNAs are key regulators of cell proliferation, migration, and differentiation during fetal CNS development and that changes in the levels of these molecules may be implicated in brain cytoarchitecture defects affecting postnatal functions such as food intake, learning, and motor skills.

# REFERENCES

Ai J., Zhang R., Li Y. et al. 2010. Circulating microRNA-1 as a potential novel biomarker for acute myocardial infarction. *Biochem Biophys Res Commun* 391:73–77.

Alvarez-Buylla A., Kohwi M., Nguyen T.M., Merkle F.T. 2008. The heterogeneity of adult neural stem cells and the emerging complexity of their niche. *Cold Spring Harb Symp Quant Biol* 73:357–365.

Babalola O., Mamalis A., Lev-Tov H., Jagdeo J. 2013. The role of microRNAs in skin fibrosis. *Arch Dermatologic Res* 305:763–776.

Barbato C., Ruberti F., Pieri M. et al. 2010. MicroRNA-92 modulates K(+) Cl(−) co-transporter KCC2 expression in cerebellar granule neurons. *J Neurochem* 113:591–600.

Barca-Mayo O., De Pietri Tonelli D. 2014. Convergent microRNA actions coordinate neocortical development. *Cell Mol Life Sci* 71(16):2975–2995.

Bartel D.P. 2004. MicroRNAs: Genomics, biogenesis, mechanism, and function. *Cell* 116:281–297.

Bernardo B.C., Charchar F.J., Lin R.C., McMullen J.R. 2012. A microRNA guide for clinicians and basic scientists: Background and experimental techniques. *Heart Lung Circ* 21:131–142.

Bernstein E., Kim S.Y., Carmell M.A. et al. 2003. Dicer is essential for mouse development. *Nat Genet* 35:215–217.

Blaesse P., Schmidt T. 2015. K–Cl cotransporter KCC2-a moonlighting protein in excitatory and inhibitory synapse development and function. *Pflugers Arch* 467(4):615–624.

Blondal T., Jensby Nielsen S., Baker A. et al. 2013. Assessing sample and miRNA profile quality in serum and plasma or other biofluids. *Methods* 59:S1–S6.

Boren T., Xiong Y., Hakam A. et al. 2008. MicroRNAs and their target messenger RNAs associated with endometrial carcinogenesis. *Gynecol Oncol* 110:206–215.

Bylund M., Andersson E., Novitch B.G., Muhr J. 2003. Vertebrate neurogenesis is counteracted by Sox1-3 activity. *Nat Neurosci* 6:1162–1168.

Chen C., Ridzon D.A., Broomer A.J. et al. 2005. Real-time quantification of microRNAs by stem-loop RT-PCR. *Nucleic Acids Res* 33:e179.

Chen X., Ba Y., Ma L. et al. 2008. Characterization of microRNAs in serum: A novel class of biomarkers for diagnosis of cancer and other diseases. *Cell Res* 18:997–1006.

Cheng L., Quek C.Y., Sun X., Bellingham S.A., Hill A.F. 2013. The detection of microRNA associated with Alzheimer's disease in biological fluids using next-generation sequencing technologies. *Front Genet* 4:150.

Cheng L.C., Pastrana E., Tavazoie M., Doetsch F. 2009. MiR-124 regulates adult neurogenesis in the subventricular zone stem cell niche. *Nat Neurosci* 12:399–408.

Chuhma N., Zhang H., Masson J. et al. 2004. Dopamine neurons mediate a fast excitatory signal via their glutamatergic synapses. *J Neurosci* 24:972–981.

Cimadamore F., Amador-Arjona A., Chen C., Huang C.T., Terskikh A.V. 2013. SOX2-LIN28/let-7 pathway regulates proliferation and neurogenesis in neural precursors. *Proc Natl Acad Sci USA* 110:E3017–E3026.

Clausen T.D., Mathiesen E.R., Hansen T. et al. 2009. Overweight and the metabolic syndrome in adult offspring of women with diet-treated gestational diabetes mellitus or type 1 diabetes. *J Clin Endocrinol Metab* 94:2464–2470.

Conaco C., Otto S., Han J.J., Mandel G. 2006. Reciprocal actions of REST and a microRNA promote neuronal identity. *Proc Natl Acad Sci USA* 103:2422–2427.

Coolen M., Bally-Cuif L. 2009. MicroRNAs in brain development and physiology. *Curr Opin Neurobiol* 19:461–470.

Cortez M.A., Bueso-Ramos C., Ferdin J. et al. 2011. MicroRNAs in body fluids—The mix of hormones and biomarkers. *Nat Rev Clin Oncol* 8:467–477.

Cortez M.A., Calin G.A. 2009. MicroRNA identification in plasma and serum: A new tool to diagnose and monitor diseases. *Expert Opin Biol Ther* 9:703–711.

Cuellar T.L., Davis T.H., Nelson P.T. et al. 2008. Dicer loss in striatal neurons produces behavioral and neuroanatomical phenotypes in the absence of neurodegeneration. *Proc Natl Acad Sci USA* 105:5614–5619.

Dajas-Bailador F., Bonev B., Garcez P. et al. 2012. MicroRNA-9 regulates axon extension and branching by targeting Map1b in mouse cortical neurons. *Nat Neurosci* 15:697–699.

Darnell D.K., Kaur S., Stanislaw S. et al. 2006. MicroRNA expression during chick embryo development. Developmental Dynamics: An Official Publication of the American Association of Anatomists. *Dev Dyn* 235:3156–3165.

Davis T.H., Cuellar T.L., Koch S.M. et al. 2008. Conditional loss of Dicer disrupts cellular and tissue morphogenesis in the cortex and hippocampus. *J Neurosci: Official J Soc Neurosci* 28:4322–4330.

De Pietri Tonelli D., Pulvers J.N., Haffner C. et al. 2008. MiRNAs are essential for survival and differentiation of newborn neurons but not for expansion of neural progenitors during early neurogenesis in the mouse embryonic neocortex. *Development* 135:3911–3921.

Delaloy C., Liu L., Lee J.A. et al. 2010. MicroRNA-9 coordinates proliferation and migration of human embryonic stem cell-derived neural progenitors. *Cell Stem Cell* 6:323–335.

Diaz N.F., Cruz-Resendiz M.S., Flores-Herrera H., Garcia-Lopez G., Molina-Hernandez A. 2014. MicroRNAs in central nervous system development. *Rev Neurosci* 25(5):675–686.

Dragatsis I., Levine M.S., Zeitlin S. 2000. Inactivation of Hdh in the brain and testis results in progressive neurodegeneration and sterility in mice. *Nat Genet* 26:300–306.

Dragatsis I., Zeitlin S. 2000. CaMKIIalpha-Cre transgene expression and recombination patterns in the mouse brain. *Genesis* 26:133–135.

Drager U.C., Olsen J.F. 1980. Origins of crossed and uncrossed retinal projections in pigmented and albino mice. *J Comp Neurol* 191:383–412.

Eixarch E., Figueras F., Hernandez-Andrade E. et al. 2009. An experimental model of fetal growth restriction based on selective ligature of uteroplacental vessels in the pregnant rabbit. *Fetal Diag Ther* 26:203–211.

Englund C., Fink A., Lau C. et al. 2005. Pax6, Tbr2, and Tbr1 are expressed sequentially by radial glia, intermediate progenitor cells, and postmitotic neurons in developing neocortex. *J Neurosci* 25:247–251.

Fineberg S.K., Kosik K.S., Davidson B.L. 2009. MicroRNAs potentiate neural development. *Neuron* 64:303–309.

Gilad S., Meiri E., Yogev Y. et al. 2008. Serum microRNAs are promising novel biomarkers. *PloS One* 3:e3148.

Giraldez A.J., Cinalli R.M., Glasner M.E. et al. 2005. MicroRNAs regulate brain morphogenesis in zebrafish. *Science* 308:833–838.

Glover V. 2011. Annual research review: Prenatal stress and the origins of psychopathology: An evolutionary perspective. *Journal of Child Psychology and Psychiatry, and Allied Disciplines* 52:356–367.

Gonzalez A. 2013. Osteoarthritis year 2013 in review: Genetics and genomics. *Osteoarthritis Cartilage/OARS, Osteoarthritis Res Soc* 21:1443–1451.

Gotz M., Huttner W.B. 2005. The cell biology of neurogenesis. *Nat Rev Mol Cell Biol* 6:777–788.

Gu H., Li H., Zhang L. et al. 2012. Diagnostic role of microRNA expression profile in the serum of pregnant women with fetuses with neural tube defects. *J Neurochem* 122:641–649.

Ha M., Pang M., Agarwal V., Chen Z.J. 2008. Interspecies regulation of microRNAs and their targets. *Biochim Biophys Acta* 1779:735–742.

Hanley B., Dijane J., Fewtrell M. et al. 2010. Metabolic imprinting, programming and epigenetics—A review of present priorities and future opportunities. *The British Journal of Nutrition* 104 (Suppl 1):S1–S25.

Hengst U., Cox L.J., Macosko E.Z., Jaffrey S.R. 2006. Functional and selective RNA interference in developing axons and growth cones. *J Neurosci* 26:5727–5732.

Huang T., Liu Y., Huang M., Zhao X., Cheng L. 2010. Wnt1-cre-mediated conditional loss of Dicer results in malformation of the midbrain and cerebellum and failure of neural crest and dopaminergic differentiation in mice. *J Mol Cell Biol* 2:152–163.

Huang Y., Hu Q., Deng Z. et al. 2014. MicroRNAs in body fluids as biomarkers for non-small cell lung cancer: A systematic review. *Technology in Cancer Research & Treatment* 13:277–287.

Jawerbaum A., White V. 2010. Animal models in diabetes and pregnancy. *Endocr Rev* 31:680–701.

Ji F., Lv X., Jiao J. 2013. The role of microRNAs in neural stem cells and neurogenesis. *J Genet Genom Yi chuan xue bao* 40:61–66.

Jin H., Li C., Ge H., Jiang Y., Li Y. 2013. Circulating microRNA: A novel potential biomarker for early diagnosis of intracranial aneurysm rupture a case–control study. *J Trans Med* 11:296.

Kapsimali M., Kloosterman W.P., de Bruijn E. et al. 2007. MicroRNAs show a wide diversity of expression profiles in the developing and mature central nervous system. *Genome Biol* 8:R173.

Kaspi H., Chapnik E., Levy M. et al. 2013. Brief report: MiR-290–295 regulate embryonic stem cell differentiation propensities by repressing Pax6. *Stem Cells* 31:2266–2272.

Kawase-Koga Y., Otaegi G., Sun T. 2009. Different timings of Dicer deletion affect neurogenesis and gliogenesis in the developing mouse central nervous system. *Dev Dyn* 238:2800–2812.

Kim J., Inoue K., Ishii J. et al. 2007. A microRNA feedback circuit in midbrain dopamine neurons. *Science* 317:1220–1224.

Kim J.H., Auerbach J.M., Rodriguez-Gomez J.A. et al. 2002. Dopamine neurons derived from embryonic stem cells function in an animal model of Parkinson's disease. *Nature* 418:50–56.

Kloosterman W.P., Steiner F.A., Berezikov E. et al. 2006. Cloning and expression of new microRNAs from zebrafish. *Nucleic Acids Res* 34:2558–2569.

Krichevsky A.M., King K.S., Donahue C.P., Khrapko K., Kosik K.S. 2003. A microRNA array reveals extensive regulation of microRNAs during brain development. *RNA* 9:1274–1281.

Krichevsky A.M., Sonntag K.C., Isacson O., Kosik K.S. 2006. Specific microRNAs modulate embryonic stem cell-derived neurogenesis. *Stem Cells* 24:857–864.

Kroh E.M., Parkin R.K., Mitchell P.S., Tewari M. 2010. Analysis of circulating microRNA biomarkers in plasma and serum using quantitative reverse transcription-PCR (qRT-PCR). *Methods* 50:298–301.

Lagos-Quintana M., Rauhut R., Lendeckel W., Tuschl T. 2001. Identification of novel genes coding for small expressed RNAs. *Science* 294:853–858.

Lagos-Quintana M., Rauhut R., Yalcin A. et al. 2002. Identification of tissue-specific microRNAs from mouse. *Curr Biol* 12:735–739.

Laneve P., Di Marcotullio L., Gioia U. et al. 2007. The interplay between microRNAs and the neurotrophin receptor tropomyosin-related kinase C controls proliferation of human neuroblastoma cells. *Proc Natl Acad Sci USA* 104:7957–7962.

Le M.T., Shyh-Chang N., Khaw S.L. et al. 2011. Conserved regulation of p53 network dosage by microRNA-125b occurs through evolving miRNA-target gene pairs. *PLoS Genetics* 7:e1002242.

Le M.T., Xie H., Zhou B. et al. 2009. MicroRNA-125b promotes neuronal differentiation in human cells by repressing multiple targets. *Mol Cell Biol* 29:5290–5305.

Lewis B.P., Shih I.H., Jones-Rhoades M.W., Bartel D.P., Burge C.B. 2003. Prediction of mammalian microRNA targets. *Cell* 115:787–798.

Li X., Jin P. 2010. Roles of small regulatory RNAs in determining neuronal identity. *Nat Rev Neurosci* 11:329–338.

Lim L.P., Lau N.C., Garrett-Engele P. et al. 2005. Microarray analysis shows that some microRNAs downregulate large numbers of target mRNAs. *Nature* 433:769–773.

Loeken M.R. 2006. Advances in understanding the molecular causes of diabetes-induced birth defects. *J Soc Gynecologic Investigation* 13:2–10.

Louvi A., Artavanis-Tsakonas S. 2006. Notch signalling in vertebrate neural development. *Nat Rev Neurosci* 7:93–102.

Maiorano N.A., Mallamaci A. 2009. Promotion of embryonic cortico-cerebral neuronogenesis by miR-124. *Neural Develop* 4:40.

Maiorano N.A., Mallamaci A. 2010. The pro-differentiating role of miR-124: Indicating the road to become a neuron. *RNA Biol* 7:528–533.

Makeyev E.V., Zhang J., Carrasco M.A., Maniatis T. 2007. The microRNA miR-124 promotes neuronal differentiation by triggering brain-specific alternative pre-mRNA splicing. *Mol Cell* 27:435–448.

Marcelis C.L., Hol F.A., Graham G.E. et al. 2008. Genotype-phenotype correlations in MYCN-related Feingold syndrome. *Hum Mutat* 29:1125–1132.

Menghini R., Casagrande V., Federici M. 2013. MicroRNAs in endothelial senescence and atherosclerosis. *Journal of Cardiovascular Transl Res* 6:924–930.

Miska E.A., Alvarez-Saavedra E., Townsend M. et al. 2004. Microarray analysis of microRNA expression in the developing mammalian brain. *Genome Biol* 5:R68.

Mogilyansky E., Rigoutsos I. 2013. The miR-17/92 cluster: A comprehensive update on its genomics, genetics, functions and increasingly important and numerous roles in health and disease. *Cell Death Differ* 20:1603–1614.

Moldovan L., Batte K.E., Trgovcich J. et al. 2014. Methodological challenges in utilizing miRNAs as circulating biomarkers. *J Cell Mol Med* 18:371–390.

Molina-Hernandez A., Diaz N.F., Arias-Montano J.A. 2012. Histamine in brain development. *J Neurochem* 122:872–882.

Murchison E.P., Partridge J.F., Tam O.H., Cheloufi S., Hannon G.J. 2005. Characterization of Dicer-deficient murine embryonic stem cells. *Proc Natl Acad Sci USA* 102:12135–12140.

Nielsen J.A., Lau P., Maric D., Barker J.L., Hudson L.D. 2009. Integrating microRNA and mRNA expression profiles of neuronal progenitors to identify regulatory networks underlying the onset of cortical neurogenesis. *BMC Neurosci* 10:98.

Nowakowski T.J., Mysiak K.S., O'Leary T. et al. 2013. Loss of functional Dicer in mouse radial glia cell-autonomously prolongs cortical neurogenesis. *Dev Biol* 382:530–537.

Nowakowski T.J., Mysiak K.S., Pratt T., Price D.J. 2011. Functional dicer is necessary for appropriate specification of radial glia during early development of mouse telencephalon. *PLoS One* 6:e23013.

Olsen P.H., Ambros V. 1999. The lin-4 regulatory RNA controls developmental timing in *Caenorhabditis elegans* by blocking LIN-14 protein synthesis after the initiation of translation. *Dev Biol* 216:671–680.

Ornoy A., Ratzon N., Greenbaum C. et al. 1998. Neurobehaviour of school age children born to diabetic mothers. *Arch Dis Child Fetal Neonatal Ed* 79:F94–F99.

Ornoy A., Ratzon N., Greenbaum C., Wolf A., Dulitzky M. 2001. School-age children born to diabetic mothers and to mothers with gestational diabetes exhibit a high rate of inattention and fine and gross motor impairment. *JPEM* 14 (Suppl 1):681–689.

Ornoy A., Wolf A., Ratzon N., Greenbaum C., Dulitzky M. 1999. Neurodevelopmental outcome at early school age of children born to mothers with gestational diabetes. *Arch Dis Child Fetal Neonatal Ed* 81:F10–14.

Packer A.N., Xing Y., Harper S.Q., Jones L., Davidson B.L. 2008. The bifunctional microRNA miR-9/miR-9* regulates REST and CoREST and is downregulated in Huntington's disease. *J Neurosci* 28:14341–14346.

Papagiannakopoulos T., Kosik K.S. 2009. MicroRNA-124: Micromanager of neurogenesis. *Cell Stem Cell* 4:375–376.

Peng C., Li N., Ng Y.K. et al. 2012. A unilateral negative feedback loop between miR-200 microRNAs and Sox2/E2F3 controls neural progenitor cell-cycle exit and differentiation. *J Neurosci* 32:13292–13308.

Petros T.J., Rebsam A., Mason C.A. 2008. Retinal axon growth at the optic chiasm: To cross or not to cross. *Annu Rev Neurosci* 31:295–315.

Pinter R., Hindges R. 2010. Perturbations of microRNA function in mouse dicer mutants produce retinal defects and lead to aberrant axon pathfinding at the optic chiasm. *PLoS One* 5:e10021.

Rees S., Inder T. 2005. Fetal and neonatal origins of altered brain development. *Early Human Dev* 81:753–761.

Rizzo T., Metzger B.E., Burns W.J., Burns K. 1991. Correlations between antepartum maternal metabolism and child intelligence. *N Engl J Med* 325:911–916.

Rodriguez-Martinez G., Velasco I., Garcia-Lopez G. et al. 2012. Histamine is required during neural stem cell proliferation to increase neuron differentiation. *Neuroscience* 216:10–17.

Sanuki R., Onishi A., Koike C. et al. 2011. MiR-124a is required for hippocampal axogenesis and retinal cone survival through Lhx2 suppression. *Nat Neurosci* 14:1125–1134.

Sarver A.L., Li L., Subramanian S. 2010. MicroRNA miR-183 functions as an oncogene by targeting the transcription factor EGR1 and promoting tumor cell migration. *Cancer Res* 70:9570–9580.

Saurat N., Andersson T., Vasistha N.A., Molnar Z., Livesey F.J. 2013. Dicer is required for neural stem cell multipotency and lineage progression during cerebral cortex development. *Neural Dev* 8:14.

Sempere L.F., Freemantle S., Pitha-Rowe I. et al. 2004. Expression profiling of mammalian microRNAs uncovers a subset of brain-expressed microRNAs with possible roles in murine and human neuronal differentiation. *Genome Biol* 5:R13.

Sheinerman K.S., Umansky S.R. 2013a. Circulating cell-free microRNA as biomarkers for screening, diagnosis and monitoring of neurodegenerative diseases and other neurologic pathologies. *Front Cell Neurosci* 7:150.

Sheinerman K.S., Umansky S.R. 2013b. Early detection of neurodegenerative diseases: Circulating brain-enriched microRNA. *Cell Cycle* 12:1–2.

Simeoni U., Barker D.J. 2009. Offspring of diabetic pregnancy: Long-term outcomes. *Semin Fetal Neonatal Med* 14:119–124.

Stappert L., Borghese L., Roese-Koerner B. et al. 2013. MicroRNA-based promotion of human neuronal differentiation and subtype specification. *PLoS One* 8:e59011.

Steculorum S.M., Bouret S.G. 2011. Maternal diabetes compromises the organization of hypothalamic feeding circuits and impairs leptin sensitivity in offspring. *Endocrinology* 152:4171–4179.

Swindell E.C., Bailey T.J., Loosli F. et al. 2006. Rx-Cre, a tool for inactivation of gene expression in the developing retina. *Genesis* 44:361–363.

Takayama C., Inoue Y. 2007. Developmental localization of potassium chloride co-transporter 2 (KCC2) in the Purkinje cells of embryonic mouse cerebellum. *Neurosci Res* 57:322–325.

Tan S.L., Ohtsuka T., Gonzalez A., Kageyama R. 2012. MicroRNA9 regulates neural stem cell differentiation by controlling Hes1 expression dynamics in the developing brain. *Genes Cells* 17:952–961.

Tortosa E., Montenegro-Venegas C., Benoist M. et al. 2011. Microtubule-associated protein 1B (MAP1B) is required for dendritic spine development and synaptic maturation. *J Biol Chem* 286:40638–40648.

van Vliet E., Eixarch E., Illa M. et al. 2013. Metabolomics reveals metabolic alterations by intrauterine growth restriction in the fetal rabbit brain. *PloS One* 8:e64545.

Visvanathan J., Lee S., Lee B., Lee J.W., Lee S.K. 2007. The microRNA miR-124 antagonizes the anti-neural REST/SCP1 pathway during embryonic CNS development. *Genes Dev* 21:744–749.

Wegner M. 2005. Secrets to a healthy Sox life: Lessons for melanocytes. *Pigment Cell Res* 18:74–85.

Weston M.D., Pierce M.L., Rocha-Sanchez S., Beisel K.W., Soukup G.A. 2006. MicroRNA gene expression in the mouse inner ear. *Brain Res* 1111:95–104.

Widera C., Gupta S.K., Lorenzen J.M. et al. 2011. Diagnostic and prognostic impact of six circulating microRNAs in acute coronary syndrome. *J Mol Cell Cardiol* 51:872–875.

Wienholds E., Kloosterman W.P., Miska E. et al. 2005. MicroRNA expression in zebrafish embryonic development. *Science* 309:310–311.

Wienholds E., Plasterk R.H. 2005. MicroRNA function in animal development. *FEBS Lett* 579:5911–5922.

Yu J.Y., Chung K.H., Deo M., Thompson R.C., Turner D.L. 2008. MicroRNA miR-124 regulates neurite outgrowth during neuronal differentiation. *Exp Cell Res* 314:2618–2633.

Zhang Z., Li S., Cheng S.Y. 2013. The miR-183 approximately 96 approximately 182 cluster promotes tumorigenesis in a mouse model of medulloblastoma. *J Biomed Res* 27:486–494.

# Index

## A

Acetylcholinesterase enzyme gene (AChE), 50
AChE, *see* Acetylcholinesterase enzyme
  gene (AChE)
ACTH, *see* Adrenocorticotropic hormone
  (ACTH)
Acyl-protein thioesterase 1 (APT1), 10
AD, *see* Alzheimer's disease (AD);
  Antidepressant (AD)
Adrenocorticotropic hormone (ACTH), 23
AEDs, *see* Antiepileptic drugs (AEDs)
Affective disorders, 21, 101; *see also* MiRNAs in
  neurogenesis/neuroplasticity
AGO, *see* Argonaute (AGO)
ALAD, *see* Aminolevulinate dehydratase
  (ALAD)
ALS, *see* Amyotrophic lateral sclerosis (ALS)
Altered miRNA expression, 106, 110
  in epilepsy, 164–165
  sites of dysfunctions, 106
Alzheimer's disease (AD), 207; *see also*
  Neurodegenerative diseases
Aminolevulinate dehydratase (ALAD), 50
amiRNAs, *see* Artificial miRNAs (amiRNAs)
Amyloid precursor protein (APP), 208, 214
Amyloid β (Aβ), 209, 214
Amyotrophic lateral sclerosis (ALS), 142, 207,
  208; *see also* Neurodegenerative
  diseases
Antidepressant (AD), 35, 105; *see also* Tricyclic
  antidepressants (TCAs)
  blood miRNA changes and, 45
  -induced changes in neurogenesis, 109
  microRNAs and, 35
  miR-16 and, 29
Antiepileptic drugs (AEDs), 155
Anti-inflammatory drugs, 159
Anxiety disorders, 22
APP, *see* Amyloid precursor protein (APP)
APT1, *see* Acyl-protein thioesterase 1 (APT1)
Argonaute (AGO), 5
Armitage protein, 11
Artificial miRNAs (amiRNAs), 147; *see also*
  Prion diseases
  combating prions with RNA interference,
  146–147
  dual targeting miRNA cassette, 148
  dual targeting of prion transcripts with, 147
  to knockdown prion protein, 146
  mouse lines expressing, 147

Astrocytes, 158
Aβ, *see* Amyloid β (Aβ)

## B

Bacteria, 140
BAF53a, 244–245
BBB, *see* Blood–brain barrier (BBB)
BD, *see* Bipolar disorder (BD)
BDNF, *see* Brain-derived neurotrophic factor
  (BDNF)
Bioinformatic tools, 4
Biological fluids, 112
Biomarkers, 210
Bipolar affective disorder (BPAD), 22; *see also*
  Mood and anxiety disorders
  clinical studies of miRNAs in, 38–39
  treatment of, 23
Bipolar disorder (BD), 111
Blood–brain barrier (BBB), 159
Blood mononuclear cells (BMC), 214
BMC, *see* Blood mononuclear cells (BMC)
Bovine spongiform encephalopathy (BSE), 140;
  *see also* Prion
BPAD, *see* Bipolar affective disorder (BPAD)
Brain, 4
  cell types in, 25
  -enriched miR-134, 171
  injury, 158, 159
  -specific miRNA, 10
Brain-derived neurotrophic factor (BDNF), 9, 24,
  102, 108, 187
BSE, *see* Bovine spongiform encephalopathy
  (BSE)

## C

cAMP response element binding (CREB), 9,
  24, 187
Catecholamines, 23
CCI, *see* Chronic constriction injury (CCI)
Cellular and molecular processes, 143
Cellular prion, 141; *see also* Prion
Central nervous system (CNS), 64, 184
Cerebellar granule neuron (CGN), 243
Cerebrospinal fluid (CSF), 12, 68, 41, 211
CFA, *see* Complete Freund's adjuvant (CFA)
CFH, *see* Complement factor H (CFH)
CGN, *see* Cerebellar granule neuron (CGN)
Chronic constriction injury (CCI), 193
Chronic unpredictable stress (CUPS), 30

Circulating cell-free microRNAs, 237
  alternated retinal ganglion cell axons, 242
  BAF53a, 244–245
  as CNS development biomarkers, 245–246
  KCC2, 243–244
  effect of loss of Dicer, 241
  miR-9, 244
  miR-92a, 243
  miR-183, 243
  miR-124a, 244, 245
  miRNAs-expression, 239
  effect of mutant Dicer mice, 240
  in neural development, 238–243
  and neurogenesis, 243–245
  neurological damage and microRNA level,
    246–248
  as regulators of CNS development, 238
  role in axon pathfinder control, 242
  and targets, 247
Circulating miRNA, 212; *see also*
    Neurodegenerative diseases
  ALS and, 223–225
  Alzheimer's disease and, 214–219
  exosomes, 212
  FTD, 225
  Huntington's disease and, 225
  Parkinson's disease and, 219–222
  techniques used, 213
CJD, *see* Creutzfeldt–Jakob disease (CJD)
Clustered regularly interspaced short palindromic
    repeats (CRISPR), 92
CNS, *see* Central nervous system (CNS)
Complement factor H (CFH), 219
Complete Freund's adjuvant (CFA), 188
Complexin 2 (CPLX2), 72
Complex regional pain syndrome (CRPS), 196
Corticotrophin-releasing factor receptor type I
    (CRFR1), 47
Corticotropin-releasing hormone (CRH), 23
CPLX2, *see* Complexin 2 (CPLX2)
CREB, *see* cAMP response element binding
    (CREB)
Creutzfeldt–Jakob disease (CJD), 140; *see also*
    Prion
CRFR1, *see* Corticotrophin-releasing factor
    receptor type I (CRFR1)
CRH, *see* Corticotropin-releasing hormone
    (CRH)
CRISPR, *see* Clustered regularly interspaced
    short palindromic repeats
    (CRISPR)
CRPS, *see* Complex regional pain syndrome
    (CRPS)
CSF, *see* Cerebrospinal fluid (CSF)
CUPS, *see* Chronic unpredictable stress
    (CUPS)
Cynomolgus macaques, 144

**D**

DA, *see* Dopamine (DA)
DBS, *see* Deep-brain stimulation (DBS)
Deep-brain stimulation (DBS), 220
Dementias, 207; *see also* Neurodegenerative
    diseases
Depression, 22; *see also* Mood and anxiety
    disorders
  clinical studies in, 40–41, 42–44
  models of, 23
  monoamine hypothesis of, 23
  preclinical studies of miRNAs in, 32–34
Depressive-like behavior, 31; *see also* Mood and
    anxiety disorders
  microRNAs and, 31, 35
  microRNAs and antidepressants, 35–36
  microRNAs and mood stabilizers, 36
DGCR8, *see* DiGeorgie syndrome critical region
    gene 8 (DGCR8)
Dicer, 163
  enzyme, 146
  knockout zebrafish, 7
  requiring cofactors, 104
DiGeorgie syndrome critical region gene
    8 (DGCR8), 5, 75, 89; *see also*
    MicroRNA dysregulation in
    schizophrenia
  haploinsufficiency mouse model, 77–80
DJ-1, 220
DNA(cytosine-5)-methyltransferase 3A
    (DNMT3a), 30, 193
DNMT3a, *see* DNA(cytosine-5)-
    methyltransferase 3A (DNMT3a)
Dopamine (DA), 239
Dopamine receptor D2 (DRD2), 30
Dopaminergic hypothesis, 83
Dorsal root ganglion (DRG), 188
Double-stranded RNAs (dsRNAs), 4
DRD2, *see* Dopamine receptor D2 (DRD2)
DRG, *see* Dorsal root ganglion (DRG)
dsRNAs, *see* Double-stranded RNAs (dsRNAs)
Dual targeting miRNA cassette, 148

**E**

Early growth response protein (EGR3), 82
ECF, *see* Extracellular fluid (ECF)
ECS, *see* Electroconvulsive stimulation (ECS)
ECT, *see* Electroconvulsive therapy (ECT)
EE, *see* Environmental enrichment (EE)
EGR3, *see* Early growth response protein
    (EGR3)
eIF2c, *see* Eukaryotic translation initiation factor
    6 (eIF2c)
Electroconvulsive stimulation (ECS), 35
Electroconvulsive therapy (ECT), 23

Endogenous genes, *see* Housekeeping genes
Environmental enrichment (EE), 49
Epigenetic mechanisms, 29
Epilepsy, 154; *see also* Injury-induced
    epileptogenic mechanisms;
    MicroRNA biogenesis pathways
  acquired, 157
  altered microRNA expression in, 164–165
  causes of, 156
  diagnosis, 155
  epigenetic control of gene expression, 160–161
  genetic mutations, 156–157
  mechanisms of, 156
  microRNA biogenesis in, 163–164
  microRNA profile, 165–170
  microRNAs regulating inflammation in,
    170–171
  nonpharmacological treatments, 156
  research on microRNA, 165
  treatment, 155
  up-and downregulated miRNAs, 168–169
Epileptogenesis, 154; *see also* Epilepsy
EPSCs, *see* Excitatory postsynaptic currents
    (EPSCs)
ERK-MAP kinase, *see* Extracellular signal-
    regulated kinase mitogen-activated
    protein (ERK-MAP kinase)
Eukaryotic translation initiation factor 6
    (eIF2c), 105
Excitatory postsynaptic currents (EPSCs), 80
Exosomes, 144–145, 212
Exp-5, *see* Exportin-5 (Exp-5)
Experience-dependent plasticity, 9
Exportin-5 (Exp-5), 5
Extracellular fluid (ECF), 219
Extracellular signal-regulated kinase mitogen-
    activated protein (ERK-MAP
    kinase), 112
Ezh2, 105

**F**

Feingold syndrome, 244
FMRP, *see* Fragile X mental retardation protein
    (FMRP)
Fragile X mental retardation protein (FMRP), 6
Frontotemporal dementia, 208; *see also*
    Neurodegenerative diseases
Frontotemporal lobar degeneration (FTLD), 208;
    *see also* Neurodegenerative diseases
FTLD, *see* Frontotemporal lobar degeneration
    (FTLD)

**G**

GABA, *see* γ-amino butyric acid (GABA)
γ-amino butyric acid (GABA), 155, 184

GCD, *see* Granule cell dispersion (GCD)
GEA, *see* Genome enrichment analysis
    (GEA)
Gene
  expression, 104
  targeting technology, 92
Genome enrichment analysis (GEA), 73
Genome-wide association studies (GWASs), 37,
    70; *see also* MicroRNA dysregulation
    in schizophrenia
  functional implications of data, 71–73
  functional investigation of SNPs, 73
  miR-198, 71
  replication successes and failures, 70–71
Glucocorticoid receptors (GRs), 24, 108
Glutamatergic signaling system, 67
Granule cell dispersion (GCD), 161
GRs, *see* Glucocorticoid receptors (GRs)

**H**

HD, *see* Huntington's disease (HD)
HDAC, *see* Histone deacetylase (HDAC)
Heterogeneous nuclear ribonucleoprotein A2B1
    (hnRNPA2B1), 148
Heterogeneous nuclear ribonucleoprotein K
    (hnRNPK), 49
Heterogeneous ribonucleoproteins
    (hnRNPs), 209
High mobility group box 1 (HMGB1), 159
Hippocampal neurogenesis, 105
Histone deacetylase (HDAC), 161
HMGB1, *see* High mobility group box 1
    (HMGB1)
hnRNPA2B1, *see* Heterogeneous nuclear
    ribonucleoprotein A2B1
    (hnRNPA2B1)
hnRNPK, *see* Heterogeneous nuclear
    ribonucleoprotein K (hnRNPK)
hnRNPs, *see* Heterogeneous ribonucleoproteins
    (hnRNPs)
Housekeeping genes, 213
HPA axis, *see* Hypothalamic–pituitary–adrenal
    axis (HPA axis)
HT, *see* 5-Hydroxytriptamine (HT)
Huntington's disease (HD), 207, 208; *see also*
    Neurodegenerative diseases
5-Hydroxytriptamine (HT), 107
Hypothalamic–pituitary–adrenal axis (HPA
    axis), 23

**I**

IL-1β, *see* Interleukin-1β (IL-1β)
Infectious agents, 140
Inflammatory pain, 193; *see also* Pain

Injury-induced epileptogenic mechanisms, 157;
      *see also* Epilepsy
   axonal/dendritic reorganization, 159–160
   extracellular matrix and wound repair, 160
   gliosis, 158–159
   inflammation, 159
   neuronal death, 157–158
Interferon, 24
Interleukin-1β (IL-1β), 159

**K**

KCC2, *see* Potassium-chloride co-transporter
      member 2 (KCC2)
Knockout (KO), 70
KO, *see* Knockout (KO)
Kuru, 140; *see also* Prion

**L**

Learned helplessness (LH), 31
Lewy bodies, 219
LH, *see* Learned helplessness (LH)
LNA, *see* Locked nucleic acid (LNA)
Locked nucleic acid (LNA), 220
Long-term potentiation (LTP), 79
LTP, *see* Long-term potentiation (LTP)

**M**

Mad cow disease, *see* Bovine spongiform
      encephalopathy (BSE)
Major depressive disorder (MDD), 31, 102;
      *see also* MiRNAs in neurogenesis/
      neuroplasticity
   miRNAs impact in, 110–112
   studies on miRNAs and, 114–123
Mammalian target of rapamycin (mTOR), 160
MAOIs, *see* Monoamine oxidase inhibitors
      (MAOIs)
MAP1B, *see* Microtubule-associated protein 1B
      (MAP1B)
MAP2K5, *see* Mitogen-activated protein kinase
      5 (MAP2K5)
Maternal deprivation (MD), 29
Maternal stress (MS), 30
Matrix metalloproteinase 9 (MMP-9), 160
MCI, *see* Mild cognitive impairment (MCI)
MD, *see* Maternal deprivation (MD)
MDD, *see* Major depressive disorder (MDD)
MeCP2, *see* Methyl CpG-binding protein 2
      (MeCP2)
Medial PFC (mPFC), 78
mESC, *see* Murine embryonic stem cells (mESC)
Methyl CpG-binding protein 2 (MeCP2), 9, 108
MFS, *see* Mossy fiber sprouting (MFS)
22q Microdeletion syndrome, 76

Microprocessor, 5
MicroRNA (MiRNA), 3, 102
   Armitage protein, 11
   association between MDD and, 110–112, 114
   biogenesis, 4–6, 61, 104–105
   -bound-RISC complex, 162
   in brain disease, 12
   brain-specific miRNA, 10
   cognitive functions, 10
   and dendritic spines, 9–10
   experimental approaches, 13
   -expression, 239
   genes, 107
   genes controlled by miR-124, 8
   impact in stress-related disorders, 109–110
   impact in suicidal behavior, 112–113
   in local protein synthesis, 107
   mimics, 226
   in nervous system, 7
   neurogenesis and synaptic plasticity, 105–107
   in neuronal development and
      differentiation, 7–9
   and neurotrophic factors, 108–109
   nomenclature of, 6
   perspectives, 12–13
   RNA-induced silencing complex, 10–12
   sites of dysfunctions, 104
   size, 5
   suicidal behavior and, 124–125
   at synapse, 9
MicroRNA biogenesis pathways; *see also*
      Epilepsy; MicroRNA profile
   in brain, 162
   Dicer, 163
   in epilepsy, 163–164
   miRNA-bound-RISC complex, 162
   in regulation of dendritic spines and neuronal
      activity, 171–172
   in seizure-induced neuronal death, 172
MicroRNA dysregulation in schizophrenia,
      60, 92–93; *see also* Genome-wide
      association studies (GWASs)
   Dgcr8 haploinsufficiency mouse model,
      77–80
   dopaminergic hypothesis, 83
   effect on target molecules, 85
   evidence for, 64
   functional narrative of, 81
   genetic, 70
   glutamatergic signaling system, 67
   investigation of, 75–77
   medication effect on microRNA expression,
      80–81
   microRNA research, 91–92
   miR-107-CHRM1, 87–88
   miR-132, 87
   miR-132-NMDAR, 86

miR-137, 73–75
miR-195-BDNF, 88–89
miR-212-PGD, 87
miR-219-NMDAR, 85
miRNAs biogenesis, 61
pathways section, 89
peripheral MicroRNA biomarkers, 68–69
postmortem findings, 64–68, 90
SCZ risk gene, 82–84
tyrosine hydroxylase, 87
MicroRNA profile; *see also* Epilepsy; MicroRNA
    biogenesis pathways
    after status epileptics, 165–166
    in temporal lobe epilepsy, 167, 170
Microtubule-associated protein 1B
        (MAP1B), 244
Microvesicles, 212
Mild cognitive impairment (MCI), 218
miR-9, 244
miR-92a, 243
miR-16 levels vs. 5-HT transporter
        expression, 112
miR-124a, 244, 245
miR-137, 73–75
miR-183, 243
miRISC complex, *see* MiRNA-induced silencing
        complex (miRISC complex)
MiRNA, *see* MicroRNA (MiRNA)
MiRNA-induced silencing complex (miRISC
        complex), 146
MiRNA Registry, 6
MiRNA responsive element (MRE), 6, 223
MiRNAs in neurogenesis/neuroplasticity, 103,
        105–107, 127–128
    altered miRNA expression, 106, 110
    antidepressant-induced changes in, 109
    gene expression, 104
    glucocorticoid receptors, 108
    limitations, 113, 126–127
    methods, 103–104
    microRNAs and neurotrophic factors,
        108–109
    microRNAs biogenesis and expression,
        104–105
    miR-16 levels vs. 5-HT transporter
        expression, 112
    miRNAs impact in MDD, 110–112, 114
    miRNAs impact in stress-related disorders,
        109–110
    miRNAs impact in suicidal behavior,
        112–113, 124–125
    miRNAs in local protein synthesis, 107
Mitogen-activated protein kinase 5
        (MAP2K5), 46
MMP-9, *see* Matrix metalloproteinase 9
        (MMP-9)
Monoamine oxidase inhibitors (MAOIs), 22

Mood and anxiety disorders, 21, 52–53
    clinical studies, 37–41, 49–50
    depressive-like behavior and therapies, 31–37
    expression studies in peripheral blood, 50, 52
    genetic influence, 22
    genotyping and rare variants, 46–47, 50
    microRNA role in brain, 24–25
    microRNAs and stress, 29–31
    miRNAs in anxiety, 48, 51
    molecular neurobiology of, 23–24
    peripheral tissue studies, 45–46
    postmortem studies, 41, 45
    preclinical studies, 25–28, 47
    review of MicroRNAs in, 25
    serotonergic signaling, 49
    treatment of, 22–23
Mossy fiber sprouting (MFS), 159
mPFC, *see* Medial PFC (mPFC)
MRE, *see* MiRNA responsive element (MRE)
MS, *see* Maternal stress (MS)
MSA, *see* Multiple system atrophy (MSA)
mTOR, *see* Mammalian target of rapamycin
        (mTOR)
Multiple system atrophy (MSA), 220
Murine embryonic stem cells (mESC), 239

**N**

NADPH oxidase 4, 194
NEFL, *see* Neurofilament light chain mRNA
        (NEFL)
Neural stem cells (NSCs), 245
NeurimmiRs, 196–197; *see also* Pain
Neurodegenerative diseases, 139, 143, 207; *see
        also* Circulating miRNA
    ALS, 208
    Alzheimer disease, 208
    biomarkers in, 210–211
    frontotemporal dementia, 208
    FUS/TLS inclusions, 209
    Huntington's disease, 208
    miRNA biogenesis and function, 211–212
    miRNA mimic, 226
    mitochondrial dysfunctions, 210
    oxidative stress, 210
    Parkinson's disease, 208
    RNA processing, 209
    TDP-43 inclusions, 209
    therapy, 226
Neurodevelopmental disorders, 108
Neurofibrillary tangles (NFTs), 214
Neurofilament light chain mRNA (NEFL), 223
Neurogenesis and synaptic plasticity, 105–107
Neurogenic inflammation, 184
Neurons, 243
Neuropathic pain, 193; *see also* Pain
Neuropeptide Y (NPY), 64

Neuroplastic changes, 24
Neuropsychiatric conditions, *see* Mood and
    anxiety disorders
Neurostimulation, 156
Neurotransmitters, 23
Neurotrophic factors, 108–109
Next generation sequencing (NGS), 92, 213
NFTs, *see* Neurofibrillary tangles (NFTs)
NGS, *see* Next generation sequencing (NGS)
NLH, *see* Nonlearned helpless (NLH)
NMDA, *see* N-methyl-D-aspartate (NMDA)
N-methyl-D-aspartate (NMDA), 10
    antagonists, 80
    interaction between miR-219 and, 85
Nonlearned helpless (NLH), 31
NPY, *see* Neuropeptide Y (NPY)
NSCs, *see* Neural stem cells (NSCs)
Nucleic acid damaging factors, 141

## O

Obsessive-compulsive disorder (OCD), 22
OCD, *see* Obsessive-compulsive disorder (OCD)

## P

PACT, *see* PKR-activating protein (PACT)
Pain, 183
    acute, 183
    alterations in miRNA expression, 196
    animal surrogate models of, 186
    burden of, 183
    chronic, 183
    coexisting conditions, 184
    future perspectives, 198–199
    human diseases characterized by miRNA
        alterations, 186
    miRNA alterations in inflammatory, 188, 191
    miRNA alterations in neuropathic, 192–194
    miRNA alterations in response to, 195
    miRNA modulation studies, 189–190
    miRNAs as potential biomarkers, 197–198
    neurimmiRs as modulators of inflammation,
        195–197
    nonneuronal components in, 185
    transcriptional changes of miRNA in
        neurons, 187–188
Pain processing, 184
    central sensitization, 184
    miRNA in, 188
    miRNAs as modulators in, 185–187
    neurogenic inflammation, 184
    pathways, 184–185
    peripheral sensitization, 184
Parasitic worms, 140
Parkinson's disease (PD), 207, 208; *see also*
    Neurodegenerative diseases

Pathogenic prion (PrP$^{Sc}$), 141, 143; *see also* Prion
PBMCs, *see* Peripheral blood mononuclear cells
    (PBMCs)
PD, *see* Parkinson's disease (PD)
Peripheral blood mononuclear cells (PBMCs),
    68, 218
Peripheral miRNA biomarkers, 68–69
PFC, *see* Prefrontal cortex (PFC)
PGC-1α, *see* Proliferator-activated receptor γ
    coactivator-1α (PGC-1α)
PGD, *see* Phosphogluconate
    dehydrogenase (PGD)
Phosphatase and tensin homolog (PTEN), 112
Phosphogluconate dehydrogenase (PGD), 66
PKC, *see* Protein kinase C (PKC)
PKR, *see* RNA-activated protein kinase (PKR)
PKR-activating protein (PACT), 104
Postpartum psychosis, 40
Posttraumatic stress disorder (PTSD), 22
Potassium-chloride co-transporter member 2
    (KCC2), 243–244
PPI, *see* Prepulse inhibition (PPI)
Precursor-miRNA (pre-miR), 162
Prefrontal cortex (PFC), 30, 62
pre-miR, *see* Precursor-miRNA (pre-miR)
Prepulse inhibition (PPI), 78
Primary miRNAs (pri-miRNA), 4, 104
pri-miRNA, *see* Primary miRNAs (pri-miRNA)
Prion, 140; *see also* Artificial miRNAs (amiRNAs)
    cellular into pathogenic prion, 141
    historic events, 140–141
    kuru, 140
    mode of replication, 141–142
    scrapie, 140
    self-seeding fibrils, 148
Prion diseases, 142, 148–149; *see also* Artificial
    miRNAs (amiRNAs)
    miRNA signature, 143–144
    mutations in miRNA-binding proteins, 148
    prion protein interactions with miRNA,
        145–146
    prions and microRNAs in exosomes, 144–145
Programming, 246
Proliferator-activated receptor γ coactivator-1α
    (PGC-1α), 223
Protein kinase C (PKC), 112
Protein-only theory, 140
PrP$^{Sc}$, *see* Pathogenic prion (PrP$^{Sc}$)
Psychoactive drugs, 111, 113
PTEN, *see* Phosphatase and tensin homolog
    (PTEN)
PTSD, *see* Posttraumatic stress disorder (PTSD)

## Q

qRT-PCR, *see* Quantitative RT-PCR (qRT-PCR)
Quantitative RT-PCR (qRT-PCR), 213

# R

Raphe nucleus (RN), 35
RBPs, *see* RNA-binding proteins (RBPs)
Real-time polymerase chain reaction (RT-PCR),
    68, 213
Repressor element 1 silencing transcription
    (REST), 8, 110, 225
Retinal ganglion cells (RGC), 242
RGC, *see* Retinal ganglion cells (RGC)
Ribonuclease (RNase), 5
Ribosomal RNAs (rRNAs), 213
RN, *see* Raphe nucleus (RN)
RNA-activated protein kinase (PKR), 104
RNA-binding proteins (RBPs), 209
RNA induced silencing complex (RISC), 4,
    162, 211
RNA interference, 146
RNA processing, 209
RNase, *see* Ribonuclease (RNase)
rRNAs, *see* Ribosomal RNAs (rRNAs)
RT-PCR, *see* Real-time polymerase chain
    reaction (RT-PCR)

# S

sALS, *see* Sporadic ALS (sALS)
SC, *see* Spinal cord (SC)
Schizophrenia (SCZ), 60, 62; *see also* MicroRNA
    dysregulation in schizophrenia
Sciatic nerve crush (SNC), 188
Scrapie, 140; *see also* Prion
SCZ, *see* Schizophrenia (SCZ)
Seizures, 154–155
Selective serotonin reuptake inhibitors
    (SSRIs), 23
Serotonergic signaling, 49
Serotonin–noradrenaline reuptake inhibitors
    (SNRIs), 23
Serotonin transporter (SERT), 35
SERT, *see* Serotonin transporter (SERT)
Short hairpin RNAs (shRNAs), 147
shRNAs, *see* Short hairpin RNAs (shRNAs)
Single nucleotide polymorphism (SNP), 37, 60
siRNAs, *see* Small interfering RNAs (siRNAs)
SIRT1, *see* Sirtuin1 (SIRT1)
Sirtuin1 (SIRT1), 11
SMA, *see* Spinal muscular atrophy (SMA)
Small interfering RNAs (siRNAs), 146
SMRI, *see* Stanley Medical Research Institute
    (SMRI)
SNC, *see* Sciatic nerve crush (SNC)
SNCA, *see* α-Synuclein (SNCA)
SNL, *see* Spinal nerve ligation (SNL)
SNP, *see* Single nucleotide polymorphism (SNP)
SNRIs, *see* Serotonin–noradrenaline reuptake
    inhibitors (SNRIs)

Spinal cord (SC), 223
Spinal muscular atrophy (SMA), 223
Spinal nerve ligation (SNL), 187
Sporadic ALS (sALS), 223
SSRIs, *see* Selective serotonin reuptake
    inhibitors (SSRIs)
Stanley Medical Research Institute (SMRI), 37
STG, *see* Superior temporal gyrus (STG)
Stress, 29; *see also* MiRNAs in neurogenesis/
    neuroplasticity
    individual's ability to cope with, 31
    miRNAs and, 29–31, 109–110
    -related proteins, 47, 49
    studies on miRNAs in, 26–28
Suicidal behavior; *see also* MiRNAs in
    neurogenesis/neuroplasticity
    miRNAs impact in, 112–113
    studies on miRNAs and, 124–125
Superior temporal gyrus (STG), 75
α-Synuclein (SNCA), 208

# T

TALENS, *see* Transcription activator-like
    effector nucleases (TALENS)
TAR, *see* Transactivation response (TAR)
TAR RNA-binding protein (TRBP), 104
TCAs, *see* Tricyclic antidepressants (TCAs)
Temporal lobe epilepsy (TLE), 157
TH, *see* Tyrosine hydroxylase (TH)
TLE, *see* Temporal lobe epilepsy (TLE)
TLR, *see* Toll-like receptor (TLR)
TNF-α, *see* Tumor necrosis factor-α (TNF-α)
Toll-like receptor (TLR), 219
Transactivation response (TAR), 104
Transcription activator-like effector nucleases
    (TALENS), 92
TRBP, *see* TAR RNA-binding protein (TRBP)
Tricyclic antidepressants (TCAs), 22
Tri-reagents, 213
Tumor necrosis factor-α (TNF-α), 219
Tyrosine hydroxylase (TH), 66

# U

Unipolar depression, 22
3′untranslated region (3′-UTR), 4, 61

# V

Valosin containing protein (VCP), 208
Vascular endothelial growth factor
    (VEGF), 24
VCP, *see* Valosin containing protein (VCP)
VEGF, *see* Vascular endothelial growth factor
    (VEGF)
Viroids, 140

Viruses, 140
Visceral inflammation, 192; *see also* Pain

## W

Walter and Eliza Hall Research Institute
    (WEHI), 92
WEHI, *see* Walter and Eliza Hall Research
    Institute (WEHI)

Wild-type (WT), 78
WM, *see* Working memory (WM)
Working memory (WM), 72
WT, *see* Wild-type (WT)

## Y

Yeasts, 140

Milton Keynes UK
Ingram Content Group UK Ltd.
UKHW040109071024
449327UK00019B/941